Memory Landscapes of the Inka Carved Outcrops

Memory Landscapes of the Inka Carved Outcrops

From Past to Present

Jessica Joyce Christie

LEXINGTON BOOKS
Lanham • Boulder • New York • London

Published by Rowman & Littlefield
A wholly owned subsidiary of The Rowman & Littlefield Publishing Group, Inc.
4501 Forbes Boulevard, Suite 200, Lanham, Maryland 20706
www.rowman.com

Unit A, Whitacre Mews, 26-34 Stannary Street, London SE11 4AB

British Library Cataloguing in Publication Information Available

Library of Congress Cataloging-in-Publication Data

Christie, Jessica Joyce, 1956-
Memory landscapes of the Inka carved outcrops / Jessica Joyce Christie.
pages cm
Includes bibliographical references and index.
ISBN 978-0-7391-9488-1 (cloth : alkaline paper — ISBN 978-0-7391-9489-8 (electronic) 1. Incas—Antiquities. 2. Incas—Religion. 3. Petroglyphs—Andes Region. 4. Outcrops (Geology)—Andes Region. 5. Ancestor worship—Andes Region. 6. Memory—Social aspects--Andes Region. 7. Andes Region—Antiquities. 8. Andes Region—Religious life and customs. 9. Social archaeology—Andes Region. 10. Landscape archaeology—Andes Region. I. Title.
F3429.C523 2016
709.01'13098—c23

2015007855

∞™ The paper used in this publication meets the minimum requirements of American National Standard for Information Sciences—Permanence of Paper for Printed Library Materials, ANSI/NISO Z39.48-1992.

Printed in the United States of America

This book is dedicated to Dr. Terry Grieder,
UT Austin, who got me started.

Contents

List of Figures

Additional color images and drawings grouped according to book chapters are available on http://christiej4.wix.com.

Foreword

The Inka represent one of the most fascinating ancient cultures of the Americas. They managed to unify the largest territorial state in the southern hemisphere, and other than the iconic tourist image of Machu Picchu, their treatment of stone is the single most recognizable cultural identifier remaining of their presence. The monumental form in which their culture embedded landscape at the heart of its being has been one of the more difficult aspects of their ethos to pass on to the Western mind.

Jessica Joyce Christie's work offers us a significant new contribution in the underexplored study of Inka sculpted rock and its place in the Andean landscape. She uses a variety of approaches and methods to lead to an enhanced understanding of Inka stonework. She uses phenomenology, which holds that reality consists of objects and events as they are perceived or understood in human consciousness and not of anything independent of this awareness; she deploys this methodology in the culturally constructed landscape which holds the sculpted stone elements which form the subject of her study.

Well-thought-out descriptions are backed up by informative illustration allowing the reader to fully understand the layout and form of the sites in their surrounding environmental setting. As we progress through the book, Inka sacred space is contextualized by our being taken through the sites along the existing access routes. This approach adds to our increasing awareness of the mountainous bedrock underlying the Inka cultural rock sculptures, as these seem in places to arise seamlessly from rough native rock to sculpted outwardly almost organic surfaces.

Thus viewing the Inka abstract stonework in a structured and responsive landscape allows us to begin to read into the sculpted rock some of the underlying reality from which it arose. The early colonial documentary source material and perceptive discussion of Inka mythology by the author have been used to good effect to delimit the sculpted stone-scapes associated with named Inka rulers. This has led to a recognition of a stone ideology created by the Inka Pachakuti Yupanqui, "the Inverter of Worlds", and his immediate successor and son, who thereby produced the largest single corpus of modified rock assembled in the Inka heartland. The Inka not only read an animated nature in the native rock, but unlike their predecessors they are the only Andean civilization to sculpt and modify the bedrock underfoot. To this nation, stone was as much alive and part of the social aspect of their consciousness as were their crops, herds, and the peoples inhabiting the land. The structured distribution of sacred shrines across the landscape allowed them to formulate an environment which could be understood by its inhabitants and which helped them anchor timing of the yearly agricultural cycle and the scheduling of agricultural task to designated groups of workers.

With the earlier work by Maarten van de Guchte, César Paternosto, and Carolyn Dean, and now Christie's illuminating discussion and interpretation of the subject, there has been a radical advance in our understanding of meaning, context, and use of a major cultural symbolic signifier deployed by the Inka state to express its presence and right to rule across its domain.

Christie progresses her study fully into the present day use and meaning of Inka carved rock sites in its existing cultural context.

Frank Meddens, BA, PhD, FSA, MIFA
Department of Geography, Royal Holloway, University of London
e-mail frank.meddens@googlemail.com

Acknowledgments

I express my deepest gratitude to my friends and assistants in Cusco for supporting and furthering my fieldwork, most importantly Raul Ccorahua Quintana and Henry Conza Leon, who have always found means to make my mountain treks possible and enjoyable. I could not have updated my work without the help of my dear friend Flor de Maria Huaycochea at the Universidad Nacional de San Antonio Abad del Cusco, who has tirelessly kept me informed about current archaeological work and events in Cusco. My thanks go out to the Ministerio de Cultura in Cusco for granting permits and to the many affiliated archaeologists and workers who have consulted with me and enriched my writing.

East Carolina University has supported my fieldwork for this book with a large research grant that fully covered the mapping season in 2001. Dean Chris Buddo of the College of Fine Arts and Communication and Director Michael Drought of the School of Art and Design added funds to increase the illustrative material as the book was about to go into production. I am greatly indebted to the technical experts who assisted me with the images, foremost to the ever-patient and never-tiring Kelly Adams and Mike Dixon. Laurie Godwin and her student workers formatted most of the images. Joyce Newman helped to improve the flow of the text by proofreading two chapters. My graduate assistant Greg Banks was instrumental in making the final production deadline by searching all of my index terms. I feel profound gratitude for this help at my local institution.

Dumbarton Oaks in Washington, D.C., generously supported the writing phase of this book with a summer fellowship awarded in 2008. The extensive literary resources, the dedicated staff support in particular from Bridget Gazzo, and the stimulating collegial networks of Joanne Pillsbury, then director of pre-Columbian studies, I encountered there have indelibly furthered my work. My thanks go out to the numerous colleagues who have reviewed this work and guided and inspired me to take it to new fronts. Some have remained anonymous, but invaluable input was provided by Patricia Netherly, Frank Meddens, Adam Herring and Maarten Van de Fuchte.

Most precious on a personal level has been the company and friendship of my son, Brian Christie Garrett, whom I dragged on his first hikes at the age of five. Later he became my project photographer and loyal assistant. Now an engineering student at East Carolina University, he still spends time with me and our godchildren in Cusco.

Notes on Orthography and Terminology

In Andean Studies, there is no consensus on the issue of the correct spelling of ancient and Quechua names. In the past decades, many Mesoamerican as well as Andean scholars have argued that we should make every effort to spell ancient and contemporary indigenous names the way Native people have spelled them in order to express respect. I am committed to make a sincere effort, but the matter is complicated for a number of reasons: first, I do not speak Quechua. Therefore, I follow the spellings of Andean names by Terence D'Altroys (2002), who has made the same commitment. Secondly, different authors, including the chroniclers who wrote down Quechua words the way they heard them as Spanish speakers, spell the same names differently, causing confusion for any reader. In direct quotations, I will maintain their spellings. Thirdly, spelling crosslinks with grammar in multiple ways. One example frequently occurring in the following text is the Quechua plural construction with the ending—*kuna*. I decided to use the English plural ending—s, for example, *wak'as* as opposed to *wak'akuna*, to render the book more reader friendly and because as non-Quechua speaker, I cannot aspire to follow all other grammar rules which may affect spellings. Finally, I use *Inka* instead of *Inca* because it is the form that is closer to Quechua, yet I use *Cusco* instead of *Cuzco* or *Qosqo* because the first is the legal spelling used in daily life in Peru today. *Carved rock, carved stone, sculpted rock, sculpted stone* are used as synonyms.

Furthermore, I am aware of the debate about the appropriateness of using terms coming from Western fine art vocabulary to describe aesthetic forms created by non-Western cultures whose languages do not carry precise translations of the Western term *art* (for example, Berlo and Phillips 1999; Dean 2006a). Carolyn Dean (2006a) argues that using Western fine art terms to refer to non-Western visual culture amounts to a contemporary act of cultural colonization. While her accusation greatly concerns me, I have never observed that Andean people would feel that way. I decided to keep using Western art terms (rock art, carved rocks, sculpted boulders, Inka artists) because the book is written in the English language, and grappling with finding the morally responsible terms—for which there is no consensus—will distract many readers.

Introduction

BACKGROUND AND CONTEXT

Of the Book: Purpose, Methodologies, and Findings

This book is about a medium of Inka art that has been sorely neglected in the literature, scholarly discussions, or public interest: the carved rocks and outcrops sculpted in situ and attended to as isolated shrines. All tourist buses stop at Q'enqo on the Cusco City Tour, and visitors marvel at the large stone on the platform, climb the outcrop on the carved steps, and crawl through the cave. Few make it down to Q'enqo Chico or walk up to Laqo. Scholarly interest in the carved rocks has not been much more active.

This is surprising since Inka sculpted outcrops are amazing monuments worked in situ usually in spectacular landscape settings. The Inka were drawn to boulders and outcrops with certain unusual features, such as aesthetic surface formations, caves, and springs, and they cut them into vertical and horizontal volumes and masses. Viewing such works today, we think of platforms, seats, steps, and altars. At times, they added relief sculptures of animals and architectural walls. Many are found around Cusco, in the Urubamba Valley, and fewer in the outlying regions of the original Inka empire.

Inka carved rocks have been discussed in conference papers, articles, and book chapters with the exception of an influential dissertation in 1990 and a recent scholarly book. Primary sources of information come from archaeology and ethnography. Many sculpted boulders have been reported on as small items in larger archaeological projects; and since about 2000, the Ministerio de Cultura in Peru has archaeologically investigated many carved outcrops in the Parque Nacional de Saqsaywaman just northeast of Cusco. Colonial Spanish writers mention sites with carved rock shrines, but the references are piecemeal. The main problem is that the information about Inka sculpted outcrops is scattered and a current archaeologically anchored analysis of their role in Inka society does not exist.

My book contributes such an analysis: it brings the diverse material together and reconstructs the role of boulder sculpture in Inka society, focusing on their religious, political, and social contexts. I discuss techniques and styles, introduce the formal categories of Inka carving, and include the man-made and natural features the rock art sites are often associated with. Many sites are illustrated by drawings which show the carvings in relationship to such features. This is another vital aspect that has been absent in most existing publications. The Inka did not carve the rocks as formal independent sculptures in the Western sense, but they have to be understood in the context of their environment, and my drawings will make that clear.

I have selected five significant site settings of major carved rock complexes. They begin in Cusco, the capital and one of the core areas in which the Inka style of rock carving was formulated. Chapters 2 and 3 move to the origin sites of the Inka dynasty on Lake Titicaca and at Pacariqtambo; chapters 4 and 5 extend to the royal estates of Machu Picchu and Chinchero to analyze how Inka rulers continued to explore rock carving and began to commission sculpted boulders in outlying regions as a political strategy to claim Inka presence. Style and formal categories will be described and available archaeological, linguistic, ethnohistoric, and ethnographic sources will be consulted. I attempt to reconstruct and interpret the function of these carved rocks, how they may have been used in specific ceremonies, and which role they may have played in the context of Inka state politics. These discussions are presented under the heading *Groundwork*. The book presents sufficient cultural context to make it useful for scholars specializing in other American societies or in landscape studies to draw cross-cultural comparisons.

The case studies identify patterns that carved rocks constituted major building blocks in the imperial strategies of the Inka state. It is also clear that the Inka were not the first Andean people to modify empowered stones in their landscape settings. In the discussions, I will refer back to pre-Inka societies and how the Inka instituted a new ideological category of stone shrines. Further, the boulder sculptures are still considered shrines (*wak'as*) by Cusquenos today: traditionalists continue to make offerings (*pagos* understood as tributes to *pachamama* [Mother Earth]) at more secluded carved rocks, and in many rural villages, ceremonies and offerings are performed at rock art sites. I will present the multiple layers of perception and physical appearances of the sculpted rock complexes under discussion over time and extract the conceptual essences which have conditioned the Andean view of stone, mountains, and their landscape setting. The Inka manipulated this worldview into a calculated political direction.

The methodology used may be likened to a mosaic work to which tessera after tessera—in the form of data from different academic disciplines—is added. I question the traditional empiricist notion that this mosaic work can reach completion or exist in a neutral and objective state. This is due to the fact that some data are simply missing and, secondly, that the information is filtered through the pen of Spanish writers and through the eyes of colonial and contemporary consultants as well as of researchers who all approach the carved rocks from their cultural perspectives and with their agendas. In this sense, the mosaic consists of layers of tesserae not fixed in time or space.

The two primary theories I have selected to apply to the discourse of Inka carved rocks are phenomenology as developed by British scholars, such as Christopher Tilley (1994; Tilley and Bennett 2004; Smith 2003:62–66) and employed by some American archaeologists (Van Dyke 2007); and the position of a political landscape structured by sets of relations between subjects and objects as advocated by T. A. Smith (2003). These selections are anchored in practice theory debating the roles of culturally conditioned structures versus individual actions in the processes of shaping human societies (for a pointed discussion see Janusek 2004:6–16). A phenomenological approach is grounded in existentialism and its debate of the subjective/objective dichotomy. In basic terms, we know the world by separating ourselves as individuals from it, but at the same time, we can understand ourselves only as part of this world. We bridge the distance between objective reality and our subjective consciousness through existence, meaning our daily lived experiences and participation in the world. Thus the human body becomes the point of dialectical mediation between consciousness and the physical world, subject and object (Van Dyke 2007:38–39). Human experience is stimulated by sensation understood as a communion between body and thing leading to perception, which always involves reciprocity between the body and the world and a continuous interchange between the two. Phenomenology treats the human body as universal in its physical configuration and assumes that all human beings share similar perceptions of spatial experiences measured by the body and senses, such as up and down, front and back, left and right. Building on other writers such as Merleau-Ponty, Tilley and Bennett (2004:24) reason that the traditional Western distinction between *Nature* and *Culture* is meaningless:[1] "The world does not exist independently of us, nor do we create it. Our interaction with the world brings it and us into being. As such it is mistaken to draw distinctions between natural and cultural landscapes and places or the material and the mental. They are intertwined in social Being."

Applying these premises and working from the starting point of contemporary bodies moving in the same spaces populated by past peoples, researchers can begin to think about how the latter experienced, perceived, and constructed landscapes. I believe a phenomenological approach is especially productive in the Andes since Andean topography so spectacularly dramatizes scale and the fundamental human spatial divisions. Although many Inka settlements associated with carved boulders lie in ruins and contemporary road access has altered the physical approach to ancient sites, my own observations walking Inka trails and exploring relations of visibility between carved outcrops and landscape features will provide one fruitful avenue of tying the physical natural and man-made landscape to the ideological landscape constructed by Andean minds (Knapp and Ashmore 1999).

Although phenomenologists emphasize that human experience is never entirely physical or exclusively derived from bodily processes but always simultaneously culturally conditioned, this latter dimension is not their primary focus. In order to contextualize Inka sculpted outcrops more forcefully in enduring Andean religious, social, and political arenas, I have chosen to employ Smith's theory of a political landscape constructed by relations. Interestingly, Smith analyzes political landscapes through the dimensions of experience, perception, and imagination. It is intriguing how his definition of spatial experience overlaps and differs from the individual bodily experience of space formulated by phenomenologists. Smith (2003:73) equals spatial experience with material practices describing "the flow of bodies and things through physical space." He lists "distribution, transport, communication, property rules, land use, resource exploitation, and administrative, economic, or cultural divisions in physical space" under spatial experience. Spatial perception is the realm of signs and codes in which observers engage their senses and generate evocative responses to spatial forms and aesthetics. According to Smith (2003:74), spatial imagination constitutes "the analytic domain of representations," the step "from maps and pictorial landscapes to spatial theory and philosophy." Smith's approach (2003:69–72) interconnects the three dimensions through human social life, and his framework of political landscapes becomes the stage on which to explore the relations between the lived, conceived, and perceived. Smith's theory of landscapes of relations will prove most useful in examining the social and ritual settings the Inka constructed for the sculpted boulders and their strategy of *stone ideology* taken up below. Secondary tactics, such as "Doubling of Sites and Essences" (chapter 2) and core and periphery dynamics (chapter 4), are aspects of the political landscape formation processes surrounding carved rocks. Following Smith's (2003:73–77) approach, I will structure the analyses in chapters 2 through 6 into sections titled Appearance, Experience, and Perception and Essence, Imagination, and Stone Ideology in order to reconstruct the relational networks of a specifically Inka political landscape.

This book presents Inka sculpted boulders from a theoretically informed perspective as social constructions of a landscape, which were culturally motivated and perceived. Rock art sites marked special places within the vast space of the Andes in which the Inka actively engaged with the world around them and remembered their past. The Inka themselves constructed their environment differently in different social contexts; anthropologists have cautioned that formal resemblances, in this case a consistent stylistic vocabulary, never hold the same meaning to all people even within the same culture and have warned not to give in to the tyranny of "culture" concept, which is used as an illusion that a homogeneous, consistent system of meaning is shared by all members of a society (Abu-Lughod, Appadurai, Friedman, Turner cited in Inomata 2006:190–191). Although a general collectively held association of the boulder sculptures with the Inka state likely existed, individual site contexts were shaped and performed through local actions and experiences. The chroniclers overlaid their perspectives on their accounts; and contemporary Quechua people and we as outsiders approach Inka carved rocks with cultural experiences of the twenty-first century. These parameters set up serious limitations to our "objective" understanding of the Inka sculpted outcrops. The implied wholeness of such an "objective" understanding will be fragmented into layers of vision in this book. All conclusions will necessarily be tilted by social frameworks and address different time- and culture-specific ways of viewing carved rocks rather than a one-fits-all model.

A secondary theory I follow in the discussion of contemporary uses of sculpted outcrops is inspired by Lila Abu-Lughod, who discusses the dilemma of how to collaborate ethically as well as productively with indigenous people and outlines strategies of how to overcome the self–other paradigm in her chapter with the provocative title "Writing against Culture" (1991). In the process of investigating non-Western cultures and explaining their cultural differences in teaching and writing, we also help to construct and maintain them—after all, they are the objects and topics of our interests, and we would not want them to go away! Anthropological as well as archaeological and art historical discourse sets up relations of hierarchy between the researcher (self) and indigenous people (the other), which by necessity assign power to the former and position the latter at an inferior level. Abu-Lughod (1991:147) reasons that the concept of "'culture,' shadowed by coherence, timelessness, and discreteness, is the prime anthropological tool for making 'other,' and difference, . . . tends to be a relationship of power" and suggests ways of going beyond this dilemma by "writing against culture." One strategy she formulates is writing "ethnographies of the particular" to be pursued as an instrument toward greater tactical humanism (Abu-Lughod 1991:158–159). Ethnographies of the particular are intended to replace generalization, which, as the accepted writing style of the authoritative discourse of professionalism, implants ideas of cultural homogeneity, coherence, and timelessness in the minds of readers and listeners. Instead, ethnographies of the particular foreground local actions by individuals living their particular lives, inscribed in their bodies and words (Abu-Lughod 1991:150). This is what I introduce throughout this book and in particular with the explanations of rock art sites at Copacabana offered by Porfirio Huanca Sanchez and Ramon Calle (chapter 2).

The analysis of the five case studies will lead to the following three arguments: *Finding 1* is that Inka carved rocks had multiple functions in the political, social, religious, and economic contexts of Inka society. Some sculpted boulders certainly marked locations where offerings to various supernatural entities were presented. This practice continues today in the form of pagos which are normally conducted at rocks further removed from the main tourist sites (for example, near Q'enqo). Other carved boulders were seats from which visual connections with specific *nevados* (Snow Mountains) were established (see example at Q'enqo Chico), and/or they overlooked important ritual spaces (Sabacurinca at Saqsaywaman). I will connect certain carved rocks forming grid-like patterns in Cusco, Copacabana, and Chinchero with yupanas and khipus as illustrated in Guaman Poma de Ayala's drawings and argue that they functioned as similar counting devices written in stone. In Cusco, such counting devices would have been useful in the context of the zeq'e system according to which offerings made at individual wak'as had to follow a ritual calendar and rules. In Copacabana, they would have counted the number of pilgrims and the offerings they brought; and in Chinchero, the carved counting device may have helped to record quantities of agricultural products since it is situated close to the central terrace sector.

Secondly, I will make the argument that the style of Inka rock carving was authored by Pachakuti Inka Yupanki and packaged into his political tool of stone ideology (*Finding 2*). Here my book succumbs to the "great man" paradigm justly criticized by Brian Bauer and Alan Covey (in Bauer 2004:71–72). I hope to demonstrate that foregrounding Pachakuti as the primary actor does not imply a literal acceptance of all his actions detailed in the chronicles and does not have to cancel out an equal emphasis on the broader social contexts and the multilayered political landscape in which he performed. The main evidence for my reasoning is the relative certainty that Machu Picchu was one of Pachakuti's royal estates and that it was commissioned and constructed in one design and in the short time span of several decades, approximately A.D. 1450 to the early 1500s. Therefore I consider the carved rocks of Machu Picchu Pachakuti's personal commissions, and similarly sculpted boulders grace his other estates. I will show that his son, Thupa Inka Yupanki, repeated the style of his father and included elaborately carved rock complexes into his estate at Chinchero. On the other hand, Wiraqocha Inka, Pachakuti's father, only had minor and simple carved rocks on his estate at Juchuy Qosqo, and Wayna Qhapaq reduced the integration of rocks to one foregrounded boulder in the middle of the plaza on his personal estate at Urubamba. However, he commissioned two beautifully sculpted outcrops at his northern settlement of Ingapirca located in what is now Ecuador. I think these data support a fairly compelling argument that Pachakuti developed and supported Inka rock sculpture, which also has ramifications for the Cusco area since it strengthens the attribution of certain carved rock wak'as on the zeq'e lines to him and to his panaqa.

Rock and stones have been sacred in the Andes since time immemorial. By about 1000 B.C. during the Early Horizon (see Burger 1992:163; Conklin and Quilter 2008:XXXII, fig. 1.2), people at Chavin de Huantar cut a vast network of ritual galleries into a mountain and excelled in sophisticated stone sculpture. Tiwanaku people during the Middle Horizon (c. 700–900 A.D.) practiced stone working techniques which may have inspired the Inka. Throughout the Andes, stone has been considered the essence of the mountain who himself is seen as a double of the human body (Bastien 1978; 2006:135) and both of them are thus alive and not inanimate objects as they are treated in Western thought. Among the Inka, rocks and caves could be the places of origin of ayllus and panaqas; and people were believed to have been transformed into stone, like some of the Ayar brothers from the emergence story in remote times. The well-known myth of the Tired Stone or Piedra Cansada relates that this huge block became tired on the transportation route to Saqsaywaman, that it wept tears of blood and refused to move any further (see chapter 4).

In addition, stone has always been a symbol of solidity and material permanence; it transcends mortal human existence because it is timeless and absolute (see Valcarcel 1964:53–54). I argue that Pachakuti capitalized on such preexisting notions and symbolic relations and imprinted upon bedrock and outcrops his own style of vertical and horizontal cuts and geometric forms. As will be shown below (see particularly the Machu Picchu chapter), his was not a "Conquest" style that would bluntly have subordinated the natural to the cultural, but rather a subtle dialogue which expressed itself in relations and degrees of modification to stone ranging from untouched but potentially sacred bedrock, to activated sculpted boulders, and finally to precisely cut, geometric building blocks used in monumental Inka constructions. I speculate that this style of Inka rock art possibly began as a personal discourse between Pachakuti and the natural environment since the carved rocks on his royal estates and at the place of his initial burial (Patallacta/Q'enqo) are found in private settings, and most do not qualify as public works of art. Such a "personal discourse" must have been quite intriguing since we know from multiple ethnographic sources that carved stone wak'as were strong-willed agents who did not always agree with the wishes of the ruler. Yet the style of rock carving also took on a political component perhaps late in Pachakuti's reign or under his son, Thupa Inka, to mark "Inka-ness" and the presence of the state at Chinchero and at outlying administrative centers, such as Copacabana and Samaipata, Ingapirca, as well as at the ceremonial center of Pumaurqu/Maukallaqta. These outlying sites with sculpted stone wak'as may have extended the zeq'e system to the four suyus and eventually to the entire empire. Thus carved rocks signaled Inka identity and the state. In a more aesthetic sense, they evoked representations of terraces as well as ritual connections with the upper- and underworlds.

I will integrate all these beliefs and concepts under the term *stone ideology*, which I postulate was instituted by Pachakuti but draws on numerous Andean practices layered in time and space. As this ideology formed and spread in the empire, it was, as I believe, readily accepted because it built upon pre-Inka pan-Andean worldviews. Ayllu members continued to construct their own stone shrines, and some of the elite may have carved rock wak'as so that not all sculpted boulders are tied to Pachakuti in person.

Finding 3 forcefully emphasizes that the general power of this ideology and the permanence of its underlying belief structures did not end in the Colonial Period. Today traditional people in the Andes continue to make offerings to pachamama at rock shrines and place Catholic sanctuaries on top of boulders. I argue that the innate life forces of stone wak'as still draw local people to bring offerings to their Mother Earth. Such ceremonies frequently ask for blessings of agricultural activities, but a growing number of ritualists petition for health, money, education, and other material possessions. All contemporary ritual offerings are private and without interference of any governmental institutions. Nevertheless, during the last decades, growing numbers of tourists and other outsiders have taken an interest in traditional Andean practices, and new economic considerations have entered the organization of ceremonies, especially large regional events. Political and legal fallout again interweaves with ritual planning as local ethnic groups (such as the Copacabana Aymara) stage their cultural independence from the Inka, as Cusco constructs its local identity as independent from the national heritage while both claim Inka roots, and as European and U.S. environmentalists align with indigenous Quechua groups to protect empowered mountains from the mining industry.

Of the Inka: A Brief Cultural History

Inka history has been reconstructed partly from archaeological data and partly from ethnohistoric and ethnographic sources, many of which include the biographies of twelve successive rulers; but the vexing problem has been that the two sets of information squarely match and that there is still no consensus on Inka chronology. During the past decades, groundbreaking archival and archaeological survey work in the Cusco region have been conducted by Brian Bauer and Alan Covey (2002) and their many colleagues to analyze the strategies of state formation. Their work concentrates on the time period from A.D. 1000 to 1400 designated as Late Intermediate Period, which followed Wari control of the Cusco area. The time span of roughly five hundred years between the beginning of the Late Intermediate Period and the arrival of the Spanish in the 1530s may approximate the lifetimes of the commonly accepted twelve generations of rulers listed by the Spanish writer Miguel Cabello Valboa in the sixteenth century. Cabello Valboa documents birth and death dates of Inka rulers and correlates them with events in the Old World (for example, he sets the death of Thupa Inka in 1493 [Cabello Valboa 1951:354]). Nevertheless, it is not clear how archaeology and ethnography may productively work together.

Raul Porras Barrenechea is one of the lone voices who includes two extensive chapters about Inka methodologies to perform history by means of storytelling, songs, khipus, and paintings (1963:103–145). These are significant mosaic tesserae in our efforts to reconstruct Inka visual culture, for example, the references to Aymara and Quechua words in early colonial dictionaries with the meaning "paper, writing" and "to write, paint" (Borras Barrenechea 1963:111). Ritual performances of Inka history were accompanied and explained by songs, and there is reasonable certainty that some were held at carved rock sites discussed in this book (chapter 3). Of course, these songs or *haylli* and stories transmitted selective manipulated histories yet they would add the indispensable Inka vision to our twenty-first century efforts toward understanding Inka history if we had surviving direct records.

I will begin by briefly contextualizing the ethnographic debate to be juxtaposed by a chart with dates derived from archaeology in the Cusco region by Bauer and Covey (Table 2). In the mid-twentieth century, the ethnographic perspective of Inka history was divided into two camps: John Rowe (1946) compared the lengths of reign of the Inka rulers recorded in the various Spanish chronicles and eventually proposed a series of absolute dates for most of them. He thereby established a chronological framework of a line of dynastic successions and wrote about Andean history as a linear series of events. This approach follows Western historicist models developed for and applied to European history (Urton 1990:5–6).

The opposing camp has been led by Tom Zuidema, who read the chronicles as mythical texts which encoded the social organization of Cusco (1964) and other local knowledge. Since we have no written indigenous precontact accounts of the Inka empire, he reasoned that we have no basis on which to measure the historicity of events as recorded by Spanish writers. He has viewed the dynastic successions of Inka rulers in the chronicles as histories of narratives which had been told and retold and in the process were modified and colored by various ideological and social motivations on the part of the indigenous consultants and the Spaniards themselves. Zuidema (1964) further proposed an alternative reading of Inka chronology challenging the dates established by Rowe: since Andean communities have always been divided into Hanan and Hurin sectors and follow leaders from each moiety, the succession of Inka rulers should be read as two parallel descent lines. This idea was further developed by Pierre Duviols (1979, 1980) and by Franklin Pease (1994:102–107; 2004:56–59, 128–132; see also Rostworowski 1999:177–181; Julien 2000:10) and implies that the Inka empire was governed by sets of corulers representing the Hanan and Hurin moieties, thus shortening Inka genealogy to six generations.

Although Zuidema's work offers an unsurpassed richness of ethnographic information, the problem brought forward with regard to his approach is that it is a rigid structuralist anthropology which does not allow room for historical processes and change (Nowack 1998; Julien 2000:6–11). Franklin Pease (1994:36–37, 44–45; 2004:162–179) has similarly argued that an Andean "history" never fully existed and that the consciousness of history was brought to the Andes by the Spaniards. He finds evidence for this reasoning in the observation that early Spanish chronicles cover the Inka past in a very cursory way; only later Spanish accounts delve into "Inka history" and thus reflect European historiographical practices to give the Inka a past by applying European canons of universal history to what were essentially Inka myths (see also Julien 2000:7). Although Zuidema's approach precisely reflects general problems with ethnography, it largely nullifies the usefulness of the Spanish chronicles for obtaining information about the Inka. More recent authors have taken a middle ground position by accepting that the Inka clearly separated the past from the present and the future but not in the detailed linear fashion narrated by the Spanish writers, for example, Bauer's and Covey's processes of state formation as evidenced in the archaeological record.

I share such a position, which is further supported by Catherine Julien (2000), who has offered a reading of Inka history which attempts to "excavate" Inka genres from the Spanish texts by comparing multiple Spanish versions of critical events. Although her ambition to penetrate to Inka source material through the many filters of time, changing cultures, and languages is admirable, her definition of Inka genres is not clear, and her conclusions remain, by necessity, subjective. She reaffirms that the Inka had a well-defined historical consciousness independent from Spanish understanding of history. For the Inka, historical genealogy was vital to document descent from Manqo Qhapaq, the dynastic ancestor, and accompanying qhapaq status. Pachakuti Inka Yupanki, the ninth ruler, did not only redesign Cusco as the imperial capital, but also altered the ritual calendar, strengthened the economic system to better support Cusco, and devised a "social engineering project" which redefined the marriage system to forge an ideology which advertised that the Inka ruling elite constituted a divine and pure social class descended from the Sun and therefore was entitled and sanctioned to govern all others. Julien presents Pachakuti as a genius leader who reformulated preexisting Andean practices and earlier settlements to frame the Inka ruling dynasty as the legitimate sovereigns of the Andean world. Her reading of Inka history supports my representation of Pachakuti, and his strategy of stone ideology discussed in this study can be added to the many tools he deployed to elevate the divinely sanctioned status of his panaqa and descent line.

In the following, I will present a simplified overview of Inka history and culture drawing on lines of evidence from ethnography and archaeology in order to set the stage for the investigation of the sculpted outcrops. My position is that the Inka had a concept of history and that it can best be approached through major themes rather than as a linear upward course leading from the creation events to empire collapse. Concurring with Julien, the themes of origins and emergence, qhapaq status, as well as the transformation of Cusco are essential to comprehending Inka cultural practices including stone ideology and thus will be the focus of attention in the summary below.

The two great Middle Horizon power players, the Wari empire and the Tiwanaku state, declined in the tenth century. In the succeeding Late Intermediate Period lasting from approximately 1000 to 1400 A.D., the Inka were one of a multitude of ethnic groups or senorios rivaling for power in what is now the Cusco region. What set them apart from other groups was that urban Cusco grew in size while the surrounding areas were in a process of depopulation and settlement shift from agricultural valley-bottom to fortified ridge-top sites (Covey 2006:84–89). D'Altroy (2002:59–61) characterizes pre-Imperial Inka society as a chiefdom. Sometime around 1400, they began to emerge as the dominant regional polity, conquering and subjugating their neighbors. The archaeological record of these neighbors is dominated by Killke ceramics which have been found in the center of Cusco, in the Cusco Valley, as well as in further outlying areas. Bauer and colleagues (2004:71–90) emphasize that state formation was not the outcome of a few successfully fought battles but the result of long-term processes. For instance, they note significant population shifts of Killke Period sites in the Cusco Valley. After 1000 A.D., several large new villages and extensive sets of terracing and canal systems were constructed on the northern slopes where few agricultural settlements had existed previously. Bauer and Covey (2004:77) interpret the ability to organize and construct large public works in order to create surpluses for the elite as an important strategy of the emerging Inka state. Valley-bottom sites southeast of Cusco along the Huatanay River were abandoned to probably establish a buffer zone between the Cusco Basin and the Lucre Basin controlled by the neighboring Pinahua and Mohina (Bauer 2004:76, map 8.3).

Contrasting with the archaeological survey results are the mythohistorical explanations of the origin of the Inka dynasty and the founding of Cusco as their capital, which must be understood as a later retrospect strategy to justify and legitimize the growing power of the Inka state (see chapters 3 and 4). Based on the two dominant stories of origin, the Inka were not locals but foreigners sent by deities from supernatural places. Scholars believe that the official myth centered on the site of Pacariqtambo (D'Altroy 2002:49–52; Bauer 1992:109–123). Near Pacariqtambo, there was a place with three caves called Tampu T'oqo. The Creator God summoned four brothers and four sisters who founded the Inka dynasty from the central cave and other groups who became the ancestral Cusco ayllus from the adjoining caves. The four brothers and sisters were paired as couples, and Manqo Qhapaq and his sister/wife Mama Oqllu assumed a leadership role. These ancestors began to hike in search of a place to settle and eventually founded Cusco (chapter 4).

The second story places the point of origin on the Island of the Sun in Lake Titicaca. Here the Creator God Wiraqocha called forth the sun, the moon, and the stars from holes in a sacred rock (Urton 1999:34–37). This rock became the destination of one of the most important pilgrimages in the Andes and was appropriated by the Inka into their state ideology (chapter 4). After setting the celestial bodies in motion, Wiraqocha fashioned humans. The Sun selected Manqo Qhapaq and Mama Oqllu as the principal couple and sent them in a northwestern direction to find a "chosen" valley for their people to settle. Here the first myth picks up the narrative, as they reemerged from Tampu T'oqo together with the ancestral ayllus and went on to found their capital (Cusco) and a new state.

It is significant that both locations of origin have been identified and have rock art and/or sacred rocks as their centers. Chapters 3 and 4 analyze each site respectively. Thus it must be noted early on that sacred boulders and carved rock wak'as were vital elements in the construction of a specifically Inka political landscape of origin.

Manqo Qhapaq supposedly became the first ruler in a line of twelve or thirteen successors. He and his followers were the first to live in what was to become the imperial capital Cusco. Ethnographic sources indicate that at the time of their arrival, the future site of Cusco was occupied by a village with the name of Acamama. This settlement comprised four sections: Quinti Cancha or the District of the Humming Bird; Chumbi Cancha or the District of the Weavers; Sairi Cancha or the District of Tobacco; and Yarambuy Cancha, which was probably a mixed district inhabited by Aymara and Quechua speakers (Rostworowski 1999:7). Acamama was also divided into upper (*Hanan*) and lower (*Hurin*) halves. It follows that the quadripartite and dual spatial divisions which the Inka interpreted politically and integrated into their state organization were already present in this pre-Inka settlement. It can be shown that concepts of a four-part order as well as binary oppositions and complementarity were part of a pan-Andean system of thinking (Rostworowski 1999:7–8).

There is still no consensus in the literature as to whether the first seven rulers were historical or mythological individuals. Beginning with the eighth ruler, Wiraqocha Inka, the ethnographic records become a lot more extensive. Much of the information focuses on conquests, and it may have been this ruler who changed Inka military strategies from raiding and alliances to territorial expansion (D'Altroy 2002:63). The stories of conquest begin with the notorious Chanka wars widely discussed in the literature (D'Altroy 2002:63–65; Rostworowski 1999:22–36; Betanzos 1996:19–43 [1557:part I, chaps. VI–X]). The Chanka were another local senorio with ambitions to enlarge their territory and inevitably had to clash with the Inka. They outnumbered the Inka army, and Wiraqocha Inka fled to his estate at Caquia Xaquixaguana (Juchuy Qosqo) together with Inka Urqon, his son and heir designate, and most of the elite. His other son, Inka Yupanki, decided to stay in Cusco and defend it or die along with a few equally committed men. The night before the battle, Inka Yupanki prayed for assistance, and the Creator God appeared to him in a dream and vision, promising additional warriors and final victory. And so it happened: during battle, the number of Inka fighters magically increased as stones turned into warriors (*pururaucas*), and the Chanka were repelled. There followed a period of tension between Inka Yupanki and his father, who remained at Caquia Xaquixaguana. But eventually Wiraqocha Inka traveled to Cusco, placed the fringe of rulership on Inka Yupanki's head, and gave his son the title "Pachacuti Ynga Yupangui Capac Yndichuri," which means "change of time, King Yupanque, son of the Sun." In the end, Pachakuti accepted his father's apologies for past offenses, including a failed assassination attempt, and the two participated in the festivities of Cusco together until the death of Wiraqocha Inka (Betanzos 1996:75–77 [1557:part I, chap. XVII]). Juan de Betanzos provides us with one of the most detailed accounts of Pachakuti's life and commissions which are central to any study of the carved rocks; therefore I will continue to paraphrase him. According to Miguel de Cabello Valboa (in D'Altroy 2002:45), the victory over the Chanka occurred in A.D. 1438 and Pachakuti ruled from 1438 to 1471.

Another notable project Pachakuti Inka Yupanki undertook was the re-building of the city of Cusco. Betanzos (1996:69–73 [1557:part I, chap. XVI]) offers a vivid account which depicts Pachakuti as a farsighted urban planner and effective organizer of labor. After making the decision to rebuild the city of Cusco, Pachakuti met with the Inka nobility to discuss how to procure stone and clay. "After this was done, Inca Yupanque outlined the city and had clay models made just as he planned to have it built" (Betanzos 1996:69 [1557:part I, chap. XVI]). After all the materials and workers were assembled, Inca Yupanque ordered everyone from the city of Cusco to leave their houses, take out everything they had in them, and go to the small towns nearby. As this was done, he ordered those houses to be torn down. With this done, cleaned up, and leveled, the Inca with his own hands, along with the rest of the lords of the city, had a cord brought; indicated and measured with the cord the lots and houses that were to be made and their foundations and structures (Betanzos 1996:71 [1557:part I, chap. XVI).

Pachakuti had mobilized a crew of fifty thousand, and they worked for twenty years from the first improvements of channelizing the Tullumayo and Saphy rivers which flow through Cusco until the completion of the building program. When the city was finished, Pachakuti held a town meeting inviting all the lords of Cusco and the lower classes of inhabitants. He ordered that the sketch of the city and the clay model that he had commissioned be brought to this meeting. With the model in front of him, he assigned houses and lots to the nobility of Cusco and all other residents. He had three lords, whose names are given as Vicaquirao, Apomayta, and Quilescache Urcoguaranca, settle between the Temple of the Sun and the point where the two rivers join. They belonged to his lineage, but they were illegitimate sons of lords and therefore held a lower social rank. Pachakuti named this section of his city Hurin Cusco or lower Cusco and its far end Pumachupa, which means lion's tail (figure I.1). The area from the Temple of the Sun on up, between the two rivers up to the hill of Saqsaywaman, he distributed among the prominent lords of his lineage and his own direct descent line. This section became Hanan or upper Cusco. Based on this description and references by other writers, the new layout of Cusco is understood as visualizing the body of a puma with its tail at Pumachupa and the head at Saqsaywaman (Gasparini and Margolies 1980:46, fig. 35).

While at a first and rather superficial glance, this may look like a unique and detailed account of a precontact Imperial urban design, the reader has to be aware that Betanzos's report was colored by his personal life. He married

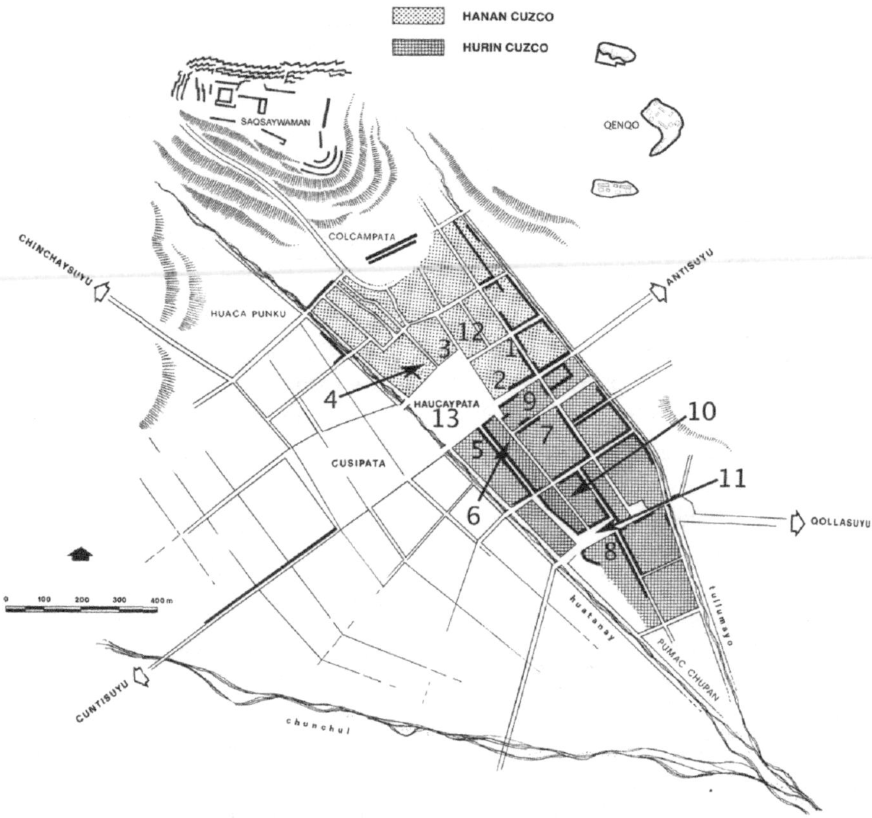

Figure I.1. Cusco. Plan of City in the Form of a Puma. Adapted from Gasparini and Margolies 1980.

a woman who was a niece of the emperor Wayna Qhapaq and had been the wife of Atawallpa in her first marriage. Thus his in-laws belonged to Pachakuti's descendant kin group, and they were his main consultants (Betanzos 1996:ix-xi [1557]). It is to be expected that they tried to enumerate the great accomplishments of their famous ancestor, and Betanzos himself likely shared their partisanship. Nevertheless, although some of the specifics may have been invented by Betanzos's family, his account together with those of other Spanish writers substantiates that Pachakuti rebuilt Cusco according to a well-deliberated design with calculated political connotations.

Archaeologists have configured a fairly concise picture of Inka Cusco (Bauer 2004:111–137). Inka Cusco grew between the Saphy/Huatanay and

Tullumayo Rivers and was centered on the main plaza, which had two sectors separated by the Saphy/Huatanay River: Awkaypata and Kusipata. Victor Angles Vargas (1988 vol. I:79–88) discusses the names and spellings of the two sections of the main plaza in Cusco. The Spanish writers spelled Awkaypata and Kusipata in a number of different ways because the Quechua language had not been written down and each author wrote indigenous words as they heard them. The two names translate as "place of crying" (Awkaypata) (Angles Vargas 1988 vol. I:81–82) and "place of rejoicing" (Kusipata) (see also Cornejo Bouroncle 1946). Angles Vargas relates the names to Inka ritual practices.

A feature of great interest is the foregrounded rock or *usnu* situated in the main plaza. Nobody knows exactly what the usnu looked like or where in the main plaza it stood because the descriptions by the chroniclers are not consistent (see, for instance, D'Altroy 2002:115, 329; Bauer and Dearborn 1995:36). Tom Zuidema (1980) reasons that there were two usnus used for solar observations. In authoritative publications, John Hyslop (1990:69–101), Zuidema (1980), and Frank M. Meddens (1997; Meddens, McEwan, and Vivanco Pomacanchari 2010) have investigated the Cusco usnus, other usnus, and the complex symbolic connotations associated with them. They conclude that usnus could assume a number of material forms: they could be stone pillars, stone seats, stone basins or fonts linked to underground channels, platforms, or truncated pyramids. Depending on their forms, usnus had a variety of functions ranging from an elevated seat of authority, which was usually accentuated by colorful canopies, to locales for libation rituals and solar observations (see also Christie 2007a, b; Meddens, Branch, Frouin, and McEwan 2010; McEwan, Meddens, Ramon, Ferreira, and Vivanco Pomacanchari 2007). Some carved rocks probably shared metaphorical aspects of the usnu.

In the greater context of the capital, the whole main plaza constituted the dividing line between the *hanan* (upper) and *hurin* (lower) divisions, and the four roads of the empire departed from it to the four *suyus* (quarters of the empire). The dualistic and quadripartite structures of Inka social and political organization may well have been reflected in the Awkaypata–Kusipata division itself.

The main plaza was fronted by major imperial buildings identified as the palaces of Inka rulers and as *kallankas*. The Qasana compound at the northwest corner of Awkaypata, beside the Saphy River, was probably Wayna Qhapaq's palace (Bauer 2004:117). Beside the Qasana stood the Coracora compound which Garcilaso de la Vega mentions as Inka Roqa's palace (Bauer 2004:122). At the northeast corner, above Awkaypata, the palace of Waskhar was situated; and near the present chapel El Triunfo, an earlier palace of Wayna Qhapaq may have existed. The cathedral likely took the place of Wiraqocha Inka's palace (Bauer 2004:122–124). The southwest corner of the plaza was occupied by the Amarukancha compound, probably built by Waskhar (Bauer 2004:124–125); and separated from it by Loreto Street stood the Aqllahuaci compound where the *aqllas*, young women selected for state and religious service, lived (Bauer 2004:128–129). Hatunkancha compound was located on the southeast corner of Awkaypata, and Garcilaso identifies it as the palace of Pachakuti Inka Yupanki (Bauer 2004:130–134).

Some of these palace compounds included kallanka buildings. A kallanka is generally defined as a great hall or long structure with a gabled roof supported by a row of pillars set the entire length of the long axis and exhibiting numerous door openings facing a plaza (Gasparini and Margolies 1980:196). Garcilaso de la Vega provides the most detailed description of the Cusco kallankas (1963:198, 260–262 [1605:Bk.6, Ch.4; Bk.7, Ch.10]). He talks about four great halls, the largest of which was the Qasana compound situated at the northwest corner of Awkaypata.

Although the urban design of Cusco was probably Pachakuti's brainchild, the brief foregoing investigation shows that the capital was built out and added to by his successors. Some of Pachakuti's political intentions may further have been connected with the carved rocks. Although it may not be possible to date these works to the reign of a particular ruler with absolute certainty, a solid argument can be made that Pachakuti included many sculpted boulders intentionally in the design of his personal sites. The evidence is twofold. First, the impressive carved outcrops of Q'enqo were likely part of the site in which Pachakuti maintained a house dedicated to his cult and where he eventually died (for a full discussion, see chapter 2). Second, carved rocks form important and integral features at Pachakuti's royal estates, such as Pisaq and Machu Picchu (see chapter 5 below). I argue that Pachakuti elevated rock art to a formal language which would express political concepts of stone and state ideology.

In terms of military conquests, Pachakuti and his son, Thupa Inka Yupanki, together accumulated a territory which reached from Cusco in the center to Tumipampa and north of Quito in modern Ecuador, and in the south beyond Santiago de Chile (figure I.2). Stretches of the central and southern coasts were not part of the empire (D'Altroy 2002:65–74; Paerssinen 1992). Pachakuti's successors continued to focus on territorial expansion. Thupa Inka Yupanki is widely seen as the conqueror par excellence. He added vast territories to the east into the lowland jungle areas, southeast, and south beyond today's Santiago de Chile into northwestern Argentina. In the north, Thupa Inka conquered Quito, and on the central coast, he annexed the Lima area. Like his father, Thupa Inka made rock art a vital element of his sites (see chapter 6.

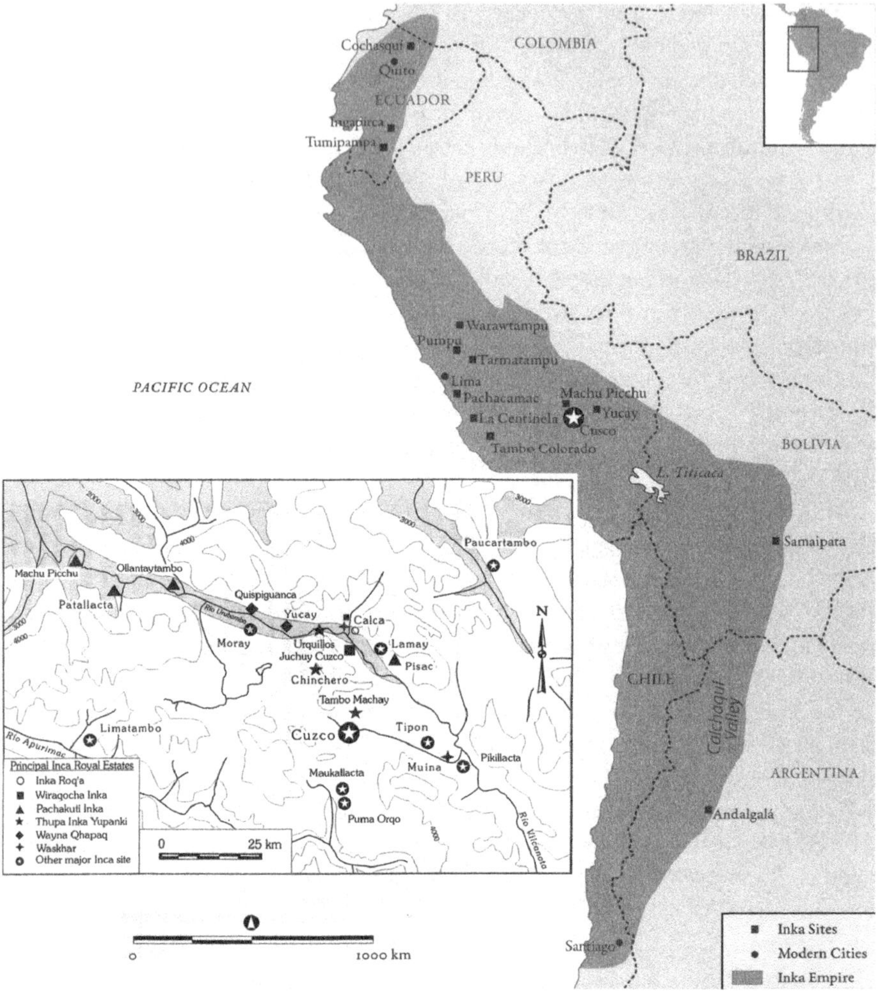

Figure I.2. Map of the Inka Empire. The insert shows the Cusco region and the Urubamba Valley with royal estates marked according to D'Altroy 2002 at an enlarged scale. Adapted from Burger, Morris, and Matos Mendieta 2007 and D'Altroy 2002.

Wayna Qhapaq succeeded Thupa Inka in 1493. He was said to have made military and administrative affairs equally important priorities of his government. Part of his military campaigns involved securing borders, but he also advanced in the north beyond Quito to a site named Pasto and in the southeast beyond Samaipata (Paerssinen 1992). The southeastern border line is of special interest to this study because it affects the role Samaipata played in relation to the border zones (see Christie 2002, 2005b, 2014; Meyers 2007:238–242). Thus the Inka Empire reached its widest territorial expansion under the reign of Wayna Qhapaq. While he was fighting in the north, an epidemic broke out, and Wayna Qhapaq himself died from it in 1528. His heir designate, Ninan Cuyuchi, experienced the same fate, which precipitated a deep conflict over succession rooted in the competition among the panaqas.

Waskhar and Atawallpa, two sons of Wayna Qhapaq by different mothers who belonged to competing panaqas, fought over the title of rulership bitterly almost until the arrival of the Spaniards in 1532. Most chroniclers agree that Waskhar became emperor in Cusco with the consent of the nobility. His half-brother Atawallpa established a base of power in the north, now Ecuador, where part of the Inka army was stationed and challenged the emperor's authority. Waskhar responded with military action. Isolated attacks turned into battles waged for years. It has been estimated that at least one-third of the population of Tawantinsuyu had been killed by war and disease just prior to the Spanish arrival, leaving the Inka state in a weakened condition (D'Altroy 2002:78–82; Titu Cusi Yupangui 2005[1570]:xii). In the end, Waskhar was captured, and Atawallpa came out of the war victorious. In Cajamarca, Atawallpa received word of the arrival of the Spanish at Tumbes on the north coast. The Spaniards under the leadership of Francisco Pizarro and Diego de Almagro quickly made their way up the Andes and arrived at Cajamarca in November 1532. Atawallpa was taken prisoner and eventually convicted of treason and executed in July 1533 (Titu Cusi Yupangui 2005[1570]:xii–xiii).

The Spaniards marched on to Cusco where they entered unopposed on November 15, 1533. Pizarro gave most of Cusco's palaces and other buildings to members of his own party, and they all set out to strip the Inka capital of its wealth. Most widely talked about were the gold and silver objects looted from the Qorikancha. In December of the same year (1533), Manqo Inka was installed as the new Inka ruler following traditional ceremonial protocol. Manqo Inka was one of Wayna Qhapaq's sons who had escaped the massacre inflicted by Atawallpa's men, and Pizarro believed he would be a compliant puppet through whom they could govern. But Manqo Inka developed higher ambitions. He managed to escape from Cusco, assembled an army, and became leader of the resistance. In 1536, his forces laid siege on Cusco while Francisco Pizarro was on a visit in Lima. During the fighting, the Spanish were far outnumbered yet managed to storm Saqsaywaman, the fortified citadel above Cusco. In the end, they held the capital, and an Inka attack on Lima (Ciudad de los Reyes) failed as well. The resistance movement did not give up but went underground into the eastern lowland jungles. After his initial escape from Cusco, Manqo Inka had founded Vilcabamba (identified today as the archaeological site of Espiritu Pampa)[2] as his new capital of the

Inkanate where he organized an indigenous counterforce to the emerging colonial states in Cusco and Lima. For thirty-six years after the unsuccessful siege of Cusco, the Inka maintained an independent state at Vilcabamba where they enjoyed their liberty, fomented plans to regain their lands, and organized campaigns against the invaders. Vilcabamba was governed by three rulers, as Manqo Inka was succeeded by Titu Cusi, who in turn was followed by Thupa Amaru. It is significant that sacred rocks were again part of the neo-imperial design. At Vilcabamba, a huge unmodified boulder sits next to the kallanka fronting the plaza. At the entrance to the Vilcabamba region, a hillside next to the palace architecture of Vitcos is activated with exquisite examples of Inka rock art. I suggest that Manqo Inka and his successors understood sculpted stone and perhaps sacred rocks in general as the material markers of the presence of the Inka state. This is why it was so important to carry this tradition to the new capital even though natural stone resources are limited in the lowlands.

For decades, the Spaniards were unsuccessful in breaking the resistance. Finally Viceroy Francisco de Toledo led an expedition authorized by Charles V and conquered Vilcabamba in 1572.

The imposition of stable Spanish rule was a slow process. After the passage of the Leyes Nuevas (New Laws) in 1542, which placed the administration of the Spanish American territories under the authority of officials appointed by the Spanish Crown and not by the conquistadores, the Crown dispatched a series of representatives titled viceroys to Peru to establish order. Civil administration was slowly built up. Most notable was Viceroy Francisco de Toledo for subjugating Vilcabamba (see above) and for his sweeping social reforms. Toledo's views were grounded in the argument that the Inka Empire was an illegitimate monarchy which had inflicted injustices upon their subjects, and therefore Spain had a moral mandate to eliminate their barbarity. His views translated into the creation of *reducciones*, which were European-style villages and towns into which Andean men and women living in dispersed hamlets were relocated (Titu Cusi Yupangui 2005[1570]:xvii). Although Toledo's policies aimed at eradicating traditional lifestyles, he and many of his contemporaries began to take a closer interest in Andean culture, partly in order to be better informed about how to subdue it.

BACKGROUND OF SOURCES

Summary of Ethnohistoric and Ethnographic Sources

The list of written colonial sources about Inka life, culture, and history is long and complex. Researchers typically resort to annotated editions of historical source material. Raul Porras Barrenechea produced some of the most ambitious compilations of sources concerning the history of Peru from pre-Inka times into the twentieth century (1962). What stands out in his work is the inclusion of Inka methodologies of transmitting history by means of khipus, writing (he references Quechua and Aymara words for writing/painting, Porras Barrenechea 1963:109–112) and the performance of *haylli* songs (1963:141–145). The most recent and invaluable help to any twenty-first century scholar in sorting through documentary sources about the Andes written between 1530 and 1900 are the three volumes edited by Joanne Pillsbury (2008). Volume 1 contains summary chapters discussing categories and topics of texts as well as the Spanish institutions which commissioned them. One essential merit of this volume is to draw the reader's attention to lesser-known documents concerning aspects of indigenous cultures not mentioned in the commonly cited chronicles of European-born writers. Volumes 2 and 3 contain entries on specific writers and documents in alphabetical order. Contributions were made by 124 scholars from nineteen countries, which were in turn reviewed by an international advisory board, reflecting the multicultural perspectives of the completed *Guide*.

In general, a major problem is that eyewitness accounts (ethnohistoric)—from about the 1530s to 1550—were written by soldiers who were part of the invading force under Francisco Pizarro and Diego de Almagro, without reflection or understanding of what they were confronting in an alien land. Their accounts are factual and lack personality (Porras Barrenechea 1963:148–149). During these first decades of Spanish presence in the Andes, the difficulties of communication were staggering since indigenous people and Europeans did not understand each other's languages; translators still had to be trained, and new words were pronounced and used as well as written down the way individuals heard them.

Writers in the mid- to late-sixteenth century broadened their interests from an account of the conquest and the Spanish experience to an investigation of Inka culture. Pedro de Cieza de Leon (1553) wrote four volumes on Inka culture; Juan de Betanzos (1557) presented Inka history from the viewpoint of Pachakuti's descendants. The Licenciado Polo de Ondegardo was a highly informed administrator in Cusco. His writings were motivated by the desire to understand how indigenous practices and institutions worked so that he could better manage the native population. He may have been the author of the original zeq'e system manuscript (see below), and in 1559, he discovered the hiding place of the royal mummies. While the early conquistador-writers generally implied that native peoples belonged to a race inferior to the European and therefore their civilization by force was justified, the authors of the later and more reflexive chronicles, for example Cieza de Leon, expressed respect for indigenous societies (Porras Barrenechea 1963:148–152). Other sixteenth-century authors, such as Cristobal de Molina and Cristobal de Albornoz, will be discussed in the context of the zeq'e system (see below). The latter prided himself to have personally destroyed over two thousand native shrines in the Huamanga (now called Ayacucho) region (D'Altroy 2002:13).

A driving force in the endeavor to increase knowledge about indigenous peoples was Viceroy Francisco de Toledo, who between 1570 and 1580 not only implemented the colonial administrative system, but also sponsored numerous authors and colonial "fieldworkers" to collect information about Andean cultural practices. Their reports have been published under the generic title "Informaciones." Toledo inspired another ideological shift projected in the chronicles authored during his time, which constructed native societies as barbaric and Inka rule as unjust (Porras Barrenechea 1963:153–154). After Toledo, chronicles show a tendency to idealize Inka life, partly because some writers were the offspring of Inka and Spanish marital unions. Franklin Pease (1995:45–48) emphasizes that Spanish authors in the early seventeenth century (ethnographic) could no longer work with eyewitnesses and had to rely on the memory of their consultants about the previous generation or on earlier written sources. This chronological distance may have been one factor that led them to frame Andean history by the universal vision of European history whose origins are anchored in the Bible.

The most important native writers are Garcilaso de la Vega and Felipe Guaman Poma de Ayala. While both had indigenous blood, they both were also highly educated according to Renaissance European and Spanish standards. Garcilaso was the son of a Spanish soldier and an Inka princess and lived in Cusco until 1560. At that time when he was still very young, he moved permanently to Spain, where he wrote his *Royal Commentaries of the Incas* which he completed in 1609. Garcilaso was naturally interested in establishing the legitimacy of his Inka ancestors and presented the Inka state as a benevolent and stable monarchy in the European sense (see D'Altroy 2002:14–15; Pease 1995:42). Guaman Poma de Ayala was also of mixed ancestry. In 1613, he finished a long letter to King Carlos V—a letter that took on the format of a book and was illustrated by hundreds of drawings of Inka history, religion, and customs as well as of Spanish abuses (1987[1615]). His drawings provide an irreplaceable source of visual detail to the text. At least two drawings depict carved rocks: in one, an Inka ruler converses with sculpted stone idols while in another, an Inka ruler is seated upon a throne or *usnu* which closely resembles a boulder with fine vertical and horizontal cuts (see below). At the same time, Guaman Poma strongly expressed the universal European vision of history that all human beings are descendants of Adam and Eve and from this starting point, he divided Christian and Andean history into a set of ages forming an oddly matching construct. In 2004, the facsimile of the Galvin manuscript attributed to Martin de Murua was published; it contains 113 ink and watercolor illustrations exhibiting striking similarities in style and detail with Guaman Poma's drawings (Murua 2004[1590]). Recently, Guaman Poma's authorship of the *Nueva crónica y buen gobierno* has been questioned, and there is an ongoing debate about the identity and relationship of the illustrators of the *Nueva crónica* and the Galvin manuscript (Dean 2006b:95–96; Ossio 2004).

Seventeenth-century chronicles further gain in descriptive details. The texts most relevant for this study are those composed by priests mentioning sacred stones or stone idols which might refer to carved rocks and which the priests aimed to destroy. One important religious writer was Alonso Ramos Gavilan. He was born in Huamanga, traveled and worked in different locations in Peru, but then stayed in Copacabana, Bolivia, for an extended period of time. His most important written work is titled *Historia del celebre Santuario de Nuestra Senora de Copacabana* and was published in Lima in 1621. While his focus is clearly on the Lake Titicaca region and the pilgrimages to the Island of the Sun, he mentions and describes idols and wak'as in other parts of the Andes, some of which were sacred stones.

One of the latest chroniclers was the Jesuit priest Bernabe Cobo. Born in Andalusia, he visited Mexico and spent most of his adult life in Peru. In 1653, he completed his multivolume work on Inka history, religion, and customs (1990[1653]). Since his writings postdate the conquest over a hundred years, he had to rely on informants who recalled the generation that preceded them. The greatest value of Cobo's publication lies in the fact that he still had access to important manuscripts which are now lost, such as the original manuscript of the Cusco zeq'e system (see chapter 2).

Further, virtually inexhaustible sources of information about the land and people of Peru rest in Andean and Spanish archives. Beginning in the early Colonial Period, the conquistadores were required to send back reports of what they discovered and saw in the new lands. Later in the sixteenth century, specific guidelines for such reports were issued, and the Spanish Crown ordered formal inspections, so-called *visitas*, of its holdings region by region. The inspectors had to use a standardized series of questions about life before and under the Inka and—of particular interest to the Crown—about the natural resources of each region (D'Altroy 2002:20; Jimenez de la Espada 1965[1881–1897]:XLVII–LII). More visitas were conducted under Viceroy Toledo's vast restructuring program in 1570–1572 (mentioned above). The final goal was total and universal: to obtain a complete knowledge of the lands and cultures of the New World directly from the indigenous people—a knowledge which would facilitate Spanish administration and optimize economic exploitation of the new provinces. Although the idea was grand, its realization was carried out as a piecemeal effort to which hundreds of individuals contributed over decades. Spanish agencies collected a vast amount of records in the form of reports, notes, maps of varying levels of detail, which were warehoused. Starting around 1570, efforts were organized to turn all this information into book format with the summary title *Relaciones de las Indias*. Many partial editions have been published over the centuries; the most useful volumes for scholars of the Andean region were the work of Don Marcos Jimenez de la Espada completed in the late nineteenth century. As a student, Jimenez de la Espada was an explorer of South America; and upon his return to Madrid, he became a professor and scholar of the natural sciences. Later in his career, he turned his attention to the history of South America. In 1881, the International Congress of Americanists charged him with editing the *Relaciones de Indias*; and in 1897, Jimenez de la Espada published four volumes of the *Relaciones Geográficas de Indias. Peru*. His work represents an exhaustive, critical, and well-documented edition, which includes annotations, appendices, and an index and remains the single edition most widely consulted in the academic literature.

Another set of useful archival sources are court records. Andean peoples quickly began to use the Spanish courts to launch a variety of claims, for example, against Spaniards for recompensation of services or against neighbors to settle grievances. Most valuable are documents referring to land tenure. Many complaints and court petitions arose when local groups tried to regain lands that had been taken by the Inka and given to colonists resettled by the Inka state as well as from competition over the inheritance of land. Such land documents have been helpful in identifying local wak'as in the Cusco zeq'e system (see the work of Brian Bauer 1998 below).

Previous Reports and Studies of Carved Rocks

Accounts by Explorers and Early Scholars

In the late nineteenth and early twentieth centuries, several European and American explorers traveled to South America, and some of them left written records and oftentimes detailed descriptions of archaeological sites including invaluable illustrations of the ancient ruins. They were driven by a mission to investigate a past so unlike their own, which often colored their accounts and drawings. One of these explorers was George Ephraim Squier who traversed Peru and the Lake Titicaca region in Bolivia in the mid-1800s. He published a detailed account of his travels divided into chapters covering distinct regions (1877). In his chapter on Saqsaywaman, Squier describes some of the stone carvings, mentions specifically the Piedra Cansada, and concludes:

> Its top, like the tops of hundreds of other rocks on the plateau of the Rodadero, is cut into what appears to be seats and reservoirs of every shape; its sides are cut into niches and stairways–the whole a maze of incomprehensible sculpture and of apparently idle although elaborate workmanship (Squier 1877:475).

His last sentence indicates that he found himself confronted by a piece of material culture to which his memory could make no connection; he had to concede to the mystery of this stone sculpture and declare it "incomprehensible" and "idle." While many of Squier's descriptions of sites go very much in depth, he was too quick in labeling and interpreting structures without much evidence.[3]

His drawings, for example, the one of the Sacred Rock near Calca with a sculpted channel terminating in the head of a serpent (1877:518), constitute important historical records; and some of his arguments, such as his explanation of the use of *Inti-huatanas* (1877:524–529), demonstrate his close familiarity with many chronicles.

Another interesting nineteenth-century explorer was the Frenchman Charles Wiener, who published his account of travels in Peru and Bolivia in 1880. Like Squier, he visited many Inka sites containing carved rock complexes: for example, Sayhuite, Ollantaytambo, Copacabana, the Island of the Sun, and Cusco. His descriptions are keen and illustrated by a large number of drawings which constitute important historical documents. I have been able to identify and verify the sculpted rocks in many of his drawings but in others not. The question arises whether those latter carved rock complexes have been modified in the 120 years or so that have elapsed between Wiener's visits and mine or whether some of his illustrations were inaccurate to begin with. I am inclined to assume the latter since some of the drawings look quite imaginary: for example, his depiction of the Inkamisana sector in Ollantaytambo (Wiener 1880:542). Unlike Squier's interpretations, Wiener's generally do not reference the chronicles and only cite local stories and legends.

In the early twentieth century, the academic community, for the first time, took an interest in the carved rocks. Max Uhle, a German professor who had conducted important fieldwork at many Andean sites including Pachacamac and Tiwanaku, presented a seminal paper at the International Congress of Americanists in Vienna in 1908. His paper, titled "Zur Deutung der Intihuatana" (1910), discussed eleven carved rock complexes located east of Saqsaywaman; and he arrived at two conclusions: first, the sculpted outcrops were physically linked to burial sites in caves and the carved shelves, platforms, and steps were used in rituals serving the ancestors; second, he noted the knob-like carvings or gnomens at many of these sites and concluded that they were part of simple altars of the sun and did not function as observatories themselves. Uhle based his first conclusion upon the form of carvings, such as niches and platforms, which could have accommodated mummies and on his finds of bones in some caves associated with or nearby certain sculpted rock complexes (for example, Q'enqo Grande, Luis Valcarcel 1935c, cited by van de Guchte 1990:146–147, chapter 2). Uhle's second conclusion is important because it speaks against the still popular interpretation of sculpted knobs as *intiwatanas* or markers of the sun first proposed by Squier (see John Rowe cited by Wright and Valencia Zegarra 2001:53–54). Uhle simply describes how complex the observation of shadows cast by the knobs is and that there are no clear patterns. Uhle's study constitutes a first academic attempt by a European scholar to make sense of an Inka past by applying twentieth-century methods of accumulating a data base of material objects, identifying patterns, and drawing generalizing conclusions.

An important investigation of the Sacred Rock on the Island of the Sun in Lake Titicaca was conducted by the Swiss-born Adolph Bandelier (1910), who spent nine months on the Islands of the Sun and Moon surveying, excavating, and consulting with local residents. Bandelier combined his field methods with an in-depth knowledge of the ethnohistoric and ethnographic literature, producing an academic book which continues to serve researchers today as a valuable resource. Bandelier drew early plans and architectural drawings of Inka sites and photographed ceramic finds. One example of his close familiarity with ethnohistoric sources is the reference to the account of the first Spanish visitors who saw the Sacred Rock in July 1534. They do not mention that the rock was covered by plates of gold and finely woven textiles as was reported by seventeenth-century writers. Further, Bandelier shares interesting data about artifacts associated with the Sacred Rock. From its surroundings came many items included in the so-called Garces collection, such as gold and silver figurines, small pins of gold and silver, as well as five stone chests containing ponchos.

Critical for this book are the numerous publications by Hiram Bingham (1922, 1979), who rediscovered Machu Picchu for the modern world in 1911 and authored studies about Machu Picchu which are available in various editions, Vitcos, Vilcabamba the Old, and Choqequirau which will be cited in the corresponding chapters of this book. Bingham was a professor at Yale University and directed several expeditions into the Machu Picchu and Vilcabamba regions. He discovered and/or was one of the first twentieth-century visitors at these remote Inka sites; he had the ruins cleared, produced accurate maps, conducted excavations, and collected artifacts.[4] The most detailed and thorough work Bingham performed was at Machu Picchu itself. His style of writing, at times, includes elements of an explorer's story and adventure accounts which make his books appealing to the general reader. Yet his descriptions of buildings and masonry techniques are always meticulous, include practical commonsense approaches, and lead to interpretations and discussions evidencing his solid familiarity with the chronicles and published sources at the time of his writing. For example, his analysis of the surface of the sacred rock in the Semicircular Temple (Torreon) leads him to conclude that an excessive amount of flaking occurred due to repeated and/or long burning fires (Bingham 1979:93). Bingham does not see the degrees to which the Inka modified stone as the material expressions of Pachakuti's state ideology as will be argued in this book. Rather, he dismisses walls that had no logical purpose as products of people's pastime: "That (the signal station) on top of Machu Picchu . . . seems to have been occupied by a number of men who passed their spare time in building walls on the very edge of one of the most stupendous precipices in the Urubamba Canyon" (Bingham 1979:41).

ACADEMIC DISCUSSIONS OF SCULPTED ROCKS USING VARIOUS METHODOLOGIES

In the 1930s, Cusco archaeologist Luis Valcarcel and colleagues conducted important reconnaissance work at a number of carved rock complexes around Cusco and in the Urubamba Valley. They photographed the rock art, measured the associated architecture, and drew sketches and maps of their location. They published descriptive reports of the following sites without considering the cultural context in which these sites might have functioned: Suchuna, which is a carved rock area within Saqsaywaman (Valcarcel 1935a), Sapantiana (Valcarcel 1935b), Q'enqo (Valcarcel 1935c), Inkamisana at Ollantaytambo and Choqueilla (Llanos 1936), Tetekaka, Kusilluchayoq, Lacco, and Lanlakuyoq (Franco Inojosa 1937), and Quillarumi (Franco Inojosa 1939). At the time of this writing, the same carved rock complexes within the zone of Saqsaywaman are being excavated by the Ministerio de Cultura and new data, especially about associated artifacts and architecture, are forthcoming. Valcarcel's work is of further importance because it constitutes the first efforts by Peruvians as opposed to foreigners to investigate the past of their country.

A remarkable study of Inka sculpted rocks was published by Walter Krickeberg in 1949. Krickeberg was not solely interested in Inka rock art but covered rock sculpture throughout the Andes as well as architecture hewn in the living rock in Mexico. Krickeberg's work is compelling by his intimate familiarity with the chronicles and the published sources of his time. He begins with the origin myths which feature sacred rocks and with the zeq'e system of Cusco which included many carved and uncarved rock wak'as. He continues to describe and discuss the sculpted rock complexes around the Inka capital and then moves to further outlying sites, such as Machu Picchu, Vitcos, Conchaca/

Sayhuite, and finally to the northern- and southernmost manifestations of Inka rock carvings at Ingapirca in Ecuador and Samaipata in Bolivia. In his discussions, Krickeberg takes up the problem of intihuatanas/intiwatanas and whether the Quechua name "intiwatana" meaning "place where the sun was tied" can be interpreted as explaining the function of this group of carved rocks as markers of the sun's movements by the way they cast shadows. Another problem Krickeberg addresses is to what extent Inka stone carving style and techniques were copied and adapted from Tiwanaku. After reading his work in detail, I think Krickeberg's excellent overview of Inka sculpted outcrops is still relevant today because no fundamentally new or different insights have been gained into the topic since then. Yet his study is outdated with regard to the methodology of analyzing carved rocks isolated from their site context. Since Krickeberg's writing, excavations have added important archaeological data from many Inka sites (for example, Machu Picchu, Sayhuite, Samaipata) exhibiting carved rocks and they can now be placed into a larger cultural context, which is the approach practiced in this book.

The methodology of Heinrich Ubbelohde-Doering (1952) is that of a formalist. He makes statements about Inka culture based upon the formal or stylistic elements of their works of art. Most useful are his high-contrast photographs of major rock art sites, stone sculpture, and aryballos. Stunning is the illustration of a human head sculpted in stone 38cm high which Ubbelohde-Doering claims was found near Q'enqo at the outskirts of Cusco. I have not seen this stone sculpture in any other source and if it truly is Inka, it would be one of two known sculpted stone heads (the other example is illustrated by van de Guchte 1990:417, fig. 15).

The art historian George Kubler briefly addresses sculpted rocks in his survey of the art and architecture of ancient America (1962:320–321). He notes the abstract way in which the carvings represent man and lists several functions of these modified boulders and outcrops: wak'as, petrified remains of the ancestors, and places where mythological events occurred. He interprets Q'enqo as the likely burial place of the ninth ruler, Pachakuti. Kubler's clear vision makes the reader aware of the complexity of roles Inka rock art assumed which scholars only now investigate in greater depth, as this book sets out to do.

The historian John Hemming and the photographer Edward Ranney collaborated to produce a work about Inka monuments which combined unconventional methodology and great aesthetic ambitions (1982). Hemming provided the rich and thoroughly documented text, and Ranney added photographs of Inka carved rocks and architecture which are intended to capture Inka spirituality, the Inka sense of visual space, and sacred landscape.

Susan Niles's study of *Callachaca* (1987) includes another important contribution to our understanding of carved rocks. She identifies the site as a royal estate owned by Amaru Thupa Inka, a brother of Thupa Inka. Her methodology is based on the study of ethnohistorical accounts and a close observation of standing architectural remains and modified landscape features, which involves mapping but no excavation. Her book contains one chapter on "Shrines and Holy Places" in which she tentatively identifies shrines on the fourth, fifth, sixth, and seventh zeq'e lines of Antisuyu, some of which are modified rocks (chapter 2). Her work can be considered one of the earliest in a series of archaeological and theoretical studies which have looked closely at the nature of the zeq'e system and how it formed an important layer of the sacred landscape the Inka formulated around Cusco.

In the late 1980s, Victor Angles Vargas (1988) published three authoritative volumes titled *Historia del Cusco Incáico* in Cusco. Using his knowledge of Quechua and oral traditions as well as his study of the Spanish chronicles, he set out to bring Inka history to a wide audience by taking the reader on a thorough tour of Cusco City and many Inka settlements in the Department of Cusco. His detailed descriptions based upon personal familiarity with the sites, his close reading of the Spanish documents, and the linguistic history of toponyms make his *Historia del Cusco* indispensable reading material for anybody studying and exploring the Inka heartland. His style of writing transmits his passion for Cusco's cultural heritage to the reader. On the other hand, Angles Vargas includes few archaeological data; his work is completely outdated when compared to the new information obtained by archaeologists through excavation and large-scale survey projects in the Cusco area.

John Hyslop (1990) followed with his impressive analysis of *Inka Settlement Planning*. He investigates all types of Inka settlements throughout the empire and focuses on their specific relation to the landscape. His book includes one chapter on "Rocks and Outcrops" and another one on "Water." This is one of the sources my work is based on; my goal is to expand his discussion of the carved boulders, analyze site contexts in greater detail, and present certain patterns which are not strictly tied to settlement planning issues.

In the same year, Maarten van de Guchte completed his dissertation titled "Carving the World": Inca Monumental Sculpture and Landscape (1990) at the University of Illinois at Urbana-Champaign under the direction of Tom Zuidema. He was the first to draw scholarly attention to Inka rock art as objects worthy of serious study, which are not simply esoteric and oftentimes crude stone modifications but constitute vital links between nature and culture in Andean worldview. His is the most thorough and comprehensive work about Inka carved rocks that has so far been written, but it was never published. My book certainly builds on van de Guchte's dissertation and will lead beyond it by, first of all, including new archaeological data; and second, by placing greater emphasis on settlements with sculpted rock complexes outside the immediate Cusco area.

At the same time, interesting studies have come out of Cusco. In 1994, Julio Palomino Diaz published a complicated argument about a grid-like carved rock complex situated in the Suchuna area of Saqsaywaman. First, he relates the carvings to pre-Inka and—what he thinks—pan-Andean measuring units as represented in Chavin de Huantar, Cumbemayo near Cajamarca, and in counting devices (*yupanas*); second, he demonstrates how this rock art site may have functioned as a sophisticated astronomical observatory that could structurally be reduced to the defining elements of an Intiwatana as exemplified by the famous example known from Machu Picchu. Diaz assumes the Inka realized that the earth is round and ties the rock art site into complicated measurements of distances between planets, calculated from the position of the sun and all multiples of the pan-Andean measuring unit. His arguments have found little support.

Brian Bauer published his influential study of the zeq'e system in 1998. The Cusco Ceque System Research Project documented shrines and reconstructed zeq'e lines through ethnographic and archival research as well as field surveys. Bauer and his colleagues redefined the zeq'e system by locating shrines on the ground and demonstrating that many lines were not straight as had previously been thought but bent and zigzagged. Of the 328 shrines or wak'as in the zeq'e system, 95 or 29 percent are standing stones, many of which were carved.

The most recent sophisticated study of Inka rockwork was authored by Carolyn Dean (2010). She analyzes Inka *culture of stone*, which complements my theme of *stone ideology* in interesting ways. She discusses rock wak'as from the visual perspectives of framing (by a rectangular masonry frame), distancing (situated in a spatial void), contouring (integrated into architecture which encircles), and carving. Her book includes a comprehensive chapter about Rock and Rule which parallels and complements my concept of stone ideology. A significant difference between our approaches is that my book focuses on case studies. Both of us reflect upon the roles of empowered rocks in colonial and contemporary Peruvian societies.

Besides the above works which are at least partly about carved rocks, more general theoretical landscapes studies have been written in the past decades, which are useful in situating the sculpted boulders within nature. These authors stress the important point that landscape and nature are not neutral and universal physical entities but are always socially constructed. They are not only physically modified by men, but they exist as mental images in people's minds derived from personal and communal experiences and emotions. In one sense, the natural setting is the passive recipient of human action, which enters into our conceptual construction of nature. This mental image of nature actively shapes our relations with it. Two authoritative edited volumes are by Wendy Ashmore and A. Bernhard Knapp, *Archaeologies of Landscape*; (1999) and by Bruno David and Meredith Wilson, *Inscribed Landscapes* (2002). Both volumes cover a range of topics with many specific case studies. Van de Guchte contributed one chapter on Inka carved rock wak'as to the Ashmore and Knapp volume. Here he (1999) develops the notion of an "aesthetic of alterity" as opposed to rigid patterns in the Inka cognition of landscape and their treatment of wak'as.

NOTES

1. Although fully recognizing the eurocentric relativity of the terms *nature* and *culture*, I will continue using them being aware that they are problematic as outlined in this introduction. The two terms are simply so ingrained in the English language that substituting them would cause more confusion.

2. In 2010, archaeological investigations at Espiritu Pampa conducted by the Ministerio de Cultura of Peru unearthed the burial of an important Wari leader. Wari presence and the identification of a likely ruler in the lowland jungle regions of Vilcabamba were wholly unexpected and have dramatically changed academic reconstructions of Wari and Inka histories and relations between them. It now appears that Manko Inqa did not establish his capital in no-man's-land but reconnected with a Wari place of power.

3. In his report on Ollantaytambo, Squier identifies isolated buildings on the slopes of mountains surrounding the village as "School of the Virgins," Horca del Hombre from which supposedly male criminals were thrown, prisons, and an Horca de Mujer solely based upon their location and architectural form (Squier 1877:502–503). Most of these buildings are now understood as storehouses.

4. The collections accumulated by Hiram Bingham were taken to Yale University and stored in the Peabody Museum. They have been the source of disputes and litigation between Yale and the Peruvian Ministerio de Cultura over the return of these objects to Peru. As part of the one hundredth anniversary celebration in 2011 of Bingham's "discovery" of Machu Picchu, a new museum has been established in Cusco which is now displaying many of the Inka artifacts.

Chapter One

Formal and Structural Analysis of Inka Carved Rocks

This part of the book will address the technical, formal, stylistic, and structural aspects of carved rocks. In the first section, I will reflect upon how the Inka may have sculpted rocks. Nobody has studied work marks on the carved rocks themselves, but Jean-Pierre Protzen has conducted thorough studies of and experiments with Inka quarrying, cutting, and stone-laying techniques (1985, 1993).[1] I will use his work and that of others to draw certain comparisons and conclusions about the sculpted rocks. Questions about techniques will lead to chronology. The Inka did not begin cutting stone in a historical vacuum but responded to stony remains of societies which preceded them. Section two will consist of a brief historical overview of stone work in the Andes focusing on Chavin de Huantar and Tiwanaku. The chronicles mention that Pachakuti chose the latter as a model when he set out to rebuild Cusco as the imperial capital. The last two sections will present the formal vocabulary and associated structural features of Inka rock art. My approach is formal, empirical, illustrative, but also selective. The goal is to give the reader a vivid idea of what Inka sculpted rocks looked like and of their variety.

Carving Techniques

In the 1980s, Protzen conducted field research in the Inka quarries of Kachiqhata and Rumiqolqa to better understand how Inka builders constructed their well-known stonemasonry of cut stones which were put together without cement and with utmost precision. The quarries of Kachiqhata are situated across the Urubamba River from Ollantaytambo and about 4km to the southwest as well as approximately 700 to 900 meters above the valley floor. They provided the rose-colored rhyolite used in the most prestigious construction of Ollantaytambo. Rumiqolqa lies about 35km southeast of

Cusco, on the left bank of the Vilcanota River (which changes its name to Urubamba in its lower course) and nearby the Inka road to Kollasuyu. Here the Inka mined andesite used in the buildings of Cusco. Protzen covers quarrying, cutting and dressing, transportation, and fitting and laying techniques. I will focus on those aspects which have relevance to the sculpted rocks.

I concur with Protzen that the choice of a particular rock type must have been of great importance to the Inka in the production of building blocks as well as of carved rock sites. Van de Guchte (1990:86–87) also comments on the selectivity of stone material in the construction of Cusco. With regard to the boulders sculpted in situ, the type of rock, of course, depends upon the geological formations they are part of. Thus many of the boulder sculptures on the outskirts of Cusco consist of limestone, and Machu Picchu is mostly built of granite. These stone materials are again different from the rhyolite and andesite Protzen studied and experimented with at Kachiqhata and Rumiqolqa. Because of the differing stone types, Protzen's observations may have limited applicability to the carved rocks. Further, scholars have attempted to correlate the distribution of sculpted outcrops throughout the Inka Empire to the local availability of certain stone materials. These issues have not been carefully studied and remain unresolved.

Protzen discusses that the extraction methods depended, again, on the type of stone material. For example, Kachiqhata lies in giant rockfall areas, and the quarrymen simply selected loose blocks that met their specifications. To part large blocks, they cut an ever-deepening groove around the blocks until they were separated (Protzen 1993:166). At Rumiqolpa, the andesite is sufficiently fractured that it can easily be broken out of the bedrock by using pry bars of bronze or wooden sticks. George Squier (1877:419) was the first to reason that the Inka cut a groove on the line of a desired fracture along which a series of holes were dug or chiseled out. He suggested that wedges of dry wood were driven into these holes, and enough water was poured into the groove to soak the wooden wedges. As their mass expanded due to the water, they helped break the rock. There is some physical evidence for this method in one block at Ollantaytambo and very good evidence in a split-rock in the quarries of Machu Picchu. However, Protzen (1985:169–170) argues that the holes in the latter were created by a metallic chisel in post-Inka times. He does not rule out the application of this technique by Inka quarrymen and sculptors, and it could well have been used to make the initial vertical and horizontal cuts in the sculpted outcrops.

Lisbet Bengtsson (1998:19–20, 44–47) surveys the literature for other possible extraction and splitting techniques which include using fire and then applying cold water, or first pounding pits into the block's surface or a V-shaped notch to create a splitting groove or to enhance natural fissures. A related proposal is that a slit was made in the rock surface, the slit was filled with some mineral or metal that expanded on heating, then it was covered up and heated, which produced a crack in the stone (E. Frank cited in Bengtsson 1998:19). Others have suggested that freezing water or urine, which is a natural organic acid, poured into cracks and fissures would have assisted in splitting rocks. The consensus is that Inka stone workers used a variety of methods to extract and split stone, which would have been applicable to quarrying building blocks and sculpting outcrops in situ. Each individual case required its site-specific method depending on the rock type and the orientation of the bedrock; for instance, when artisans worked to cut a niche out of a vertical rock wall, it would not have been practical to fill fissures with any kind of solid or liquid substances because they would have drained out. In such cases, the channel and wedge technique would have been the most effective.

The process of sculpting rock outcrops in situ stands in a reverse relationship to the process of quarrying building blocks: in the former, the end product is what is left standing on site, while in the latter; the end product is the stone block which is transported to a construction site. Indeed, nobody has directly asked whether the stone material extracted from the carved rock complexes was turned into building blocks. It clearly did not remain near the rock art and was taken away; nor have any stone working tools been found next to carved rocks.[2] It is conceivable that the Inka created and used some carved rock complexes as wak'as and as quarries all in one. Susan Niles (1987:102–105, 183–188) discusses the quarries of Choquequirau as one such example. As I will address throughout this book, the thought pattern of clearly defined black and white categories is a Western construct which developed in ancient Greek philosophy and was refined during the Enlightenment in the eighteenth century. Andean people do not generally separate categories; in their minds they overlap and intertwine, and the use of the same sculpted rock complex as wak'a and quarry would be very possible.

Once Inka stone workers had completed the basic vertical and horizontal structural cuts to sculpt an outcrop by any or a combination of the techniques explained above and had removed the excess stone material, they straightened and oftentimes polished the resulting surfaces. Stone surfaces were typically worked with round edge hammerstones. Very large hammerstones could be used for splitting and squaring stone blocks, medium-sized ones for preparing the surfaces, and the smallest for flaking and finishing the edges. Protzen (1993:170–172) describes and illustrates the work marks as roughly circular contiguous cups, approximately square-shaped adjoining pans, and adjacent parallel troughs. It is important not to mistake such work marks for stylistic carving designs. Bengtsson (1998:91) emphasizes that the work marks left on roughly prepared blocks are similar to those on finely prepared blocks.

She reasons that the same type of tools, hammerstones, was used during both phases except that workmen employed smaller tools and less force during fine preparation. Further, Inka artisans had available edge tools and chisels made out of a tin-bronze alloy. The final stage of polishing was accomplished by rubbing the surface with certain stones as hard as or harder than the block prepared. Hide or wool was also employed for rubbing. Protzen's (1993:194) experiments showed that a glossy surface could be obtained by using a flat piece of andesite together with a mixture of soil and water for rubbing and surface treatment.

Transportation is rarely an issue with regard to carved rocks since most of them were sculpted in situ. The question has come up in reference to a number of large blocks which do not seem to be part of bedrock protrusions, such as the Ceremonial Rock at Machu Picchu and the Sayhuite Monolith, and/or feature in mythological narratives, for example, the Piedra Cansada. There is no clear consensus in the literature as to whether these huge blocks were transported to the sites they now occupy. They will be discussed more fully in their site-specific contexts in chapters 5 and 2 respectively.

Fitting and lying only come up in carved rock complexes where building blocks or entire walls are set upon bedrock which occurs as part of rock art sites. Protzen (1985:177–179) has found that, as a general rule, the bedding joints of every new course are cut into the top face of the course already laid below it. It follows that squared-off work marks equating the size of medium building blocks and arranged in rows on bedrock surfaces are bedding joints for wall foundations and not examples of a peculiar rock art style (figure 1.1.).

Such bedding joints are typically found along the edges of some carved outcrops. Depending upon the final height of the walls, they would have concealed the sculptures from view.

Pre-Inka Roots of Stone Carving and Rock Art

In the Andes, the Inka were clearly not the first to quarry, dress, sculpt, and transport stones and use stone masonry. Human beings link their activities to things they are familiar with, by accessing personal and collective memories. In this sense, the Inka constructed a memoryscape by sculpting selected boulders in a specific style format. Questions arise, such as, from whom did the Inka copy styles and technologies and what did they innovate? The following is a brief chronological overview of the use of dressed stone in architectural construction and stone sculpture (following Protzen and Nair 1997:146).

The practice of selecting a stone block, shaping it into an upright slab, and setting it into public space has now been documented as early as roughly 2500 B.C. at Caral. Other early stone slabs, probably related to ethnographic *wanka*

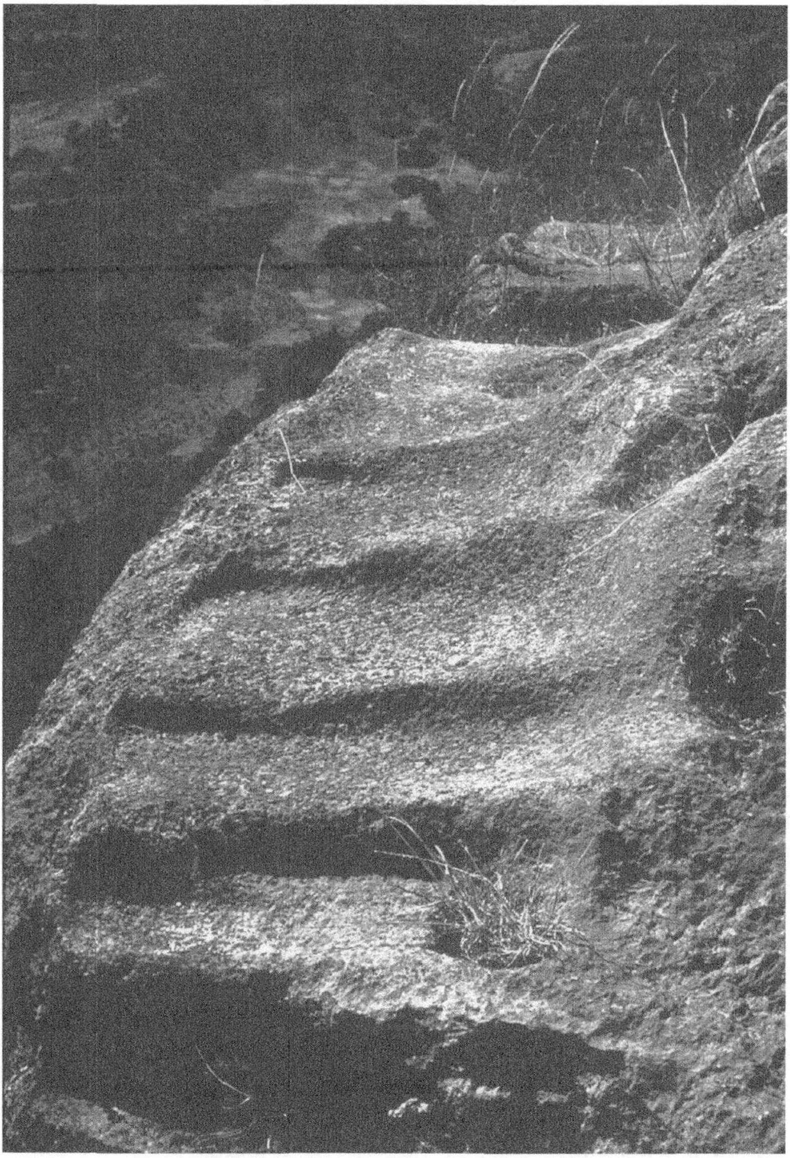

Figure 1.1. Bedding Joints for a Wall on Bedrock Surface at Pumaurqu. Photograph by Brian Garrett.

(petrified owners of land), have been reported from Huaricanga and Caballete, both located in the Fortaleza Valley in the Norte Chico region and dated in the Late Archaic (3000–1800 B.C.) (Matthew Piscitelli, personal communication, 2013). A history of stone images begins with Cerro Sechin on the North Coast. Here sometime before 1000 B.C., a monumental enclosure wall composed of large dressed and loosely fitted granite stones was built. These stones exhibit the well-known bas-reliefs of sacrificial victims. Tightly fitted cut-stone masonry is firmly established at Chavin de Huantar in the Old and New Temple complexes between about 1000 and 500 B.C. Chavin artisans also produced a large corpus of relief carvings and certain three-dimensional sculptures, such as the Lanzon and the tenon heads. In the southern Andes, sophisticated cut-stone masonry and carved stelae are associated with Pukara, a site with regional influence at the beginning of the Early Intermediate Period or about 200 B.C. In the Middle Horizon (c. 550–900 A.D.) centers of Huari and Tiwanaku, stone was worked in many different techniques. At the still inadequately investigated site of Huari, stone constructions were made of crude field stones in some sectors and of carefully cut and polished plates at Cheqo Wasi. East of Cheqo Wasi, there is a cave named Infiernillo above which petroglyphs showing curvy lines and small holes cut in the stone have been recorded (Lumbreras 1974:151–165). Further, eleven carved stone statues are known from Huari site but rarely talked about. Tiwanaku excelled in stone sculpture and techniques of cut-stone construction during the Middle Horizon while some construction activity began roughly 200 B.C. and continued through A.D. 1000.

Numerous authors have stated that the direct inspiration and model for Inka stone masonry and sculpture was Tiwanaku (for example, Gasparini and Margolies 1980:7–25). Such connections are based upon citations from the chronicles and—as we will see—superficial visual resemblances. The most specific reference comes from Bernabe Cobo. Book 12 of his *Historia del Nuevo Mundo* contains accounts of Pachakuti's military conquests and his order to copy Tiwanaku stone masonry styles in Cusco.[3] Cieza de Leon notes that he heard Indians say that the Inka made the buildings of Cusco in the form they had seen at Tiaguanaco; Sarmiento de Gamboa remarks that Pachakuti commissioned the splendid buildings of Ollantaytambo and that their stone masonry was made by captives from Collao (which is the southeastern region of the Inka empire with Lake Titicaca, near which Tiwanaku is situated) (both cited by Protzen and Nair 1997:146). Thus Cobo, Cieza de Leon, and Sarmiento de Gamboa name Tiwanaku as the model for Inka stonework. The latter two writers published their books several decades after the Spanish entered Cusco (Cieza de Leon in 1553 and Sarmiento de Gamboa in 1572), and Cobo had earlier manuscripts at his disposal. We can therefore assume that some of their information came from consultants who had lived under Inka rule.

It is significant that their consultants referred to Tiwanaku as the major source of inspiration for Inka stone masonry and sculpting styles. This important Middle Horizon center lies far away close to the southeastern shores of Lake Titicaca and near the Islands of the Sun and the Moon. The Cusco Valley contained the remains of numerous older settlements the Inka could have selected as physical and conceptual models for their capital (see Bauer 2004:31–54 for an archaeological investigation of the entire Cusco Valley). The ruins of Pikillaqta, a large Wari administrative center, would have been most noticeable. Pikillaqta (McEwan 2005) stands out because of its grid-like plan and is located only a short distance southeast of Cusco. Thus the reasons why the Inka imitated distant and not nearby places were not practical but must have been based upon ideological considerations. Lake Titicaca and its islands were pan-Andean origin places charged with wide-reaching powers, where the main celestial bodies, the Sun and the Moon, were said to have first risen from a sacred rock outcrop. At Tiwanaku, the creator god Wiraqocha fashioned an early human race out of stone. The Inka organized visitation of these sites into an important state-sponsored pilgrimage. Pan-Andean and Inka origin places and creation narratives will be discussed thoroughly in chapters 3 and 4 in their site-specific contexts. Here it suffices to state that the Inka citation of Tiwanaku was primarily political: the Inka started out as one ethnic group among many in the Cusco Valley. They managed to dominate it and they began conquering an ever-growing empire. They were successful military strategists and they were pragmatic. At the same time, they were lacking a specific history and a clearly defined cultural identity, which would make them stand out above neighboring groups. By "appropriating" visual material from Tiwanaku, they gave themselves this history and link with Andean origin places. It follows that by appropriating Tiwanaku stone-working styles, Pachakuti lent his rebuilding project of Cusco a façade of historical authority and legitimacy (see also van de Guchte 1990:46–49).

A technical comparison (Protzen and Nair 1997) of Tiwanaku and Inka stone-working techniques and styles in architecture and sculpture quickly teaches that the perceived similarities only make up an outer superficial façade of resemblance. Protzen and Nair (1997:147–153) discuss and illustrate types of Tiwanaku and Inka masonry bonds, demonstrating that they have little in common. The most obvious visual difference is that most Inka masonry bonds are executed with beveled, sunken joints which give the walls the characteristically Inka chiaroscuro effect. Tiwanaku masonry joints are in one plane and completely two-dimensional (figure 1.2).

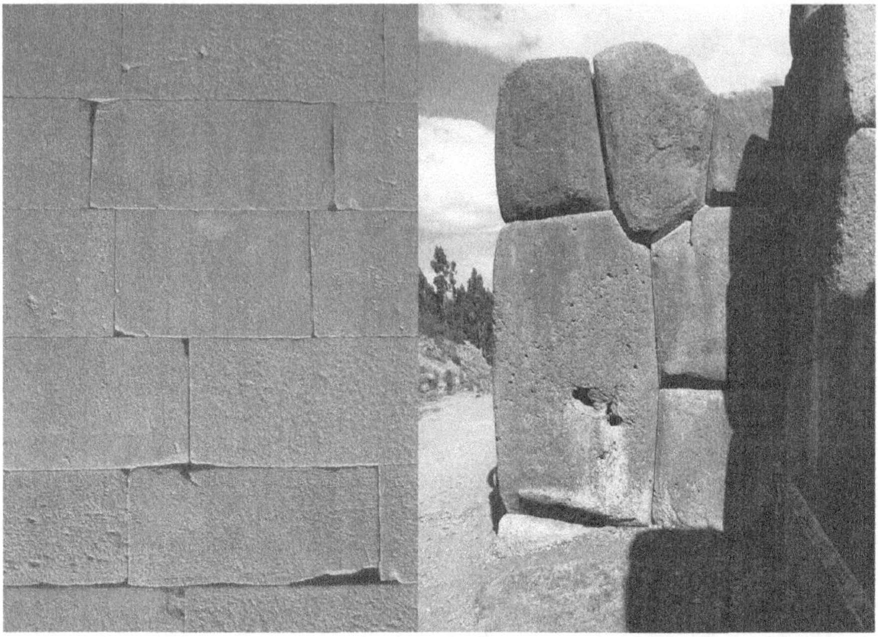

Figure 1.2. Comparison of Tiwanaku Stone Masonry Joints at Akapana, East Side, and Inka Stone Masonry Joints in Cusco, Saqsaywaman. Both Photographs by Brian Garrett.

Closely related to the differing masonry bonds are the stone-fitting techniques. Protzen and Nair (1997:158–160) emphasize that the Inka typically left the upper faces of already set stones uncut until the next stone was ready. The new stone was first carved on its lower side or bedding face and on one lateral face. The faces of the already set stones were then shaped to match those of the new one. In this manner, each stone was individually fitted into its place. At Tiwanaku, the procedure was opposite: the great majority of stones meet at right angles, have planar faces, and have the same height. Building stones were interchangeable, and Protzen and Nair (1997:159) suggest that they could have been prefabricated and mass produced. These are fundamentally different techniques and evidence of disparate conceptions as well as aesthetics of working with stone. With regard to tools, Tiwanaku stone masons probably used hammerstones like their Inka colleagues, leaving similar work marks. However, hammerstones could not produce the precise 90 degree interior angles commonly seen on Tiwanaku stone blocks. According to Protzen's and Nair's analysis, comparable cuts in Inka masonry and rock sculpture have rounded interior angles as a result of the pounding technique (with hammerstones), which is better documented for the Inka than it is for Tiwanaku (Protzen and Nair 1997:156–157). Similarly Protzen's experiments (1985:174) demonstrated that hammerstones produce dihedral angles between two adjacent faces, which are greater than 90 degrees. It follows that Tiwanaku artisans may have possessed additional tools which facilitated the creation of exact geometric cuts and forms and of which we have no record at this time.

Thus we may conclude that the similarities between Tiwanaku and Inka stonework perceived by many visitors are indeed visual and superficial. What impressed the Inka was the human feat of moving, dressing, and assembling large and smoothly finished stone blocks into walls to design spaces marked by man. Gasparini and Margolies (1980:8–12) document that Lupaqa masons from the Titicaca Basin in Kollasuyu were sent to Cusco to fulfill their obligatory labor tribute *mit'a*. At the same time, they warn of the time gap of approximately five hundred years which lay between the Tiwanaku monuments and Inka constructions and that high quality cut-stone masonry was not used during the Late Intermediate Period separating Tiwanaku and the Inka Late Horizon. I concur with Protzen and Nair's (1997:166) hypothesis that—at least the technical aspects of—Inka architecture and stone masonry are authentic inventions and not Tiwanaku derivatives. The same can be said of the carved rocks.

We will now explore the formal vocabulary of Inka rock art by first looking at the formal elements of carved rocks and, second, at associated structural features.

The Formal Elements of Inka Carved Rocks

Geometric Style. The most common style of Inka rock carving is characterized by straight cuts which the viewer perceives as vertical and horizontal lines. Protzen's (1985:175) measurements have shown that what the human eye registers as ninety-degree angles formed by verticals and horizontals is not always mathematically exact; for example, the steps of the well-known "Throne of the Inka" at Saqsaywaman have concave edges. Many of the sculpted outcrops in and around Cusco are of limestone which has severely eroded into serrated surfaces with sharp ridges. As a result, measurements taken today cannot reconstruct the degree of accuracy—here understood as a Western standard—in Inka geometry.

What seems to have mattered to the Inka was the creation of a visual geometry. A similar sense of geometry has been observed in Inka architecture (Gasparini and Margolies 1980:68–75, 187–191, 234–238; Protzen 2000; Christie 2006a), spatial layout (Duviols 1976:24–30), and a rigid application of geometry has been noted for highland textiles (Conklin 1996:325–326).

In art historical methodology, vertical and horizontal lines, ninety-degree angles, and geometric forms are usually interpreted as expressing order and control often associated with a higher power. I will argue below in the discussion of *stone ideology* that to a certain point, Inka geometry was a visual marker of the state but that it was flexible. Protzen (2000:200, 210–211, 216–217) thinks that Inka architecture was more strongly shaped by the features of the land than by a strict geometric overlay. Similar reasoning is applicable to stonework: the Inka modified stone to varying degrees, from natural boulders integrated into architectural settings to quarried cube-like building blocks which were precisely fitted into masonry walls. My argument is that Inka visual culture showcased gradated levels of control over natural elements reflected in stonework as well as in architecture and ultimately in levels of administrative control (see below and Duviols 1976).

The many examples of the style of rock sculpture using straight vertical and horizontal cuts can be grouped into potential seats (for instance, the Inka Throne Sabacurinca at Saqsaywaman (figure 1.3), and the Ceremonial Stone at Machu Picchu, chapter 5), niches (Laqo), the common steps and platforms (see, for example, Q'enqo), and possible altars or tables (Laqo). Perhaps to defy this perceived geometry, some carved rocks curve following rounded visual lines, for example at Q'enqo Chico (figure 1.4), and on the Piedra Cansada just north of Saqsaywaman as well as a fully rounded carved rock of modest size sitting on one of the *andenes* (terraces) leading up to Nusta Hispana at Vitcos.

Channels and Canals. Another type of carving consists of linear or zigzag channels. Some of the channels have two interlacing, zigzagging arms forming rhomboids and terminating in a catchment basin: for example, at Q'enqo Grande (figure 1.5), Vilcashuaman, and at Samaipata. Zigzag canals in sculpted rocks have been cross-linked with the zigzag channels in carved wooden *pacchas*, which are ceremonial drinking devices composed of a cup connected to a long stem through which the liquid flows in zigzags to the mouthpiece (Allen 2002; Carrion Cachet 1955). As Catherine Allen (2002) specifically shows, pacchas were and continue to be used in traditional ceremonies which facilitate the circulation of vital liquids through the cosmic layers. Their relatives in stone likely fulfilled a similar purpose.

Postholes. More specialized carving examples are four small postholes on the arm and backrest of a potential seat. So far, I have come across four examples: three at Q'enqo Chico (figure 1.6), and the fourth is situated in the Urubamba River at Ollantaytambo. The latter has been referred to as "Water temple" and has been discussed by van de Guchte (1990:195–200). In the ethnography, one can find beautiful examples of Inka officials seated upon thrones or litters surmounted by colorful canopies (Guaman Poma de Ayala 1987:407 [1615:398]; Jimenez de la Espada 1965:218–219 [1881–1897]). The carved seats with postholes may well be the physical leftovers of similar ceremonies.

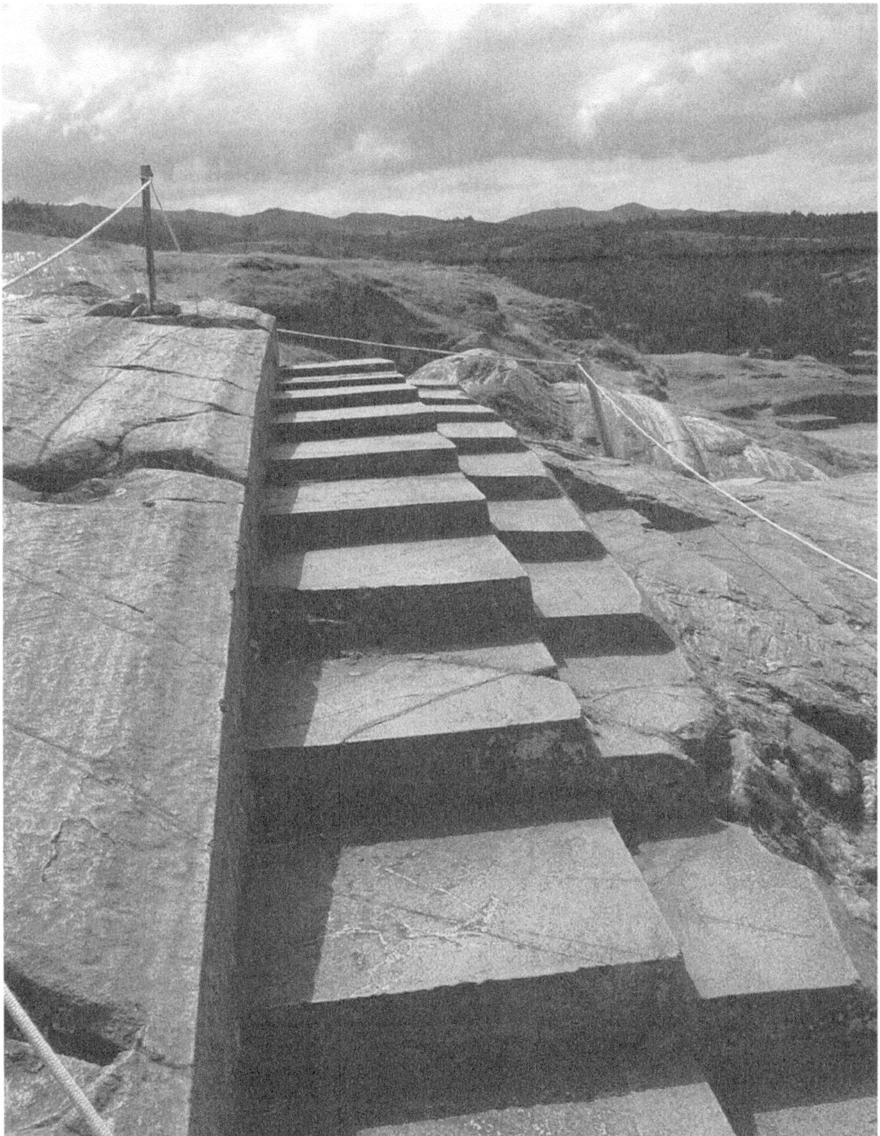

Figure 1.3. Seat-like Rock Sculpture Known as Inka Throne, or Wak'a Sabacurinca, Saqsaywaman, Cusco. Photograph by Brian Garrett.

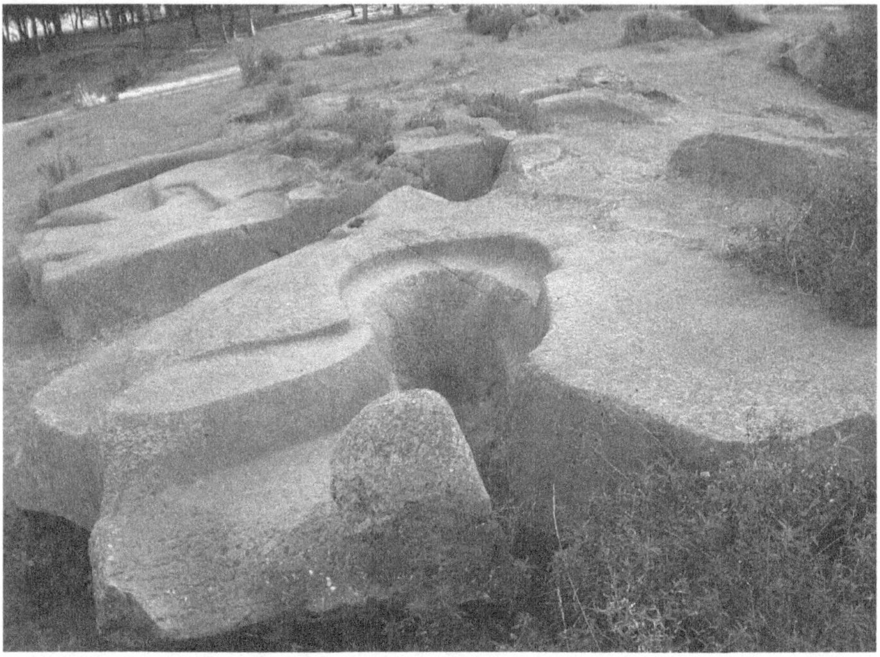

Figure 1.4. Curving Sculptures at Q'enqo Chico, Cusco. Photograph by Brian Garrett.

Cup-marked rocks are visually similar to postholes. Frank Meddens (2006) has documented a large number of stones, some of them pocketed by cup-marks, in the Chicha/Soras Valley in the southern highlands, which appear to have formed an organized system. They are dated to the Late Horizon by association with local sites. Meddens argues for multiple ritual functions of these carved rocks and their locations. I concur that they were most likely used in agricultural rituals and perhaps calendric aspects. But I view the cup-marks as a widely used and here locally organized carving style distinct from the Inka state style of geometric forms.

Gnomens or Knobs. A fifth carving style is materialized by the common gnomen or knobs, which are rounded or angular projections from smooth surfaces. They are generally interpreted as *Intiwatanas* and sundials, but nobody knows exactly how such sundials would have worked. This reasoning is based upon the superficial formal resemblance of all gnomen/knobs with the famous Intiwatana of Machu Picchu, which was so named by Bingham (1979:52, 1997:200). Intiwatana is a Quechua word meaning "the place to which the sun was tied." Besides the Intiwatana stone at Machu Picchu

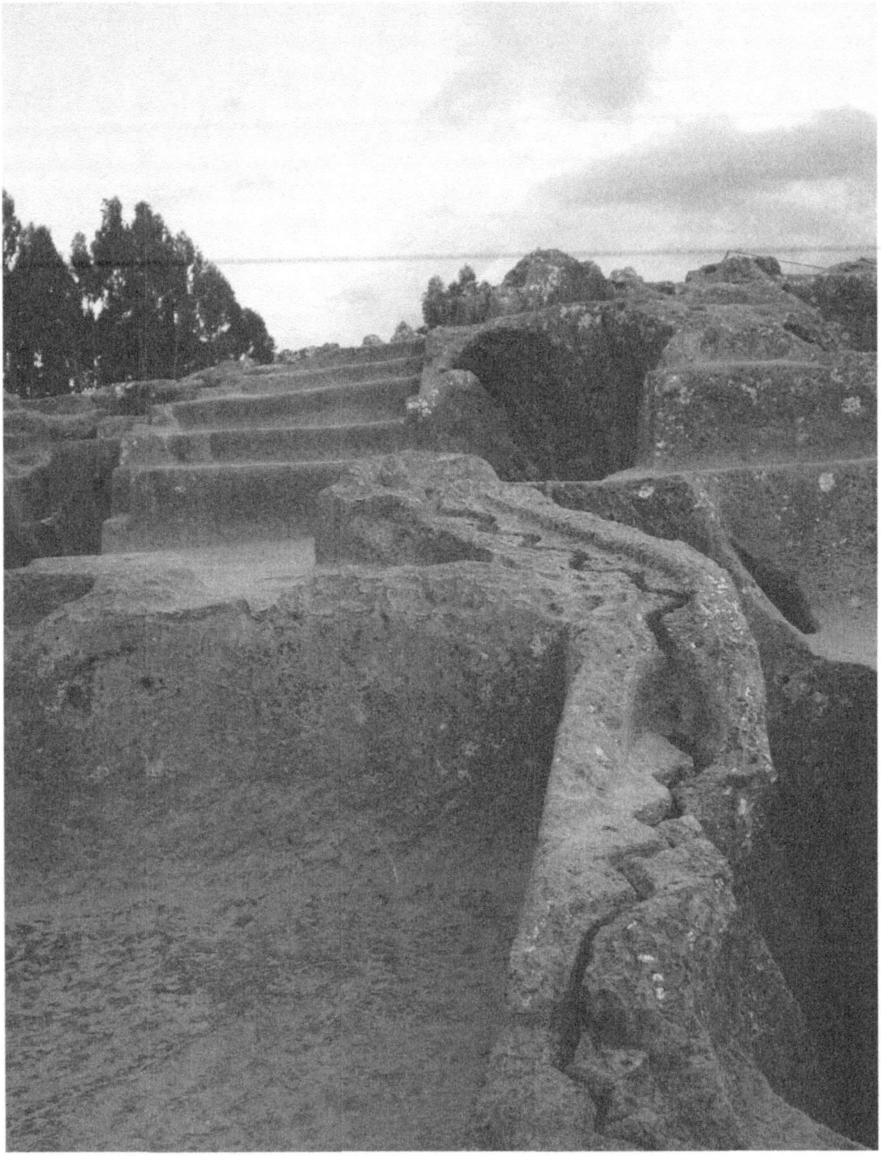

Figure 1.5. Zigzag Channels at Q'enqo Grande, Cusco. Photograph by Brian Garrett.

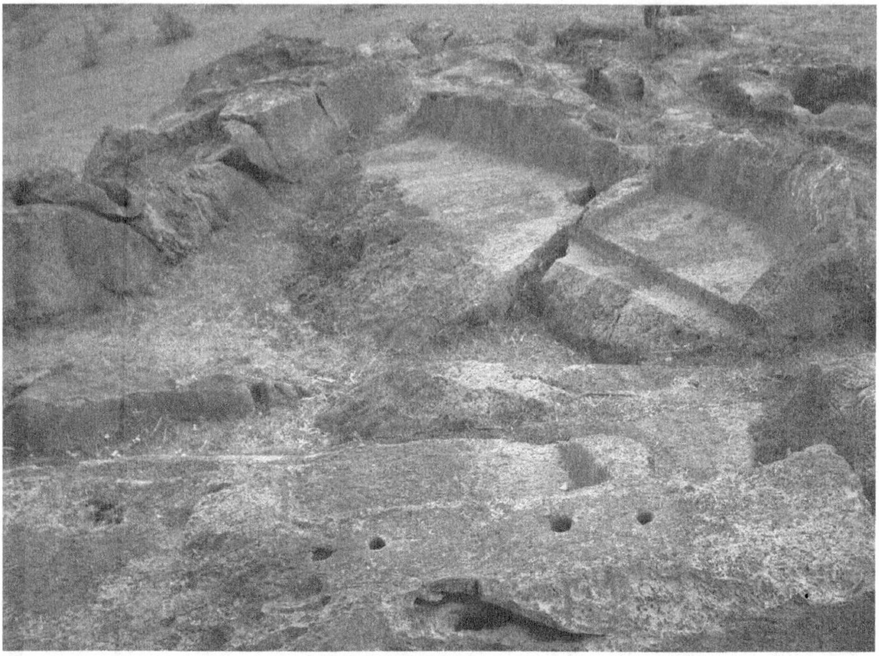

Figure 1.6. Q'enqo Chico, Postholes. Photograph by Brian Garrett.

(chapter 5), noteworthy examples of gnomens and knobs are found at the Sun Temple of Pisaq, at Laqo (figure 1.7), and at Ollantaytambo in the Inkamisana sector.

Figurative High Relief Images. A very different carving style consists of figurative images in high relief. This style is not very common; the few known examples primarily portray animals and secondarily architectural structures and some plants. The most impressive case is the monolith from Sayhuite (see Christie 2003, 2013). A small number of animal figures were carved at Laqo (figure 1.8), Chinchero (chapter 6) and Pumaurqu (chapter 4). The most frequently depicted animal, which occurs at all these sites, is the feline. Further, camelids and snakes were carved in low relief in some architectural building blocks in walls in Cusco (near Choqechaca Street) and at the Intiwatana site at Pomacocha lake just outside of Vilcashuaman. Stylized angular camelids were inlaid with white stones in terrace walls at Choqequirau.

From an art historical perspective, the Inka demonstrate a decisive preference for abstracted and geometric forms over figurative elements. They never showed much interest in copying nature in a Western Aristotelian sense. The Museo del Inka in Cusco houses the head of the one known fully three-dimensional figure sculpture of an Inka official. Rather, the Inka appear to

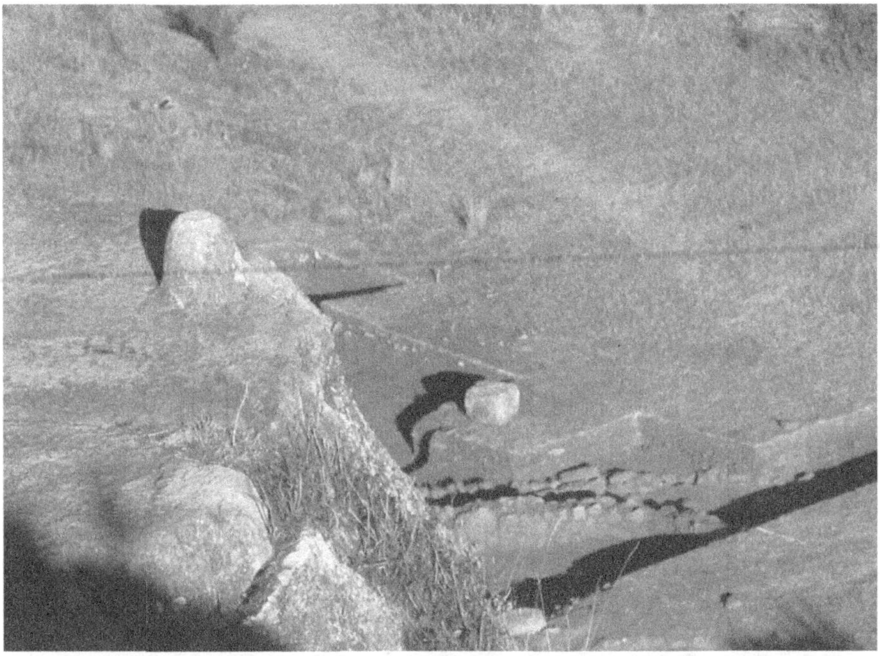

Figure 1.7. Carved Knob at Laqo, Cusco. Photograph by Brian Garrett.

have been more interested in the essence (Stone-Miller 2002:15–16) of the figure or object depicted, which was hinted at by abstracted formal elements (Paternosto 1996). Abstract figurines of humans and llamas were produced in large numbers in gold and silver metallurgy and used in sacrificial offerings (*qhapaq ucha* events). This inclination toward abstraction is related to the Inka concept and application of geometry discussed above.

Unmodified Boulders. Finally, uncarved natural boulders were integrated into architectural and sculptural settings. The Machu Picchu Sanctuary area offers numerous examples. On the trail from Machu Picchu to Intipunku, a large rock about fifty feet in height slightly leans toward the trail. A terrace was constructed in front and steps cut into the bedrock (chapter 5). This location was one of the cemeteries Hiram Bingham discovered and it contained the burial of an elite woman (Wright and Valencia Zegarra 2001:119). Thus the unsculpted boulder marked the trail and the burial site and was also a stone wak'a. Another case in point can be found at the lower entrance to Winay Wayna. Where the trail from Chachabamba enters the settlement, the Inka singled out a large rock by constructing a platform around it. As the prior example, it was not formally sculpted, and the intention of the Inka appears to have been to leave the visual impression of a natural rock. This wak'a was

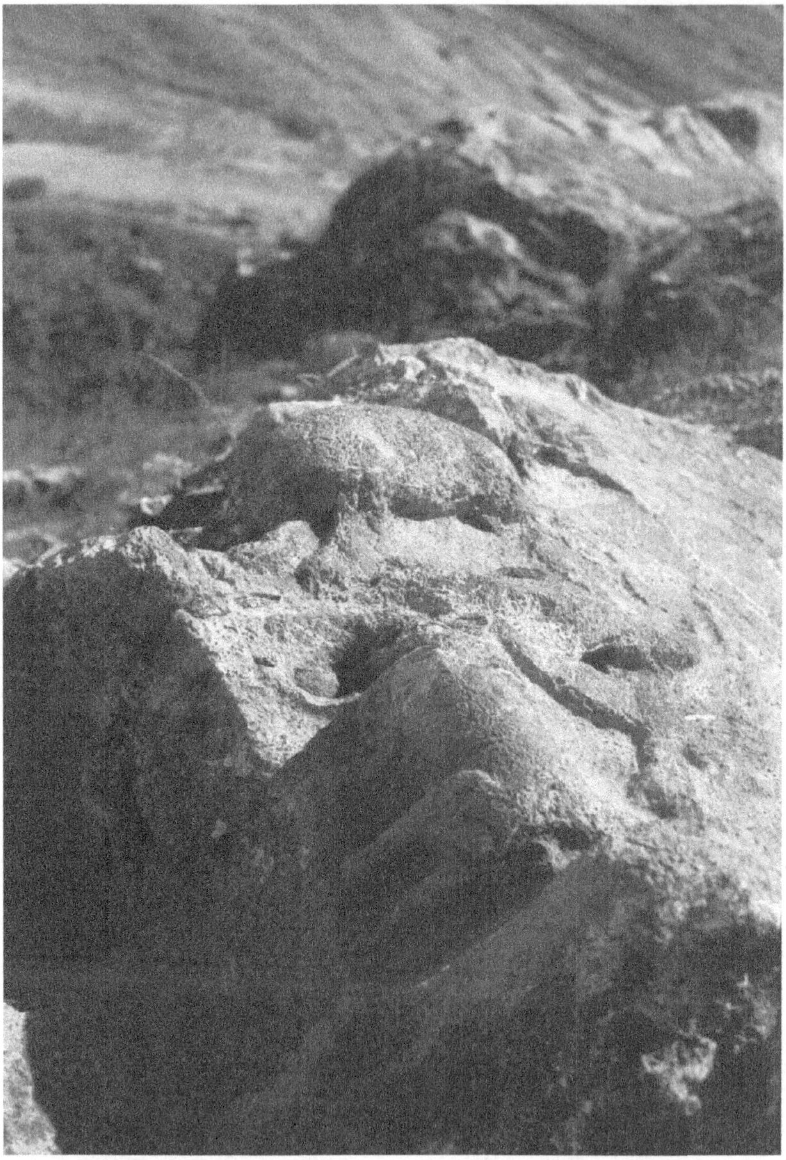

Figure 1.8. Laqo, High Relief Carvings of Animal Figures. Photograph by Brian Garrett.

most likely a *huanca chacrayoc* (Duviols 1979b:8) and as such falls outside Pachakuti's state-sponsored stone ideology.

Structural Features Associated with Carved Rocks

Bodies of Water. Many sculpted rock complexes are situated next to water sources, or the Inka channeled water through them. Examples include the previously mentioned Water temple at Ollantaytambo and a large carved boulder on the north side of Laqo which exhibits steps, platforms, as well as a snake design encircling an opening in the rock. The Water temple is immersed in the waters of the Urubamba during the rainy season (van de Guchte 1990:195–200), and a stream flows around the base of the carved boulder north of Laqo soaking the ground.

Caves. Caves constitute another vital structural feature which is often a part of carved rock complexes, for example, the well-known cave at Q'enqo Grande with the sculpted altar-like stone (figure 1.9), at Laqo, and at Amaromarcaguaci. Water and caves were building blocks of the Andean cosmogram or view of the world, as will be demonstrated throughout this book. In traditional Andean thinking, the Pacific Ocean to the west functions as the

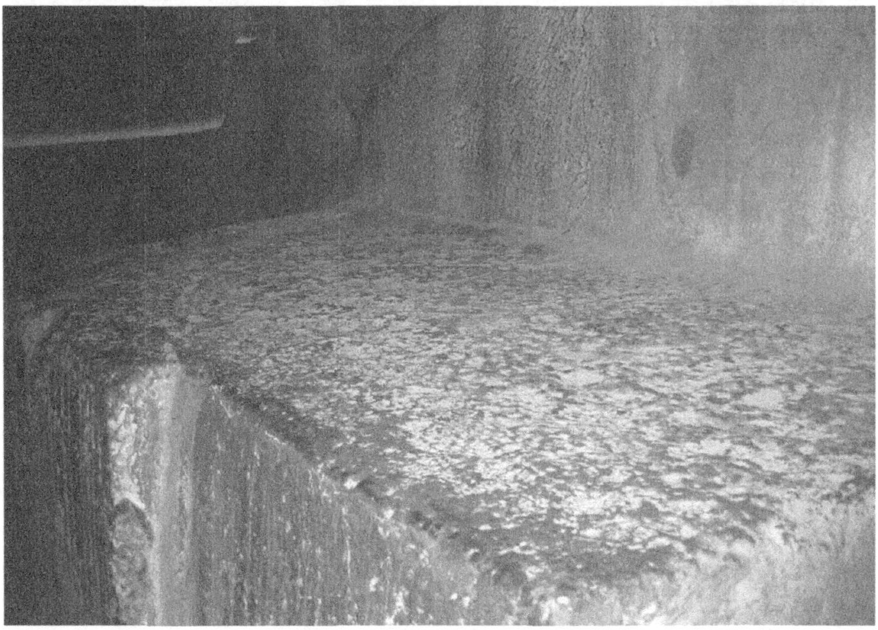

Figure 1.9. Cave with Sculpted Stone at Q'enqo Grande, Cusco. Photograph by Brian Garrett.

ultimate source of water. From it, water rises to the sky, and it is visualized as the Milky Way, which is explained as a celestial river. This metaphorical river transfers water to the nevados, the snowcapped mountaintops whose cave openings give birth to most rivers, which, in turn, transport the water to lower elevations, fill lakes and streams, and finally return water to the ocean. Thus the Andean world becomes a gigantic stony mountain formation which may be entered through caves and around which water circulates. Inka rock art assumes its place in this conceptual setting.

Architectural Wall Formations. To further enhance the context of the natural setting, the Inka added man-made architectural features to many carved rocks. Cases to illustrate this point are the Pulpituyoq rock at Patallacta and Chachabamba, both located between Ollantaytambo and Machu Picchu. Pulpituyoq rock is a modified rock outcrop which sits next to the Kusichaca River and below the main sectors of Pachakuti's estate of Patallacta. A rounded enclosure wall spirals around the top of this tall outcrop. The uncarved top of the rock is the center of an open space partially encircled by niches with cordholders. The presence of cordholders indicates that the interior space of these niches could be opened or concealed from public eye by hangings of cloth or leather. This would likely have been done during ceremonies when sacred objects were displayed in the niches and the entire open space of the structure became the locus of rituals. Chachabamba is a small settlement at the Urubamba River, where an Inka trail ascends toward Machu Picchu (chapter 5). A large boulder displaying a finely sculpted long seat occupies one side of a *kancha* (architectural group of four one-room buildings framing a courtyard) which is surrounded by a fountain system laid out in a symmetrical fashion. As will be further discussed below, the Inka modified stone and their environment in general in degrees in an ongoing dialogue. The technique of sculpture in the form of carving bedrock outcrops constituted one level of stone modification; quarrying stone, cutting and dressing it into building blocks which were assembled into walls at a new location was a further level of modification and more forceful coercion of nature. Rock art sites combining sculpture and architecture gather two layers of modifications which I interpret as a specifically Inka negotiation over control of natural forces.

Celestial Alignments. Some carved rocks may purposefully have been aligned to the cardinal directions and to the cyclical movement of the sun. There has been much speculation and little demonstration of how exactly such alignments would have worked. One well-published debate is over how the Sun Temple at Machu Picchu would have signaled the solstices by the way sunlight illuminates the top of the modified outcrop (chapter 5). At Ollantaytambo at the entrance to the Inkamisana sector, five carved knobs project from a polished wall surface. Below are steps and ledges with two pronounced cut notches. Around midday on December 22, the day of the summer solstice, the shadows cast by two sets of two knobs form two straight lines falling exactly into the two notches (Elorrieta Salazar and Elorrieta Salazar 1996: figs. 33–36). Another carved outcrop lies downstream of Ollantaytambo where the Urubamba River makes a noticeable bend. It is graced with niches, shelves, seat-like carvings, and a large flattened circular knob or seat which faces east. Local consultants call this outcrop *Chingana* (a name commonly used in the Cusco region for rock wak'as with intricate carvings and/or passageways) and explain that the first sunlight appearing over the eastern mountain range illuminates the carvings. Here lies the problem surrounding most interpretations of sculpted rocks as astronomical observatories: the shadow cast by any geometric carving leaves crisp noticeable lines which, theoretically, anyone may observe on a daily basis throughout the year. This theoretical observer could mark the movements of the shadow and the extreme points where it changes direction. The shadow would reach the two extreme points on the days of the solstices. It is my understanding that such observations could be conducted at most carved rocks; on which carved rocks and how exactly the Inka traced shadows remains, for the most part, speculation.

By providing a historical and technical overview of Andean stone working techniques, chapter 1 has set the stage for the following chapters which will now rigorously mine the site settings and layers of understanding of five selected groups of Inka carved rock complexes.

NOTES

1. He does mention some sculpted rock complexes, such as the Throne of the Inka at Saqsaywaman and the Inkamisana sector at Ollantaytambo.

2. One possible exception might be the river cobbles strewn around the well-known Ceremonial Rock outside the Main Gate to Machu Picchu. Ruth Wright and Alfredo Valencia Zegarra (2001:8–10) interpret them as symbolic features bringing the Urubamba River into presence. Gary Ziegler (personal communication, 2005) has suggested that they were brought in and placed at the carved rock by visitors to refer to mountains in a similar way as *apachetas* (heaps of rocks left at mountain passes) do. But do we, at times, get too caught up in our fascination with symbolism and search for spirituality in Inka places? Could some of these river cobbles have been used as hammerstones to work the Ceremonial Rock? Without a professional lithic analysis of these stones, this question will remain open.

3. Pachacutic came to Tiaguanaco [spelling in original text] to see their superb buildings whose cut stone masonry he very much admired because he had never seen buildings of this kind; and he ordered his people to study this construction technique very well because he wanted that the projects underway in Cuzco should be of this style (Cobo 1956:191–192 [1653:Bk.12,Ch. 12], translation by the author).

Chapter Two

Carved Rock Complexes or Groups of Carved Rocks in Specific Contexts

This and the following chapters will introduce major carved rock complexes and groups of carved rocks which constitute dominant features at specific Inka settlements and in certain spatial contexts. Chapter 2 investigates Cusco, the capital, where Pachakuti and his descent line began to institute the Inka style of rock carving on their shrines (*wak'as*) of the zeq'e lines. Chapters 3 and 4 take the reader to the origin sites on Lake Titicaca and at Pacariqtambo where the Inka ingeniously constructed their new history, which linked the elevated status of their dynasty to sacred stone formations. Chapters 5 and 6 move to the royal estates of Machu Picchu and Chinchero by the rulers Pachakuti Inka Yupanki and Thupa Inka Yupanki. At Machu Picchu, Pachakuti fully formulated Inka rock art by placing it into a discourse with the surrounding mountainscape. At Chinchero, Thupa Inka adopted his father's style of rock carving and initiated processes of exporting it to outlying regions. Under the subheading *Groundwork*, I will assemble and summarize the information provided in the existing literature: I will analyze form, reference archaeological data where they are available, and cite applicable ethnography and historical documents in order to reconstruct how the Inka may have used these sculpted outcrops. Site-specific reconstructions constitute one piece in the dynamic mosaic my book sets out to assemble, and the reader can easily find them in the sections titled *Groundwork*. The scope of the study extends further to the roles carved rocks must have played in the broader contexts of Inka political landscape, spatial divisions of the empire, state politics, and Inka identity. What sets this book apart from other publications is an emphasis on questions such as, what kind of links may have existed between carved rocks and the seat of the Inka state? Did the Inka pursue specific strategies in placing the boulder sculptures where they have been found? And in which ways and forms does rock art construct an identity of Inkaness that differentiates itself from pan-Andean consciousness? Such questions will be taken up under the subheadings: *Inka Landscape and Carved Rocks—Appearance, Experience, and Perception* and *Inka Landscape and Carved Rocks—Essence, Imagination, and Stone Ideology*, implementing T. Adam Smith's approach toward investigating political landscapes (see above).

As explained earlier, the methodology resembles a mosaic work to which tessera of insight after tessera are being added although this mosaic cannot reach completion because of changes in perspectives and interpretations through time. Therefore any methodology has to be flexible, agent-based taking into consideration layers of seeing and understanding the sculpted outcrops by individual subjects and through their bodies as well as minds in changing time periods. I have found that Smith's perspective of relational landscapes which acknowledge layers of relationships between people, settlements, and features of the land in space and over time is best suited to pursue this type of study. I do not postulate that it is possible to reconstruct a purely Inka context. The Inka themselves responded to stonework from people who preceded them, and the sculpted rock sites have been physically reconstructed and mentally reimagined throughout historical time periods to such an extent that they have fused the past with the present. I will address multiple layers of contexts as themes separated by subheadings, bringing to life relational networks in the spirit of Smith. Although the *Groundwork* section is written as a conventional narrative, the format of themes and subheadings will emphasize by means of writing style that we are only seeing insights, glimpses, and slices of the rich social dynamics in which carved rocks have acted. In Tilley and Bennett's (2004:220) words, the aim of this book cannot be "to control or fix knowledge but rather, through our practice, to open up perspectives and create opportunities for further understanding." In this spirit, five chapters will explore selected site settings.

CARVED ROCKS ON THE CUSCO ZEQ'E LINES

This chapter investigates those carved rock complexes which can be identified as wak'as on the Cusco zeq'e lines with some degree of certainty. This group of rocks is of extraordinary importance because accounts and narratives by the Spanish provide insights and help us visualize how they were used during Inka times. One has to stay alert, however, that the Spanish texts do not place neutral histories in front of us but information filtered through the values of colonial society and the personal agendas of their authors. In the *Groundwork* portion, I will critically compare statements made by several chroniclers and assemble the pertinent archaeological material as published by Brian Bauer (1998) as well as the larger ethnographic context from Tom Zuidema's (1964) initial and still authoritative study of the zeq'e system. The first part of the *Groundwork* section will explain the fundamental structure of the zeq'e system. The reader unfamiliar with Cusco can consult Bauer (1998) and Zuidema (1964) for the more complex aspects. This will lead into the identification of those wak'as which corresponded with carved rock complexes. The *Groundwork* will be followed by an exploration of these boulder sculptures in time and space to produce the creative act of writing a *present past* (Tilley and Bennett 2004:225) which correlates the tesserae pieces of multidisciplinary data with my own bodily experiences.

GROUNDWORK

The Zeq'e System and its Ethnohistoric Sources

The zeq'e system was one of three spatial divisions the Inka employed in Cusco and probably in other parts of the empire, and it can only be understood in relation to the other two. Most Spanish writers agree that the spatial organization of Cusco followed three major systems which geographically and conceptually overlapped: first, and most important, the Inka divided their empire into four regions, or *suyu*, which originated in the center of Cusco. Four major roads departed from the Main Plaza (Awkaypata) in Cusco in the directions of these four quarters: Chinchaysuyu in the northwest, Antisuyu in the northeast, Kollasuyu in the southeast, and Kuntisuyu in the southwest. These groupings were extended throughout the entire empire so that each region belonged to one suyu. The second division echoed the topography of the Cusco valley and grouped the city into two halves and moieties: the upper half or Hanan Cusco was made up of Chinchaysuyu and Antisuyu, while the lower half or Hurin Cusco consisted of Kollasuyu and Kuntisuyu. The Inka associated both moieties with specific rulers: the first five rulers in the dynastic list, Manqo Qhapaq through Qhapaq Yupanki, were affiliated with Hurin Cusco, while the later five Inkas, Inka Roq'a through Thupa Inka Yupanki, were associated with Hanan Cusco.

Overlaid onto these two divisions was the zeq'e system. Zeq'es were imagined abstract lines departing from the Temple of the Sun in the center of Cusco in a radial fashion. Altogether there were forth-one or forty-two zeq'es, and their directions were determined by the locations of numerous shrines or wak'as totaling more than 328. Wak'as were complex sacred places endowed with spirits which required offerings and regular maintenance (see, for example, van de Guchte 1990:237–329; Bray 2014). Brian Bauer's field investigations documented that of the 328 wak'as, ninety-six

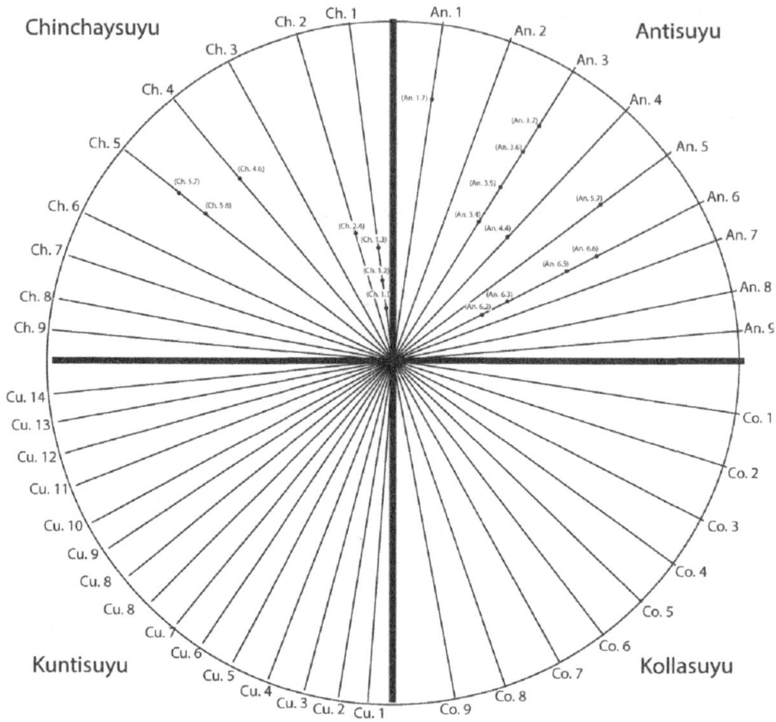

Figure 2.1. Diagram of the Zeq'e System in Cusco. Courtesy of Brian Bauer 1998; modified to indicate the locations of the carved rock wak'as.

(29 percent) were springs or sources of water and—what is most relevant to the present study—approximately 95 (29 percent) were standing stones. Together these springs and stones made up nearly 60 percent of the known shrines. Other shrine categories included hills and mountain passes (32, or 10 percent), palaces of the royal Inkas and temples (28, or 9 percent), fields and flat places (28, or 9 percent), tombs (10, or 3 percent), and ravines (7, or 2 percent). Less frequent wak'as were caves (3, or 1 percent), quarries (3, or 1 percent), stone seats (3, or 1 percent), sunset markers (3, or 1 percent), trees (2, or 1 percent), and roads (2, or 1 percent) (Bauer 1998:23); see figure 2.1:

> From the Temple of the Sun, as from the center, there went out certain lines which the Indians call ceques; they formed four parts corresponding to the four royal roads which went out from Cuzco. On each one of those ceques were arranged in order the guacas and shrines which there were in Cuzco and its region like stations of holy places, the veneration of which was common to all (Cobo 1990:51 [1653]).

These explanations were given by Father Bernabe Cobo, who was born in Spain in 1580 but spent much of his adult life in Peru and Mexico (Hamilton 2008). He was trained by the Jesuit Order and became well known for his historical writings. In 1653, four years before his death, he completed his *Historia del Nuevo Mundo* containing forty-three books divided into three parts from which the quotation above was taken. Most significant in the context of this study is Book 13, which contains four chapters describing and naming the zeq'e lines and their associated wak'as in each suyu. According to Cobo's account, Chinchaysuyu, Antisuyu, and Kollasuyu contained nine zeq'es each while Kuntisuyu is said to have had fourteen.

Most importantly, Cobo establishes a relationship between the zeq'es, their wak'as, and social groups within Cusco:

> Each ceque was the responsibility of the kinship units and families of the city of Cuzco, from within which came the attendants and servants who cared for the guacas of their ceque and saw to offering the established sacrifices at the proper times (Cobo 1990:51 [1653]).

Occasionally, Cobo is more specific and notes which kin group was responsible for a particular zeq'e. For example, for the fourth zeq'e of Antisuyu (An-4) he writes: "The fourth ceque of this road was called Collana; it was the responsibility of the ayllo and family of Aucailli Panaca and had seven guacas" (Cobo 1990:66 [1653]).

An analysis of the various kin group names included in Cobo's description of the zeq'e lines shows that the obligation of maintaining and worshipping the shrines along the zeq'es was divided between two broadly defined social groups: the panaqas of the ruling Inkas and the nonroyal ayllus (localized descent groups) of Cusco (Bauer 1998:39). The panaqas were the royal ayllus and consisted of the direct descendants of Inka kings while citizens of lower social status made up the nonroyal ayllus of Cusco.

The names of panaqas and nonroyal ayllus listed by Cobo also appear in various other colonial documents. The most important source of data is Molina's description of the Citua festival in Cusco (see Bauer 1998:40–45). During this celebration, the capital city was ritually cleansed. At its close, four groups of one hundred warriors carrying ashes from the rituals ran out from the main plaza along the four roads of the empire. At the edge of the Cusco Valley, they were met by representatives of other groups who received the ashes and threw them into the major rivers of the four regions of the empire. Molina lists the Cusco participants in the Citua festival according to their social rank as members of the panaqas and nonroyal ayllus of the imperial city, and there is evidence that each panaqa was paired with an ayllu.[1]

Molina's division and pairing of the panaqas and nonroyal ayllus in suyus is mirrored in Cobo's description of the zeq'e system where he identifies which kin groups were responsible for specific zeq'es. There is evidence that a similar pairing of royal and nonroyal ayllus may have existed in specific zeq'e clusters. In at least one case, Cobo documents that a panaqa and nonroyal ayllu maintained adjacent zeq'es (Ch-1 and Ch-2; Bauer 1998:45). Making the assumption that in each zeq'e cluster (collana, payan, cayao, see below) in which a panaqa is represented, a nonroyal ayllu must also be present (Zuidema 1964:8), and matching the data furnished by Molina and Cobo, an incomplete list can be extrapolated which connects specific zeq'es to certain royal and nonroyal ayllus (Bauer 1998:46–47). While such a list remains speculative to some degree, it is helpful to this study because it offers at least a possibility to place certain carved rock wak'as under the control of historically identifiable families.

It must further be pointed out that Cobo clustered the zeq'es in each suyu into three different groups by naming them collana, payan, or cayao. While Cobo never provided a definition of these terms, they are also contained within a number of early colonial Quechua/Spanish vocabularies, liturgical material, and administrative documents. A review of these materials by Bauer clarifies that collana, payan, and cayao are ranking terms that begin with collana and end with cayao in a descending order. The Inka used them to divide groups of individuals into subsets with different degrees of prestige, and they applied them to the zeq'e system in the same manner. Collana, payan, and cayao established a hierarchical order within sets of three lines and their associated Cusco kin groups (Bauer 1998:35–37).

Cobo was not always consistent in his presentation of zeq'e information. The Chinchaysuyu zeq'es are described in a counterclockwise order, while those of the other quarters are numbered in a clockwise direction. Another problem is that Cobo's book was written relatively late, in 1653, which was more than one hundred years after the Spanish entry in Cusco. He must have extracted much of his information on the history and religious practices of the Inka from earlier documents which writers in the sixteenth and seventeenth centuries not always referenced. Researchers have postulated that Cobo must have used an original zeq'e system manuscript which has never been found. Given the precision of Cobo's account, it has been suggested that the original information on the system must have come from an individual who kept track of the offerings to be presented to the Cusco shrines (Bauer 1998:7–8). Such individuals were known as *kamayuq* (specialist, master), *wak'a kamayuq* (shrine specialist), and *vilca kamayuq* (sacred object specialist).

Some researchers, including Brian Bauer, think that Juan Polo de Ondegardo was the author of the original zeq'e system manuscript (Bauer

1998:18–22; Presta and Julien 2008). Polo de Ondegardo was a lawyer who held a number of public positions in what are now Bolivia and Peru during the mid- to late-sixteenth century. He wrote various reports and historical treatises about shrines and religious practices of the Inka which were based upon his own observations and interviews with surviving Inka. Some of his writings make mention of the Cusco zeq'e system as well as of others. For example, in a document dated 1571, Polo de Ondegardo refers several times to a map he drew of the Cusco zeq'e system. He further reports that he personally investigated other zeq'e systems in different parts of the highlands.

The ethnohistoric literature contains many references to wak'a worship and related zeq'e systems (for an annotated list of sources, see Bauer 1998:15–21).[2] In the mid-twentieth century, the Cusco zeq'e system appeared again as a topic of great scholarly interest. Early studies were written by Paul Kirchhoff (1949) and by Cusco archaeologist Manuel Chavez-Ballon (1970). In the mid-1950s, R. Tom Zuidema conducted an extensive investigation of the zeq'e system which was published in 1964 under the title *The Ceque System of Cuzco: The Social Organization of the Capital of the Inca*. Zuidema was primarily interested in broader issues of social organization rather than in individual wak'as, which he reconstructed in a sophisticated model divided into a First, Second, and Third Representation. His arguments advance from overlapping systems of tripartition and quadripartition to quinquepartition, which organize space as well as marital, other social and economic relations and which he extrapolates from historical documents, the chronicles, and a stunning number of ethnographic sources. Zuidema's work was widely influential, but many aspects have not been generally accepted and have been revised—some by Zuidema himself. More recent studies continue to focus on organizational aspects of the Cusco zeq'es and how they may reflect the social and political dimensions of Inka society (John H. Rowe 1980, 1985; John Hyslop 1990).

Brian Bauer's project attempted to identify and verify wak'a locations on the ground and to reconstruct the exact course of zeq'e lines based upon known wak'as. Field data were complemented by archival research, and the results of his investigations were published in 1998 with the title *The Sacred Landscape of the Inca: The Cusco Ceque System*. Bauer's work is most insightful for the context of the present study because he identified specific carved rock outcrops as wak'as on zeq'e lines.

The following discussion of individual sculpted rock shrines will move from suyu to suyu beginning in Chinchaysuyu and ending in Kuntisuyu. It should be noted from the outset that the carved rock wak'as are concentrated in Chinchaysuyu and Antisuyu or the Hanan moiety of Cusco.

The Sculpted Rock Wak'as in Chinchaysuyu

Cobo describes the first wak'a on the first zeq'e line of Chinchaysuyu as follows:

> [Ch-1:1] The first was named Michosamaro; it was located up against the slope of Totocache, and they said it was one of those who they fancied had emerged with the first Inca Manco Capac from the cave of Pacaritampu. They relate that one of the women who came out of the cave with them killed him because of an act of disrespect toward her which he committed. He turned to stone, and his spirit appeared in this same place and ordered that they make sacrifices to him there. Thus the sacrifice at this *guaca* was very ancient. It always consisted of gold, clothing, seashells, and other things, and it used to be made for good rains (Cobo 1990:51 [1653:Bk.13, Ch.13]).

Albornoz names Ch-1:1 "Luchus amaro was a stone in Tococache" (Bauer 1998:140). It has been suggested that Ch-1:1 may correspond with Mesa Redonda, a circular carved stone which is part of a larger group (Muriel n.d.:3). This identification is based on the toponym Totocache or Tococache, which refers to the present Cusco district of San Blas at the eastern periphery of which Mesa Redonda is located. The place name is mentioned by Cobo, Albornoz, and Sarmiento de Gamboa.

The coordinates of this carved rock are 71 degrees 58 feet 14 inches W and 13 degrees 30 feet 33 inches S at an altitude of 3480 masl. It measures 4.90 by 4.40 meters and is 1.50 meters high (van de Guchte 1990:93). The name "Mesa Redonda" or Round Table is most certainly recent and is based on its circular form and flat surface. The most important carved features are a number of cutouts forming a stepped pattern on its western side. Van de Guchte interprets them as "bedding" to locate stones (1990:94), but they project out very little and may as well be purely aesthetic. Similar stepped patterns have been found on other carved outcrops. Mesa Redonda forms a group with three other carved rocks, two canals, and three terrace walls. If Ch-1:1 was indeed an anthropomorphized stone, then the tallest boulder called Cinca (Nose) measuring 15.50 meters by 9.60 meters would be the most likely candidate. From about 2002 to 2005, the Instituto Nacional de Cultura (now Ministerio de Cultura) conducted excavations, and they uncovered and restored at least one house foundation, demonstrating that Mesa Redonda constitutes one feature within a larger complex integrating boulder sculpture with architecture. The site has been turned into a protected area.

Ch-1:2 and Ch-1:3 were possibly among the most politically potent shrines and carved rock complexes within the entire zeq'e system. Cobo describes them as

> [Ch-1:2] The second *guaca* of the *ceque* was called Patallacta. It was a house which Inca Yupanqui designated for his sacrifices, and he died in it. The Incas who succeeded him thereafter made ordinary sacrifice here. In general, all the things which they consumed in sacrifice were offered for the health and prosperity of the Inca.
>
> [Ch-1:3] The third *guaca* was named Pilcopuquio. It is a fountain next to the house just mentioned from which an irrigation ditch issues. The Indians relate that when Inca Yupanqui had made that house for the sacrifices, he ordered that water to emerge there and afterward decreed that ordinary sacrifice should be made to it (Cobo 1990:51–54 [1653:Bk.13, Ch.13]).

Albornoz described Ch-1:2 as "Ancas pata is a cliff where the house of Viracocha Inca was" and Ch-1:3 as "Pilco puquio is a spring" (see Bauer 1998:140). His idea of Ch-1:2 is obviously very different from that of Cobo.

Bauer (1998:50–51) and Jeanette Sherbondy (1982:49, 189 n. 78) suggest that Ch-1:2 was Q'enqo Grande and Q'enqo Chico (for further discussion, see Christie 2012). Their reasoning is that several other writers in addition to Cobo link Patallacta with Pachakuti Inka Yupanki's body and that Patallacta is the name of the large hill and hacienda immediately to the north of the city, on which Q'enqo Grande and Q'enqo Chico are located.[3] Betanzos gives a very detailed account of what happened at Patallacta, which is worth quoting:

> After he was dead, he was taken to a town named Patallacta, where he had ordered some houses built in which his body was to be entombed. He was buried by putting his body in the earth in a large new clay urn, with him very well dressed. Inca Yupanque ordered that a golden image made to resemble him be placed on top of his tomb. And it was to be worshiped in place of him by the people who went there. Soon it was placed there (Betanzos 1996 [1557]:138 [part I, chap. XXXII]).

Pachakuti's statues were taken out of Cusco by Manqo Inka when he organized his revolt against the Spanish but later ended up in the hands of Francisco Pizarro. At the time of Betanzos's writing in the 1550s, only the body remained at Patallacta (Betanzos 1996 [1557]:139 [part I, chap. XXXII]). Pedro Sarmiento de Gamboa also reports that on his deathbed, Pachakuti told his son Thupa Inka Yupanki to care for his body and place it in his houses at Patallacta. Thupa Inka should have his father's image made in gold, and all the provinces he subjected should perform solemn sacrifices as well as the *purucaya* festival (see below). Unfortunately, Sarmiento de Gamboa does not describe the latter further (1942:140). In addition, he states that Pachakuti's body together with his golden portrait statue, estates, and servants were found by Polo de Ondegardo in the early 1570s in Tococache, which corresponds to the present district of San Blas in Cusco (1942:141–142). It is of further interest that Sarmiento de Gamboa mentions Patallacta in a listing of royal estates: "He [Pachakuti] built near Cuzco on a hill called Patallacta sumptuous houses and many more around Cuzco. He altered many streams so that they provided beneficial water and served recreational purposes; . . ." (1942:128 translation by the author). This reference is important since it implies that Q'enqo should also be seen in the context of royal estates (see below).

The third group of Q'enqo, named Kenko West by van de Guchte, now has to be integrated into any discussion of the site since it has been under investigation by the Ministerio de Cultura, and important new data have come to light. Ministerio de Cultura archaeologists refer to it as sector Qochapata. It encompasses the limestone rock outcrop just to the northwest of Q'enqo Grande and on the opposite side of the road to Pisaq. The Ministerio de Cultura has been excavating walls abutting the outcrop, small architectural enclosures, and a section of an Inka trail. Archaeologists have further cleared a set of terraces descending toward Q'enqo Grande. The terraces constitute an important feature which physically and linguistically correlates Qochapata, Patallacta, and Ancas pata, the name Albornoz gave to the Ch.1:2 wak'a. *Pata* is the Quechua term for anden or terrace and in the following, I will assume that Patallacta included Q'enqo Grande, Q'enqo Chico as well as Qochapata.

In the opinion of Sherbondy and Bauer, Pilcopuquio or Ch-1:3 is the now covered spring situated in the small ravine that runs between the ruins of Q'enqo Grande and Q'enqo Chico (Sherbondy 1982:49, 189 n. 78; Bauer 1998:51).

The site of Q'enqo is best known for some of the most extensive carved rock complexes in the Cusco area described in detail by van de Guchte (1990:143–151). They are located east of Saqsaywaman on the modern road from Cusco to Pisaq. As the name implies, the largest and most important group is Q'enqo Grande found at 71 degrees 58 feet 16 inches W and 13 degrees 30 feet 19 inches S at an altitude of 3,590 masl. Q'enqo Grande was explored in 1935 by Luis Valcarcel (1935a, b, c) and in the 1980s by van de Guchte (1990).

To the north of the carved outcrop extends a flat area which is enclosed by a concave, semicircular wall on its northern side. This wall was constructed in fine Imperial Cusco masonry style and contains nineteen niches and one staircase. Valcarcel assumed that the concave wall was originally two meters higher (1935a:223–233). To the north of the semicircular plaza stands a square structure which incorporates a basin or bath in the inside. This plaza was sealed off by a zigzag wall on the east side and by a rectangular structure on the west, however, leaving a narrow entrance between the structure and the outcrop (figure 2.2).

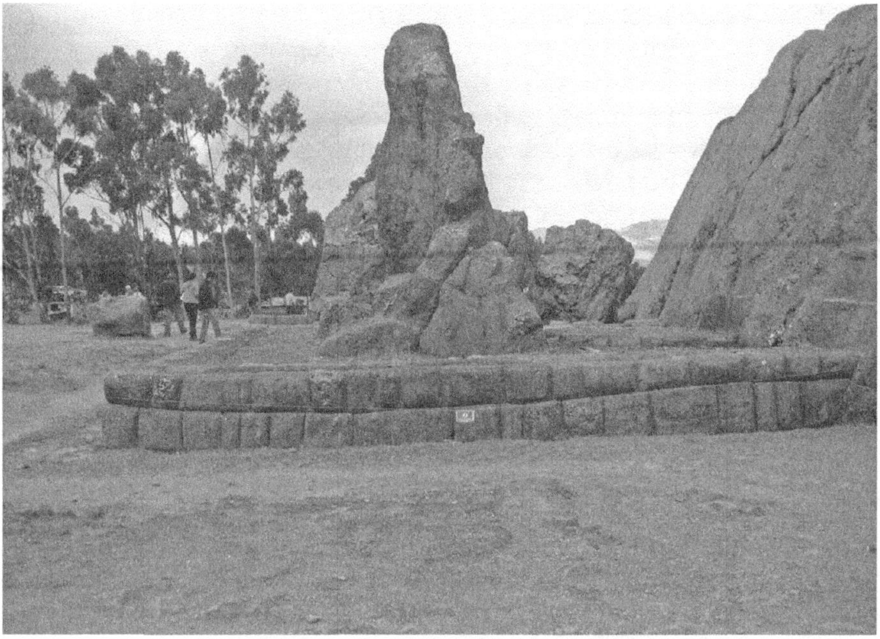

Figure 2.2. Q'enqo Grande, Semi-circular Plaza with Elevated Rock. Photograph by Brian Garrett.

The focal point of the concave wall and the semicircular plaza is a tall piece of rock placed on a masonry platform against the mass of the outcrop (figure 2.2). Although modified, this stone was never carved in the geometric style and has been variously interpreted as a phallic element (Terence Grieder, personal communication, 1982) and as a wak'a in the form of a seated puma, which is the most popular explanation. It is well possible that the Inka perceived this gigantic rock as Pachakuti's huanca related to his brother statues or *huauques* (*wauq'es*) on his tomb (see above and Duviols 1979b:17–18; and see below). The identification of this rock with a seated feline would further strengthen such a link since Pachakuti is often presented displaying puma elements. In any case, the rock was most likely mutilated in colonial times so that its interpretation remains guesswork. Van de Guchte argues that the platform walls were intended to be approximately two meters high because traces of cuts for the accommodation of building blocks can still be seen on the west side of the platform (1990:144). More traces for the "bedding" of stones are visible on top of the carved outcrop at its northernmost edge. Van de Guchte postulates the presence of another wall which would have effectively shielded the rock shrine from the eyes of viewers (1990:144–145).

It follows that the plaza with the elevated rock was a much more enclosed, secluded, and private area during Inka times than it is today. The twenty apertures in the concave wall could have been used as seats by selected individuals who would have gathered to observe certain performances and speeches held around the platform in the semicircular plaza. Some of such events may have been related to the rite of Purucaya, a funerary ceremony, which is mentioned in several chronicles. Betanzos describes this ceremony in chapter XXXI, a ritual which Pachakuti apparently had devised and ordained that it be performed for himself one year after his death and ever after for other noble Inkas (1996:134–137 [1557:part I, chap. XXXI]). Fifteen days before the actual Purucaya ceremonies, men and women of Cusco left the city, crying and carrying in their hands clothes Pachakuti had worn, his personal adornments, and the weapons with which he had fought. They visited places where he had stood and sat while alive and locations where he had accomplished his deeds and successes. At these places, they called out for him, showed him the personal items, and explained and reminded him what he had done with them while alive. Niles (1999:35–37, 45–48) discusses how such activities were part of an Inka strategy to ritually perform history and make it visible and audible in order to validate oral narratives. If Niles is correct, there is no doubt that many of the carved rock wak'as on the zeq'e lines and clearly those belonging to specified ayllus and panaqas functioned as places where personal histories were performed and ancestors remembered. Further, Betanzos's descriptions visualize and dramatize how Inka wak'as turned into places of ancestor worship (see also Glowacki and Malpass 2003:435–437). I find it reasonable to suggest that Q'enqo was one such place where elements of Pachakuti's biography were reenacted and where he might have been contacted.

Betanzos's (1996:134–136 [1557:part I, chap. XXXI]) description of the actual Purucaya ritual becomes more complicated and involved four men, each bringing ten women as well as a boy and a girl, all very well dressed. The men had long cords made of gold and fine wool tied to their belts; the women grasped and pulled the cords and thus ritually controlled them. Later, two squadrons of warriors came out from opposing sides of the plaza, representing Hanan and Hurin Cusco. They battled, and the people from Hurin Cusco acted like losers. Betanzos's description makes it quite clear that the main events just described were public and took place in the main plaza Awkaypata of Cusco. However, it is equally clear that the Purucaya was a month long, drawn-out ceremony which entailed numerous major and minor, more private activities, such as washing away the mourning and sacrificing sheep throughout the city in places where Pachakuti had spent time (Betanzos 1996:136–137 [1557:part I, chap. XXXI]). Some of the sacrifices may well have been performed at private locations, such as the semicircular enclosed plaza, within the Q'enqo sectors. Evidence that the Purucaya (or *Puroca*) became fairly common and was continued to be performed after Pachakuti's death comes from Ramos Gavilan (1988:138–139 [1621:Bk.1, Ch.22].

The plaza area of Q'enqo Grande has to be understood in connection with the immense sculpted outcrop which descends to its south. Access to the upper part is possible by means of a series of steps which were cut into the west side of this mass of limestone rock. The carvings are concentrated in the northern section while the southern section displays very few. The great majority of the carvings consist of vertical and horizontal cuts creating flat and smooth areas which form steps, platforms, and perhaps seats. Of particular interest are a recessed area which may have been used as a bath or basin, two gnomons or circular knob-like carvings, and a zigzag channel cut into the rock surface (figure 1.5). The latter has given Q'enqo its name. The Quechua word to describe such a zigzag is *q'enqo*, but it functioned as a *paccha*, which is a tool used for channeling liquids during divination rituals. An authoritative study of pacchas was conducted by Rebecca Carrion Cachet in which she illustrates several portable wooden pouring devices with zigzag channels (1955). At Q'enqo Grande, a liquid would have flowed down the zigzag, which makes a slight turn and eventually splits in two ends. It may have been important which direction and which arm of the zigzag this fluid would take. I will return to the q'enqo zigzag canal later.

Another notable feature of Q'enqo Grande is its interior chamber. It is a natural crevice which has been sculpted into a cave-like interior chamber crossing the rock mass in an east-west direction. It is here that a rectangular block has been carved into multiple horizontal planes with finely polished surfaces and sides (figure 1.9). Steps on its west side indicate that one could ascend the top platform. Behind this main sculptural work, a blind window or niche was cut into the rock wall, and a trench descends about 1.60 meters deep. Given the subterranean location of the main carving, it seems to have established a metaphorical access point to the interior of the earth. The symbolically charged position of this altar is further intensified by light patterns which originate from an opening close by the western exit of the cave. Valcarcel retrieved human remains and a plate from this interior chamber (1935a:232).

Van de Guchte proposes that part of a report by Pedro Pizarro may refer to the interior passage of Q'enqo:

> In one cave they discovered twelve sentries of gold and silver, of the size and appearance of those of this country, extraordinarily realistic. There were pitchers half of pottery and half of gold, with the gold so well set into the pottery that no drop of water escaped when they were filled, and beautifully made. A golden

> effigy was also discovered. This greatly distressed the Indians for they said that it was a figure of the first lord who conquered this land. They found shoes made of gold, of the type the women wore, like half-boots. They found golden cray-fish such as live in the sea, and many vases, on which were sculpted in relief all the birds and snakes that they knew, even down to spiders, caterpillars and other insects. All this was found in a large cave that was between some outcrops of rock outside Cuzco. They had not been buried because they were such delicate objects (Pizarro cited by van de Guchte 1990:147).

Van de Guchte concludes that the gold effigy mentioned in this text could have been one *huauque*, which means brother or double, of Pachakuti Inka Yupanki (1990:147).[4] Based on the available dates, this is possible but somewhat questionable.[5]

Q'enqo Chico is situated 25masl lower than Q'enqo Grande. The two are separated by the above mentioned ravine and a distance of approximately 40 meters. As at Q'enqo Grande, the nucleus of Q'enqo Chico is a mass of limestone; but unlike Q'enqo Grande, the Inka built a wall around it made of finely cut and irregularly shaped stones. The space between limestone and wall is packed with earth. The entrance to this structure was evidently by means of a staircase on the north side. This staircase was built roughly in the center of the north façade, but it curiously ends about 2.40 meters above present ground level. Perhaps there would have been a wooden ladder leading all the way down, or the ground may have been higher during Inka times.

On top of the outcrop of Q'enqo Chico, most of the carvings are vertical and horizontal cuts forming seat-like sculptures. They all face in different directions. At two or three of these potential seats, postholes are visible along the elevated edge (figure 1.6). Similar postholes are found at the so-called Water Temple of Ollantaytambo (see also van de Guchte 1990:195–200). I speculate that such holes may have held wooden posts which supported a perishable roof, perhaps made of cloth. If this was so, these particular seats may have functioned as true thrones foregrounded by a canopy which provided shade and was possibly decorated, upon which the most distinguished individuals presented themselves. Guaman Poma de Ayala (1987:407 [1615:398]) provides a drawing of such a throne-like seat (figure 2.3).

Other carvings are arranged around curving rock walls. These curving and highly sculptural walls are not natural but appear to be man made (figure 1.4). I recognize that this is a stylistic element peculiar to Q'enqo Chico and a few other sites, which deviates from the general linear vertical and horizontal cuts in Inka sculpted boulders. Van de Guchte noticed the foundation wall of a structure of finely cut stones which enclosed other cut stones (1990:149)—a feature which I could not locate. Nearby and possibly associated with Q'enqo are terraces (to the south) and more carved rocks as well as the foundation

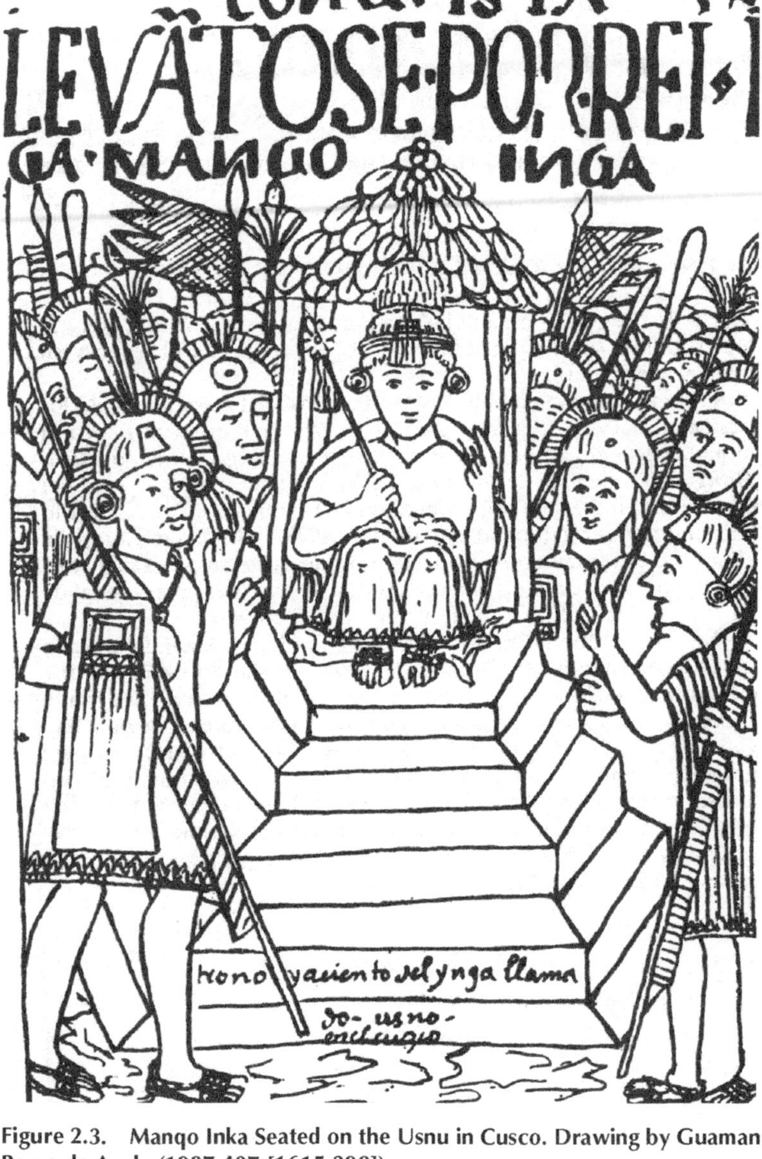

Figure 2.3. Manqo Inka Seated on the Usnu in Cusco. Drawing by Guaman Poma de Ayala (1987:407 [1615:398]).

walls of additional buildings (to the north, see Qochapata below) which were under investigation by the Ministerio de Cultura in 2007. This may indicate that the site of Q'enqo was indeed much larger than the two carved rock complexes and resembled a royal estate, which Sarmiento (cited above) implies.

A comparison of the formal features of Q'enqo Grande and Q'enqo Chico may offer one fruitful methodological avenue to deduce function of this site. At Q'enqo Grande, the spatial focus is the semicircular plaza framed by the walls in high-quality Cusco-style masonry as well as by the sculpted outcrop. The open space of the plaza implies use for gatherings of moderately sized audiences who may have turned toward the stone wak'a elevated on its platform. The carved zigzag canal on top of the outcrop suggests that it was used in ceremonies during which liquids were poured which ran down the channel and metaphorically the entire outcrop to be received by the earth. Thus the setting of open space, architecture, and sculpture seems to point toward a ritual use. The formal language of Q'enqo Chico is quite different: the architectural retaining wall of irregularly shaped but precisely fitted stone blocks elevates the carvings. The carving style emphasizes seats with postholes to support canopies; another foregrounded seat with stepped symmetrically sculpted side panels looks toward the nevado Ausangate. Therefore form indicates a realm for presentation of elite individuals or special objects although open spaces for an audience are lacking. Based upon the formal analysis of the carvings, architecture, and open space, I interpret Q'enqo Grande as a locale where semipublic ceremonies were held. Semipublic events would have included audiences of moderate size who were probably members of Pachakuti's court and after his death of his panaqa. Q'enqo Chico, on the other hand, was a private place where Pachakuti may have spent time alone or accompanied by only a few assistants. Further, due to the orientation of some sculpted seats, it was a location from which visual and metaphorical contact with the surrounding mountain landscape was established. Thus the orientation of Q'enqo Chico could be described as outward looking while Q'enqo Grande was clearly oriented inward: at the latter, visual and functional relations interacted between the architectural seats and the elevated wak'a and on top of the outcrop, attention was drawn downward toward the carvings.

The public–private dichotomy could be extended one level further. The most spacious and open area within Cusco was the main plaza Awkaypata where the more public performances of the Purucaya with the largest number of participants and the greatest audiences took place. Semipublic rites of the Purucaya, such as the sacrifices, centered upon Pachakuti's court, and his panaqa would have been conducted at Q'enqo Grande while Q'enqo Chico was used as a private sector. The staggered opposition public–private between Awkaypata–Q'enqo Grande and Q'enqo Grande–Q'enqo Chico might possibly be translated into Hanan–Hurin (see below under *Inka Landscape and Carved Rocks*).

The third group which may now be added to Q'enqo is the sector Qochapata within the Archaeological Park of Saqsaywaman (Rodriguez Limache 2007) referenced as Kenko West on van de Guchte's plan.[6] It is situated on the northwest side of the road to Pisaq opposite from Q'enqo Grande. The area of this massive limestone outcrop has long been known as a location where New Age spiritualists come to leave their offerings. In 2007, the Ministerio de Cultura began a multiyear project of excavation, conservation, and restoration (Rodriguez Limache 2007). They have shown that a rectangular perimeter wall abuts the outcrop to the west. The south and west sides of this wall are lined by irregular structures. Further to the west sits a north-south alignment of four almost symmetrical one-room enclosures with central openings. A set of terraces reinforces the slope and accommodates additional small buildings immediately southeast of the outcrop. Oscar Rodriguez Limache (2007:211) has determined that these terraces functioned as retaining walls and had no agricultural use. Below the terraces and following the direction of the road, a straight section of an Inka trail has been exposed. The sector of Qochapata thus centered on the rock outcrop which exhibits carvings in the style of vertical and horizontal cuts and was linked with the two architectural groups to the west.

Besides unearthing wall foundations, archaeologists have discovered huge quantities of associated artifacts (Rodriguez Limache 2007:206–208). One aribalo and three Spondylus shells were found in the northwest part of the sector, the great majority of the lithic items came from the east side of the outcrop, six metal objects were located in the southern part, and worked bone fragments were encountered throughout the site. Rodriguez Limache and colleagues (2007:207) counted a total of 7,200 ceramic fragments, 99.85 percent of which are Inka. The overwhelming ceramic evidence and a majority of architectural features establish that Qochapata was built and used by the Inka.

Based upon its close vicinity and similar material features of a carved rock outcrop connected with architecture, I argue that Qochapata was part of Q'enqo and thus of Patallaqta as described by Cobo. The present state of knowledge suggests that Qochapata was linked to an architectural complex more extensive than the architecture integrated with Q'enqo Grande or Q'enqo Chico.[7] Therefore it seems reasonable to argue that support personnel of Pachakuti's shrine might have been housed and objects and offerings used in the rituals at Q'enqo Grande and Q'enqo Chico would have been stored here. The Spondylus shells were clearly intended as ritual offerings.[8] In reference to the ceramics, 1,887 sherds were Inka Polychrome versus 5,021 in Inka Monochrome (Rodriguez Limache 2007:207). These numbers may imply that more than half of the ceramics did not come from elite ware but from simpler vessels used for storage or domestic activities. I reason that Qochapata was likely the support sector for the semipublic ceremonies of Pachakuti's court held at Q'enqo Grande and for Pachakuti's private space at Q'enqo Chico.

Sherbondy (1982:49, 189 n. 78) and Bauer (1998:51) identify Ch-1:3, the spring named Pilcopuquio, with the spring in the small ravine that runs between the ruins of Q'enqo Grande and Q'enqo Chico. Bauer found references to a spring called Pilcopuquiu in a number of documents concerning the lands immediately north of Cusco. In addition, his surveyors found recently made offerings beside the spring capped with cement for decades, as recently as 1991 (1998:51). Cobo relates that Pachakuti Inka Yupanki created this spring and "afterward decreed that ordinary sacrifice should be made to it" (1990:54 [1653:Bk.13, Ch.13]). Thus it appears that this tradition is still being followed in some way.

To conclude, existing evidence supports that Q'enqo Grande, Q'enqo Chico, Qochapata, and the spring Pilcopuquio(u) were the materializations of Ch-1:2 and Ch-1:3. They formed a large ceremonial complex which harmonized the artistic media of sculpture and architecture with the natural landscape and water. The complex of Q'enqo was the design of Pachakuti Inka Yupanki and reflected his vision of organizing principles for Cusco (Christie 2012). It was most likely a locale where he spent time while alive and where he was mourned and summoned after death.

At the same time, Cobo claims that the first zeq'e of Chinchaysuyu "was the responsibility of the kinship unit and *ayollo* of Goacaytaqui" (1990:51 [1653:Bk.13, Ch.13]). According to Bauer, Goacaytaqui was a nonroyal ayllu and not connected with the Inaca (or Hatun) Panaca of Pachakuti Inka Yupanki (1998:46, table 4.8). One explanation for the discrepancy that Ch-1:2 and Ch-1:3 were monuments so important to Pachakuti although his own panaqa was not in charge of this zeq'e line may be that the associations of kinship units with zeq'es changed and shifted over time. In other words, the zeq'e system and its inherent responsibilities may have functioned as a living agent-driven system in which nonroyal ayllus and royal panaqas could assume and transfer responsibilities for specific zeq'e lines and wak'as as needs arose and circumstances changed.

The second zeq'e was under the care of the Vicaquirao Panaqa, the descent group of Inka Roq'a, the sixth ruler. Ch-2 thus constitutes the royal zeq'e in the cluster Ch-1, Ch-2, and Ch-3. This connection to the imperial government is significant in the interpretation of Ch-2:4, which Cobo describes as

> [Ch-2:4] The fourth guaca was called Viroypacha; it is a conduit of fairly good water which was declared a guaca by Inca Yupanqui. It was prayed to for the tranquillity of the Inca (Cobo 1990:54 [1653:Bk.13, Ch.13]).

It is interesting to note that although the second zeq'e was the responsibility of the panaqa of the sixth ruler, Inka Roq'a, its fourth shrine was established by Inca Yupanqui/Inka Yupanki, a name which probably refers to Pachakuti Inka Yupanki, the ninth ruler, or to his son, Thupa Inka Yupanki. This supports the fluid structure of the zeq'e system as suggested above. Further, a vital point of Zuidema's (1964) arguments is that zeq'es and suyus ordered marriage classes, which underscores that the Inka zeq'e system functioned on multiple levels.

While Muriel locates Ch-2:4 on the east side of Q'enqo (n.d.:5), Sherbondy (1982:49) thinks that it was a segment of the Choquechaca/Tullumayo River that flows from Saqsaywaman through Cusco to join the Huatanay and Bauer proposes more specifically that Ch-2:4 was a waterfall in this river and an adjacent carved rock complex which is now called Sapantiana (1998:53). He refers to two of Guaman Poma de Ayala's drawings which contain locations labeled Uiruypaccha which Guaman Poma describes as a pond and a fountain in one of the palaces of the Inka (Poma de Ayala in Bauer 1998:53) and to local land documents which mention Viroypacha.

Sapantiana is located at 3,570 masl and its coordinates are 71 degrees 58 feet 53 inches W and 13 degrees 30 feet 28 inches S. Next to it flows the Tullumayo River, also referred to as Choquechaca, which comes down from the hills of Saqsaywaman, cascades over a colonial construction, and continues under Choquechaca Street to join the Huatanay River. The rock complex measures approximately 21.50m from north to south and 16.80m from east to west. Sapantiana is a limestone outcropping with carvings that consist primarily of vertical and horizontal cuts forming platforms, planes, and steps (figure 2.4). The finest carvings can be seen on the northwestern edge of the rock. The rock complex was surrounded by containing walls, and two or three single one-room structures were integrated with it. Valcarcel (1935c) recorded a southeastern and a western building. I noticed the faint outlines of the bedding stones composing the western structure, and I recorded the eastern wall of the southeastern structure as well as a chamber-like construction added to the rock on its northeastern side. The latter was also documented by van de Guchte (1990:92). Valcarcel (1935c:5) found some human remains but no artifacts associated with Sapantiana.

Accepting Bauer's identification of Ch-2:4 with the waterfall in the Choquechaca River and the adjacent carved rock complex Sapantiana (1998:53), it is possible to reconstruct and visualize Sapantiana during Inka times further by looking more closely at the drawings and text by Guaman Poma de Ayala. The first drawing referenced by Bauer is number 316 titled *Canciones y Mucica*. It illustrates two young men playing flutes on a hill and two women swimming in the *uatanay mayo* or Huatanay River at the base of this hill. In the background, there are the drawings and names of three other major water

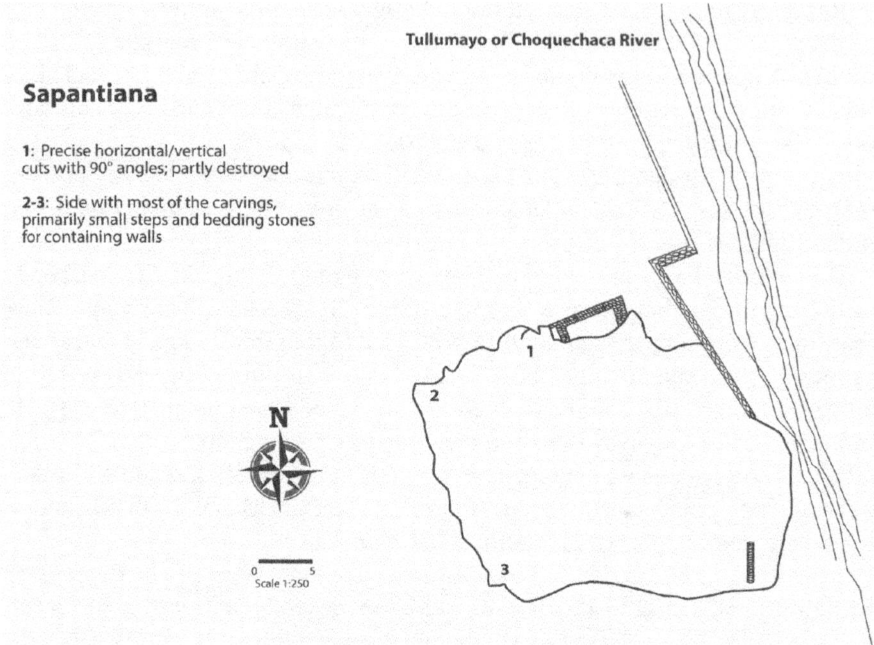

Figure 2.4. Sapantiana and Associated Features. Drawing by the Jessica Christie, Brian Garrett, Martin Carillo, and Royer Carillo.

sources for Cusco, one of which is *uiroypaccha*. The second drawing is number 1051, Guaman Poma de Ayala's map of Cusco. *Uiroypaccha* is marked in the upper right-hand corner next to a building group labeled *s. blas*. Although Sapantiana borders the present neighborhood of San Blas, their spatial relationship is reversed on Guaman Poma de Ayala's map.

Two passages in Guaman Poma de Ayala's text provide additional insights. In his chapter on the administration of the Inka, there is a paragraph on *regalos* (presents) in which he describes the things the Inka emperor had and did in his leisure time.[9] The second reference comes from the chapter about the history of the queens, captains, and ladies. Guaman Poma de Ayala reports of the eleventh queen, *Raua Ocllo.*[10] In the first reference, Guaman Poma inventories "water fountains *Uiruy Paccha*" as features of an Inka leisure palace. As noted above for Q'enqo, *paccha* signifies a device for channeling liquids, such as a fountain. In the second reference, I am assuming that *Uiroy Pata* referred to the same locale as *Viroypacha*. Gonzalez Holguin (1989:268 [1608]) lists *pacha* as "tiempo [time], suelo [ground], lugar [place]." In modern Quechua, *pacha* means "the earth, the world, our planet" and *pata* may signify "shore, sideline, high elevation, step and terrace" (see above). *Uiruy,*

Viroy-, and *Uiroy* share the same root; in general terms, both *pacha* and *pata* are spatial designations, although *pacha* emphasizes the open universal space and *pata* seems to imply specific individual places, probably the terraced shoreline of the Choquechaca/Tullumayo River on which Sapantiana sits (see Tilley 1994:8).

Thus Guaman Poma de Ayala's textual passages describe Uiruy Paccha/ Viroypacha as connected with a royal pleasure palace where exotic non-native animals were kept and music was played for the qoya/queen. The insistent association with water makes Sapantiana certainly a likely candidate for this location. The remaining inconsistency with chronology—that is, Cobo attributes the build-out of Sapantiana to Pachakuti or Thupa Inka, but Raua Ocllo was the eleventh queen or wife of Wayna Qhapaq—may be explained by the suggestion that the leisure palace of Sapantiana served several generations of rulers. Still, in 1935, Valcarcel (1935b:3) remarks that the houses in the environs of Sapantiana were residences of the nobility who wanted to live here because the climate is warmer than in the rest of the capital.

The sixth shrine on the fourth zeq'e of Chinchaysuyu is an important carved rock shrouded in a number of mythological tales. Cobo introduces it as follows:

> [Ch-4:6] The sixth guaca was a large stone named Collaconcho which was in the fortress. They declared that, bringing it for that structure, it fell three times and killed some Indians. The sorcerers said that it had replied to questions they put to it that, if they persisted in wanting to put it in the structure, all would have a bad end, apart from the fact that they would not be able to do it. From that time on it was considered a general guaca to which they made offerings for the strength of the Inca (Cobo 1990:56–57 [1653:Bk.13, Ch.13]).

According to oral tradition, the so-called Piedra Cansada or Tired Stone situated some one hundred meters to the north of the circular depression within Saqsaywaman has been associated with Ch-4:6 (see van de Guchte 1990:130–139 and Bauer 1998:56–58, who support this identification; and Paternosto 1996:83–85, who does not concur). It is a mass of bedrock limestone measuring approximately twenty-five meters across. Three carved staircases provided access to the top where most of the carvings form steps, platforms, and rectangular horizontal planes of varying sizes. Of particular interest are two carved oval-shaped depressions and a stepped ornamentation executed in high relief in which the steps outline empty negative space. Possible motives of Inka artisans to use formal inversions of positive 3-D mass and negative void space will be taken up in chapter 7. This latter design is found on the northeast side of the outcrop where a large rock section with part of the design has broken off.

During the writing of this book, the Ministerio de Cultura has excavated around the perimeter of Piedra Cansada known as Chincana Grande among Cusco archaeologists (Socualaya Davila 2007). To the north of the rock, four terraces were cleared. On the east side was a rectangular building with two niches executed in coarse masonry. The rock section split off from the main boulder fell on one side of this structure, and because of this and prior undocumented exacavations, the cultural context has been considerably disturbed. Abutting the south side of Piedra Cansada, a small architectural compound has been uncovered which terminates at another terrace wall. The differences in wall construction between the precisely cut and fitted stones of the compound and the irregular blocks in the terrace are striking. Besides the fine masonry associated with high status, the structure displays a carved canal linked to a potential fountain basin. To the southwest of Piedra Cansada, the Ministerio de Cultura has identified more water channels leading to the reservoir of Saqsaywaman as well as the location of a storage building. This storage building contained grinding instruments and large ceramic vessels leading Socualaya Davila to conclude that it was used as a support structure for rituals held at Piedra Cansada. The ceremonial character of Piedra Cansada is further corroborated by finds of tupus, small idol figurines executed in metal, and fragments of Spondylus shells reported by Wiener near the rock (cited by Socualaya Davila 2007).

The Tired Stone appears in the chronicles of Pedro Cieza de Leon (1553), Gutierrez de Santa Clara (1590), Garcilaso de la Vega (1605), Fray Martin de Murua (1613), and Cobo (see above). Guaman Poma de Ayala includes a drawing of the—uncarved!—stone in transport, which is titled The Ninth Captain Ynga Urcon and accompanied by the following explanation:

> He was the son of Topa Ynga Yupanqui and was in charge of bringing stones from Cuzco to Guanoco. They say that the stone got tired and did not want to move and that this stone cried [tears of] blood. And it stayed there until today, that his son Guayna Capac Ynga had the stone taken to Quito, Tomi [Pampa], Nobo Reyno from the city of Cuzco, Yucay, so many thousand miles (Guaman Poma 1987:154 [1615:162, 159]).

Guaman Poma and the other accounts narrate that the stone came from a distant place or was on its way to a distant location; at a certain point the stone wept blood, talked, and did not want to go any further. Therefore it was called the Piedra Cansada, the Tired Stone. The new archaeological evidence speaks strongly against the nomadic character of the carved rock complex and shows that the Inka embedded it into a permanent ritual context.

The fifth zeq'e was possibly controlled by the Inaca (or Hatun) panaqa of Pachakuti Inka Yupanki (Bauer 1998:46) and as such constituted one of the most significant lines within the zeq'e system. Cobo's description leaves little doubt that the sixth shrine on this zeq'e corresponded with a well-known carved rock.

> [Ch-5:6] The sixth guaca was named Sabacurinca; it was a well-carved seat where the Incas sat. It was very venerated, and solemn sacrifices were made to it. On account of this seat, the whole fortress was worshipped; for the seat must have been inside or next to the fortress (Cobo 1990:57 [1653:Bk.13, Ch.13]).

Most researchers (Bauer 1998:61; Muriel n.d.:14; van de Guchte 1990:10, 72, 123–126) identify Ch-5:6 with the so-called Throne of the Inka, a magnificently carved series of stepped planes (figure 1.3). It is located north of the zigzag walls of Saqsaywaman and overlooking the esplanade. The throne was sculpted out of a large outcrop composed of diorite, a stone which is extremely hard to carve, and the entire hillside is known as Rodadero or by its Quechua name Suchuna. Van de Guchte includes in the identification of Ch-5:6 a second seat-like sculpture located approximately three meters up to the west of the stepped platforms. The steps are designed in such a manner that one plane stands out as being the highest, and three planes descend on its northern side and nine planes on its southern side. Van de Guchte took precise measurements which demonstrated that the dimensions of these planes are remarkably uniform, imprinting an impression of rigid geometry in the eye of the observer. In front of this series of stepped planes are another series in a smaller scale as well as additional carved seats. Formally these planes have the contour of a stepped design. Stepped designs constitute a special category of carvings, and other examples occur at the Chingana Chica to the east of the circular depression of Saqsaywaman, at Quilla Rumi, and at the Temple of the Three Windows in Machu Picchu. Cases of formal inversion of the stepped design in which the steps are projected upside down into negative space can be found at Piedra Cansada described above, in Quespehuara, and in Chinchero.

If we take Cobo's information literally, then the Inka, including Pachakuti, would probably have been seated on the highest planes, and offerings would have been presented and sacrifices would have been made here. The centralized and most elevated seating areas were the top plane as well as the sculpted seat higher up. From these viewpoints, one enjoys a "VIP" vision of the ancient and present performance space of the esplanade and a magnificent panoramic view of the landscape surrounding Cusco. Carolyn Dean (2010:108) makes the important connection that rulers became personified *wank'a*, a type of empowered stone understood as owner of territories, by sitting on stone and overseeing space.[11]

There is a second role Sabacurinca and other, similarly carved boulders possibly played. Its stepped planes might have been used as a counting device similar to *yupanas* and structurally parallel to *khipus*. Yupanas are grid-like arrangements of square concave boxes[12] used to count objects; the sculpted planes, platforms, and steps project out in a convex fashion yet share the geometric design. A khipu is a system of colored strings consisting of a main cord and multiple knotted pendant cords. Each pendant cord documented the count of objects in one category in such a manner that groups of knots marked decimal positions with the highest positions closest to the main cord. Thus the act of counting was conducted with the yupana, and the established quantities were then recorded on the khipu. Cobo stated that "solemn sacrifices were made to it," and the individual sacrifices might have been counted before consumption on the different planes of Ch-5:6, taking on the value of decimal units. Each descending row of platforms could have functioned like a string in a khipu. A supervisor or *khipu kamayuq* might have been positioned on the top platform, and an assistant would have counted the objects or offerings. Each time the assistant counted a unit of ten, the khipu kamayuq would have placed a small stone on the second plane below him. Once the count reached ten units of ten, he would have positioned one stone on the third plane and cleared the second. The counting process may have continued like this along both rows of platforms, each calculating one object category. This argument was fully explained elsewhere (Christie 2005c, 2009). I suggest that Sabacurinca held the double function of a distinguished seating area and of a locale to present and count offerings.

The following seventh shrine is described by Cobo:

> [Ch-5:7] The seventh guaca was named Chacaguanacauri. It is a small hill which is on the way to Yucay, where the young men who were preparing themselves to be orejones went for a certain grass which they carried on their lances (Cobo 1990:58 [1653:Bk.13, Ch.13]).

Albornoz lists his shrine #29 as "Chaca guanacauri, a stone above the fortress" (in Bauer 1998:61, 139).

Sherbondy (1982:81), Zuidema (1986:192–193), van de Guchte (1990:108–110, 371), and Bauer (1998:61–62) concur that the archaeological complex of Chacan should be linked with Ch-5:7. Bauer observes that Cobo mentions a hill called Chacaguanacauri as an important place of activities during the opening days of the Capac Raymi festival and the concurrent Warachikoy ritual as well as a reference to a landmark in this area as Chacan-Guanacaure in a 1578 land document (1998:61–62).[13] Bejar Luksic, Durand Caceres, and Vranich (2010) speculate that the Wanakauri idol would have been transported to this site to preside over elite ceremonies.

The main feature of Chacan, locally known as Balcon del Diablo, is a tall limestone outcrop. The Chacan (or Saphi) River has cut a huge cavern through its base, and a stone-lined canal crosses the river on the upper level of the outcrop. This feature of a canal feeding off and then crossing the same river over a natural bridge is unique in Inka hydrological planning and must have held a special place in the Inka discourse with their natural surroundings. The upper portions of the outcrop were adorned with Inka stonework.

Bejar Luksic, Durand Caceres, and Vranich (2010) excavated around two main carved rocks and inside a small rectangular structure associated with Chacan. They recorded the remains of a channel attached to one carved rock and draining down to the Chacan River. Within the building, they found red and yellow floors and large-scale offerings of maize kernels, fruit, as well as remains of young camelids, ceramic vessels, metal artifacts made of silver alloy, and a possible tomb. These types of evidence point to an active Inka ritual site to which elites contributed offerings and where ceremonies involving the flow of liquids were staged.

The Sculpted Rock Wak'as in Antisuyu

The first zeq'e line in Antisuyu was under the care of the Socso panaqa of Wiraqocha Inka, the eighth ruler. The connection with Wiraqocha is of relevance in understanding the context of the seventh wak'a, which Cobo introduces in the following manner:

> [An-1:7] The seventh guaca was called Amaromarcaguaci; this was a house of Amaro Tupa Inca, which was on the road of the Andes [i.e., the montana] (Cobo 1990:63 [1653:Bk.13, Ch.13]).

Sherbondy (1982:83–84, 193–194 n. 86), van de Guchte (1990:72, 152–165), Rowe (1997:286), and Bauer (1998:75–76) think that a large carved rock outcrop currently called Llanllakuyoq may represent Amaromarcaguaci (An-1:7). Bauer substantiates this identification by referring to a 1636 document which states that an area with the name Amarumarca was near the Curimarca road. The road which passes Llanllakuyoq today leads from Q'enqo to Curimarca via Tambomachay and continues on to Pisaq. On the other hand, Cusco archaeologists excavating for the Ministerio de Cultura reason that An-1:7 must be Laqo (Guillen Naveros 2007).[14] In the following, I will take the position that Llanllakuyoq most likely was Amaromarcaguaci.

The rock Llanllakuyoq or Amaromarcaguaci is located just to the west of the Pisaq road at an altitude of 3,725 masl and 71 degrees 58 feet 28 inches W and 13 degrees 29 feet 32 inches S. It is cradled by two branches of an important canal in the Amaru Drainage System (see Sherbondy 1982:37–41, 64–69; Van de Guchte 1990:374–376 and illus. 23, 24). These branches unite in the proximity of a reservoir situated a short distance to the southwest of An-1:7 and then split again. The rock complex stuns the visitor through its monumental scale and its bewildering complexity of carvings, caverns, walls, and associated carved rocks. I follow van de Guchte's ordering of the site into a rock-architectural unit to the south, the zigzag wall, and the carved bedrock of Amaromarcaguaci itself with the accompanying buildings (Van de Guchte, ill. 62).

The rock-architectural unit consists of an overhanging wall of bedrock framed by stone masonry walls on three sides. A trapezoidal platform with a polished top sits against the back wall. The form of this platform and the closed space evoke the concept of an altar upon which offerings may have been laid. In between this unit and the outcrop of Amaromarcaguaci stand a long retaining wall and an intriguing zigzag wall which echoes the three-tiered zigzag wall of Saqsaywaman.

The outcrop presents itself as a large mass of bedrock which is penetrated by a variety of passages and corridors. Four major corridors enter the rock and continue to a considerable depth. Most of them are natural fissures which were modified and "embellished" by Inka artisans. They polished the sides of the corridors and lined them with shallow shelf-like projections.

The west side of the rock exhibits the most extensive carvings. They primarily consist of vertical and horizontal cuts sculpting planes, niches, and blind windows. A section of this side is fronted by a wall repeating the contour of the rock. In the northwest corner of the outcrop, the walls of two structures remain standing. They are linked to the bedrock by one additional wall and by the fact that their doorways are oriented toward the entrance of a major corridor. On top of the outcrop and approximately in its center, there are finely smoothed vertical planes with a projecting knob as well as a horizontal plane, both accessed by a stairway.

Amaromarcaguaci is surrounded by numerous smaller rocks located to the north and east. Van de Guchte describes these features in depth and elaborates on their relations via sight lines with the main outcrop (1990:155–158). His discussion further underscores the importance of visibility between carved rocks, sites, and landscape features.

The question to be explored is if there would indeed have been a historical connection between the material features just described and Amaromarcaguaci, meaning a "house of Amaru Thupa Inka," as Cobo states? The chronicles provide useful background information about Amaru Thupa Inka (summarized by Rostworowski 1953:235–242; Niles 1987:18–20; and van de Guchte 1990:159–160).[15] According to Gonzalez Holguin (1952 [1608]), *Amaromarcaguaci* may be translated as "storage house of Amaro" (see also van de Guchte 1990:158). Although this linguistic evidence leaves open whether the storage house belonged to Amaru Thupa Inka, the high official and "Segunda Persona" in Thupa Inka's government, or to Amaro/u, the serpent dragon, the first attribution has more support. Historical documents make it clear that the Amaru canals irrigated lands which belonged to Amaru Thupa Inka (Sherbondy 1982:37–41, 64–69). The architectural features of the two structures, such as coarse masonry and doorways opening toward one of the cave-like passageways, make them unlikely candidates for any kind of elite residence or ritual building. A storage function in association with the cave opening would appear more probable.

The second possible attribution could be to Amaro/u understood as the serpent dragon. Van de Guchte (1990:158–162) and Sherbondy (1982:65–68) explore the complex symbolism of the amaro/u concept and conclude that it is linked with ancestors, rivers, rains, caves, and procreation. I suggest that amaro/u is an unlikely literal house owner but surely part of the metaphorical language Amaru Thupa Inka employed in the articulation of his house Amaromarcaguaci.

The third zeq'e line contains two wak'as which seem to correspond with important carved rock complexes to the northeast of Cusco. Cobo describes these wak'as as follows:

> [An-3:0] The third ceque was named Cayao, and it had ten guacas.
>
> [An-3:4] The fourth was called Chuquimarca; it was a temple of the Sun on the hill of Mantocalla, in which they said that the Sun descended many times to sleep. For this reason, in addition to everything else, they offered it children.
>
> [An-3:5] The fifth guaca was called Mantocallaspa; it was a fountain of good water which is on the above-mentioned hill where the Indians bathed.
>
> [An-3:6] The sixth was called Mantocallas, which was a hill held in great veneration on which, at the time of shelling maize, they made certain sacrifices. For these [sacrifices], they placed on the said hill many bundles of carved firewood dressed as men and women and a great quantity of maize ears made of wood. After great drunken feasts, they burned many sheep with the said firewood and killed some children.
>
> [An-3:7] The seventh guaca was named Caripuquiu; it was a fountain which is on the slope of the said hill. They offered shells to it (Cobo 1990:65 [1653:Bk.13, Ch. 14]).

At this time, the third zeq'e of Antisuyu cannot be associated with a particular ayllu. However, it was the cayao zeq'e in the cluster An-1, An-2, An-3. An-1, collana or the zeq'e with the highest rank, was under the supervision of the panaqa of Wiraqocha Inka (the eighth ruler). It is reasonable to suggest that An-3 was maintained by a nonroyal ayllu.[16]

The carved rock complexes of Laqo and Kusilluchayoq are candidates for An-3:4 and An-3:6. Laqo is the largest carved rock complex in the northern part of the valley of Cusco and is situated at 71 degrees 57 feet 53 inches W and 13 degrees 30 feet 11 inches S at an altitude of 3,640 masl. A narrow natural corridor divides Laqo into a western and eastern sector. The western sector is by far the larger one and exhibits elaborate carvings on its south side. These carvings consist mostly of vertical and horizontal cuts forming steps, platforms, and niches. There is a cave entrance leading to a longitudinal, carved interior chamber. Walking around Laqo in a clockwise direction, the visitor passes the west and arrives at the northern side. From the northwestern corner, two sets of steps ascend to the top of the outcrop. The lower set consists of five steps carved in an unusual manner so that the western half of each horizontal plane is consistently higher than its eastern counterpart. It appears that aesthetic considerations merged with the functional aspects of the steps.

Climbing the steps just described, the visitor reaches the upper platform, which stands out for the complexity of its carvings. There are many vertical and horizontal cuts forming seat-like sculptures. Of special interest are a low and finely carved rectangular platform, a projecting knob or gnomen located at the southern edge of the top platform (figure 1.7), as well as the high relief carvings of five animals also found at this outer southern edge (figure 1.8). Two stylized four-legged animals, possibly felines with missing heads, seem to frame two bird figures with outstretched wings. The heads of these birds have never been executed, but in their stead, a circular hole was carved. Less than half a meter away from this group, a relief in the form of a snake has been documented.

The eastern sector of Laqo is much smaller than the one to the west, and it extends east of the natural crevice. Most carvings occur again on the southern side. This side may have functioned as a public façade from which the outcrop was meant to be approached in Inka times. A broad stairway running parallel to the façade or in an east-west direction ascends to a large niche or blind window which impresses through its high quality of craftsmanship. From the stairway, the visitor may enter the eastern cave. On both sides of its entrance, three animals—two snakes and one feline—have been carved in high relief. The cave consists of two connecting chambers hewn out from bedrock. At the far end, a low round platform projects from the wall below a rock crevice which opens to the outside. Natural light from the sun or moon falls through it and creates light patterns on this platform.

After the year 2000, the Ministerio de Cultura conducted major excavations around Laqo which they call *Salonniyoq* or *Templo de la Luna* (Guillen Naveros 2006, 2007). Jorge W. Guillen Naveros and colleagues divided the site into four sectors corresponding to the quadrants northwest, northeast, southeast, and southwest. In the northeastern sector, two rectangular structures were excavated, facing east. In the interior of one of them, a platform runs along the rear wall in the base of which twelve small openings were exposed (figure 2.5). Burnt maiz kernels were documented near these openings. Such architectural features are characteristic of one type of Inka storehouses or *qolqas* (Huaycochea Nunez de la Torre 1994:129–226). Guillen Naveros interprets the structures as storehouses for agricultural products and goods which belonged to and were offered to the Sun.

In the southeastern quadrant, two more rectangular buildings were investigated. They contained ceramic fragments, burnt maiz kernels, and some domestic implements leading researchers to conclude that they most likely functioned as the residences of ayllu members in charge of the care and maintenance of the gigantic wak'a.

In the southwest sector, high-status buildings and potential elite residences were identified. A large rectangular structure was erected upon a man-made platform. Its two door openings face northeast, and its interior exhibits remnants of a cream-colored wall plaster. Next to the exterior door openings,

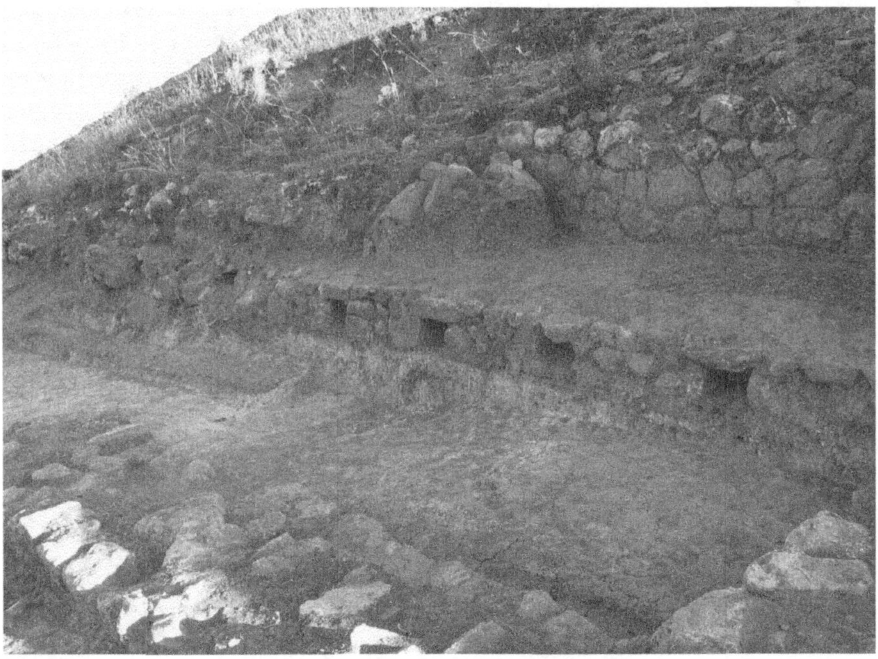

Figure 2.5. Laqo, Excavations Conducted after 2000, Northeast Sector, Storehouses. Photograph by Brian Garrett.

two sculpted stone blocks—one in semicircular form and the other rectangular—were located. Inside this building, another unique stone sculpture was encountered: it is a large circular stone vessel with wide walls and a feline stretched out in high relief on the rim. Guillen Naveros suspects that they may not have found the stone vessel in its original setting. Notwithstanding, he proposes an explanation for the feline on the stone vessel as well as the carved animal figures on top of Laqo as representations of the fauna of Antisuyu, the jungle habitat of most felines, a connection that is reinforced by the location of the outcrop near the Antisuyu road.

North of this building, near a wall and one meter below the surface, mural paintings were encountered. This wall displays a series of small double-jamb niches in low relief painted in red, blue, white, orange, and green. In another excavation, eight burials containing more than fifteen bodies belonging to high- and low-status members of Amaru Thupa Inka's court were investigated. Guillen Naveros (2006, 2007?) concludes that the elite structures have to coincide with the house of Amaru Thupa Inka listed by Cobo as *Amaromarcaguaci* in his description of the seventh wak'a on the first zeq'e line of Antisuyu. In his view, Laqo is An-1:7 (note the alternate reading above).

Kusilluchayoq is situated at a distance of about 300 meters south of Laqo at 71 degrees 57 feet 59 inches W and 13 degrees 30 feet 21 inches S, and at an altitude of 3,600 masl (figure 2.6). The site consists of two large limestone outcroppings lower than those of Laqo and separated in the center by a wide corridor. The northern half contains numerous geometric carvings formed by vertical and horizontal cuts. Of special interest are the remains of carved canals, a low disk, and a circular bowl. The southern half as well is characterized by seat-like carvings, steps, and platforms, but they seem to be condensed toward the central corridor.

The most striking sculptural elements are found in or adjacent to this corridor. One limestone block has been carved in the fully three-dimensional form of a feline positioned in the direction of the corridor. To the southeast of this statue, the narrow entrance to a cave is visible which continues underneath the southern sector. Most of the rocks facing this central passage have been smoothed and made more angular. A second rock with figurative images sits on the south side elevated above the corridor. Three monkeys and one serpent are depicted in high relief with the addition of channel-like cuts. The animals are distributed over three sides of the rock and today are in a deteriorated condition. Perhaps this elaborately sculpted rock functioned as a shrine in the channels of which libations were poured. It has also given the whole site its name: the Quechua term *Kusilluchayoq* means *the one with the monkey*. This is a recent designation, and its Inka name remains unknown.

The western section of the site consists of a number of architectural units (figure 2.6). A terrace wall was built first to retain the earth surrounding the

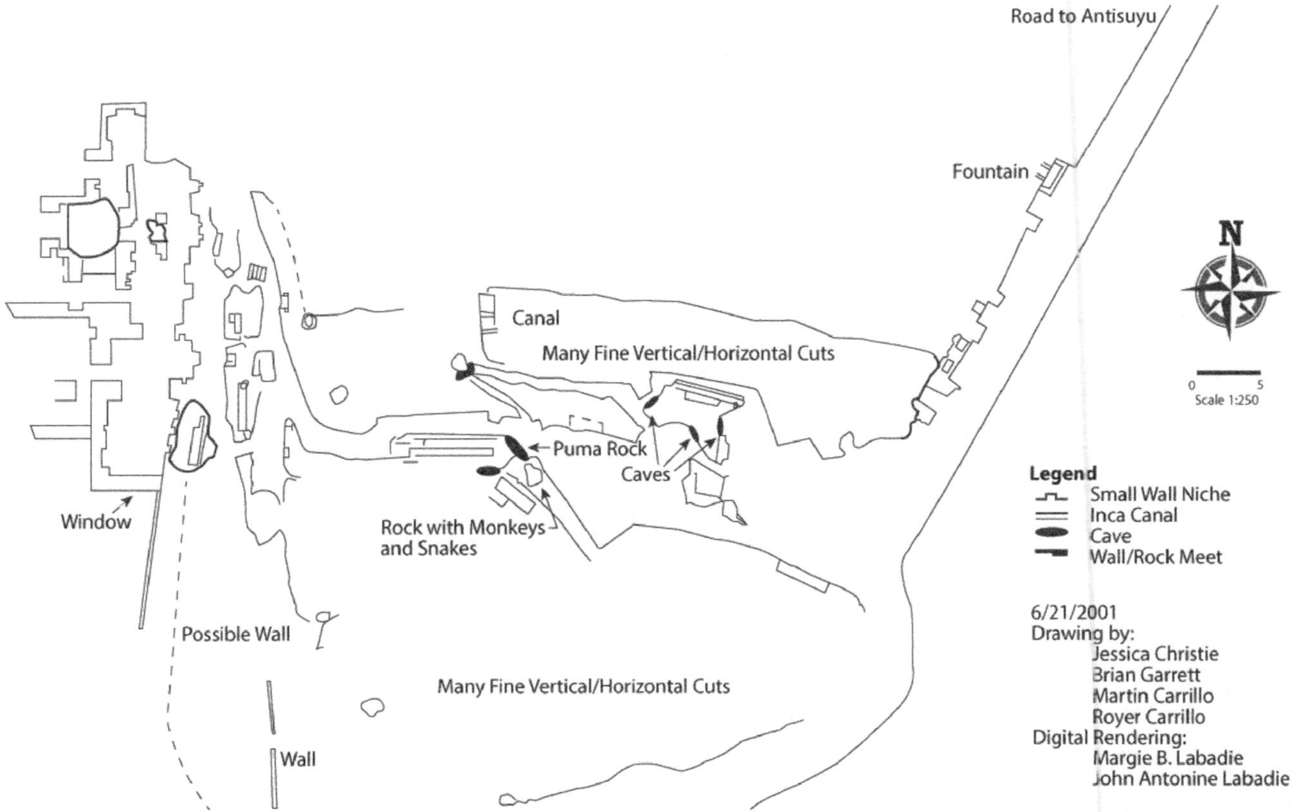

Figure 2.6. Kusilluchayoq. Plan Drawing by the author.

two carved limestone sections. A series of small rooms were attached to this wall. All walls at Kusilluchayoq are made of fieldstones, and a boulder has been integrated into the construction. Furthermore, the road from Cusco to Antisuyu passes on the east side of Kusilluchayoq and continues toward Laqo. Upon leaving the former site, a fountain outlet was set in the western side of this road. Since this fountain with two canals had no apparent agricultural purpose, its function may well have been related to the road to provide water for thirsty travelers or for ritual cleansing before entering and after leaving the capital.

Tom Zuidema (1977) and Anthony Aveni (1980:303; 1981:309, fig. 25.3) propose that Laqo corresponded with An-3:4. In their opinion, Laqo was the Sun Temple called Chuquimarca which Cobo describes as the fourth wak'a on the third zeq'e of Antisuyu. Their argument is connected with the ninth wak'a on the sixth zeq'e of Chinchaysuyu which Cobo describes as "a hill named Quiangalla, which is on the Yucay Road. On it were two markers or pillars which they thought denoted the beginning of summer when the sun reached there" (Cobo 1990:59 [1653:Bk.13, Ch.13]). Although today there is no location in the Cusco region with the name Quiangalla, this toponym does show up in several sixteenth-century land documents as being northwest of Cusco in an area crossed by the Inka road to Yucay (Bauer and Dearborn 1995:80–82). By means of astronomical calculations, Aveni and Zuidema have located Quiangalla on a large ridge descending from the summit of Huaynacorcor mountain which corresponds with the description given in the land documents (Aveni 1981:309–310; Zuidema 1981:168–169). Zuidema's and Aveni's astronomical measurements demonstrated that June solstice sunset between such pillars could not have been observed from the Qorikancha nor from anywhere else in the central Cusco area. The closest point is Laqo. Viewed from Laqo, the sun would have appeared to set in between the two pillars on the winter solstice on June 21. Based on their astronomical calculations and on Cobo's description, these two scholars identify Laqo with the Sun Temple of Chuquimarca. However, survey work conducted by Bauer and Dearborn did not find any archaeological evidence to support the suggestion that the actual pillars of Quiangalla once stood in the area of the ridge indicated by Aveni and Zuidema (Bauer and Dearborn 1995:86).

Van de Guchte points out that there were actually three high points in the area: the hill at present known as Ocururochayoc, the modified natural hill Erapata/Cruzpata/Tambillo, and the carved rock Laqo (1990:180). All three elevations may have been part of the hill of Mantocalla mentioned by Cobo. It is thus possible that the Sun Temple of Chuquimarca (An-3:4) was situated within this larger area, which makes a connection between the sculpted outcrop of Laqo and the Sun Temple less direct. Bauer adds that Kusilluchayoq could be another possible candidate for Chuquimarca (1998:81). John Rowe expressed his opinion that Chuquimarca/An-3:4 or Mantocalla may have been several large rectangular compounds, the foundations of which can still be seen on a hill to the north of Laqo (Bauer and Dearborn 1995:86–87).

Bauer (1998:81–83) identifies An-3:4 with Kusilluchayoq and An-3:6 with Laqo. In his description of festivals and sacrifices that were performed during the twelve months of the year, Cobo provides information about the Inti Raymi ceremony, held in June, which or part of which took place "on the hill of Manturcalla" which is expected to be identical with Mantocalla (Cobo 1990:142–143 [1653:Bk.13, Ch.28]).

Cristobal de Molina left a very similar account (Molina in van de Guchte 1990:180–181). D'Altroy (2002:154–155) notes that the cleric Bartolome de Segovia witnessed the Inti Raymi held in Cusco in 1535 and described it in his *Relación* (Segovia 1968:81–82 [1553]). Segovia offers a vivid and colorful account of assembled nobles in fine regalia and mummies of deceased rulers lined up in two rows and Manqo Inka seated upon a throne; they all sang to the Sun as it rose into the sky; meat was burned as offering, and many other sacrifices were made; women brought vessels with chicha and coca leaves and offered them to the Sun. In general terms, Segovia's account is consistent with Molina's and Cobo's descriptions of the Inti Raymi. With regard to location, Segovia places the events "in a plain near the road leaving Cusco toward sunrise" and "in another plain where a tree stood" (1968:82 [1553]). The first location could refer to the Antisuyu road and thus support the argument that Laqo and Kusilluchayoq were two of several locations where the Inti Raymi was staged. In reference to time, Segovia (1968:81 [1553]) gives the month of April and not June, the month of the Inti Raymi.

Bauer's identification of An-3:6 with Laqo is strengthened by a document dating to 1649 that describes the area of Mantocalla as being crossed by the Uru Uru canal as well as by the road to Pisaq (1998:82). A major canal, which originates at Ucu Ucu/Uku Uku approximately three kilometers away (this is one of the Amaru canals, see Amaromarcaguaci), runs along the southern base of Laqo, and the Inka road to Pisaq and Antisuyu passes just to the south of it.

Bauer names the stone fountain next to Kusilluchayoq on the Inka road to Pisaq (which is the Antisuyu road) as one possible candidate for An-3:5. This proposition seems logical because Cobo describes An-3:5 as "a fountain of good water" on the Mantocalla hill "where the Indians bathed" (see above). An-3:7 was a fountain named Caripuquiu situated on the slope of the Mantocalla hill (see above). Bauer suggests that Caripuquiu corresponded with a small spring on the northern slope of Laqo that is reached by means of a stone stairway (1998:83). I have not been able to find this spring.

To conclude the discussion of An-3:4, An-3:5, An-3:6, and An-3:7, it would appear that the ethnohistoric material and previous studies allow certain cautious insights into the use and function of Laqo and Kusilluchayoq in the context of the Inti Raymi festival. I suggest that the extravagant feasting may have taken place at Kusilluchayoq. This site is a low-lying carved outcrop which can be accessed without much climbing, marking it as an inviting, open, and somewhat public space. Kusilluchayoq contains many vertical and horizontal cuts, and it is conceivable that the Inka lords spread fine cloth over these carved seats as the chroniclers have reported for other rock wak'as at special occasions. There they may have sat comfortably—drinking, eating, singing, and rejoicing. Perhaps someone rose occasionally to pour chicha down the channels on the rock sculpted with monkeys and a snake to share the special liquid with Pachamama.

Feasting of the extent and length described by Molina would require an area set aside where the drink was prepared, special rooms to store the golden vases and vessels and to house the mamacunas. Both authors mention that the wives of the Inka stayed in a courtyard or patio. Some of the known buildings in the Mantocalla area are the rooms associated with Kusilluchayoq and the newly exposed structures along the northeast and south sides of Laqo, and it is likely that they functioned as the support and storage structures for the Inti Raymi.

The top of Laqo constitutes a very different type of space. This setting is highly elevated, and access is restricted to the north side. Visibility into far distances can be enjoyed by any viewer positioned in the summit area, while at the same time, the visibility of this viewer from below the outcrop is highly limited. Such a space is well suited to conduct some of the June solstice observations discussed above, and it could also have been used as a location for sacrifices which Molina and Cobo emphasize so vividly. The finely cut, low, freestanding platform might have provided an altar setting where some of the animals listed by Cobo were offered to the Sun. The top of Laqo provided room for a small and selected audience only. This elevated elite space may also have been the proper environment where the statues from carved quishuar wood mentioned by Cobo were displayed. If this was indeed so, only chosen individuals who had permission to climb up could have seen them. The larger general audience would have remained below. I view the whole hillside as a monumental stage set for the Inti Raymi celebration within which Laqo and Kusilluchayoq played the role of vital ritual stations.

The fourth zeq'e line was under the care of the Aucailli panaqa of Yawar Waqaq, the seventh ruler. The fourth shrine consisted of carved stones.

> [An-4:4] The fourth was a quarry which is near there named Viracocha. In it there was a stone which resembled a person. They say that when they were cut-

> ting stone from there for a house of the Inca, it came out so, and the Inca ordered that it should be a guaca (Cobo 1990:66 [1653:Bk.13, Ch.13]).

In her investigation of Callachaca, an area which had been developed as a royal estate by Amaru Thupa Inka (see above under An-1:7, Amaromarcaguaci), Susan Niles identifies several shrines on the fourth, fifth, and sixth zeq'es of Antisuyu which passed across the site (1987:179–206). She suggests that An-4:4 was the quarry area of Choquequirau, which is one section of the much larger site of Callachaca (1987:fig. 1.3, 183–190). Bauer (1998:84) concurs with Niles. In this quarry, carefully worked and then abandoned stone blocks can still be seen. Furthermore, three small bedrock outcrops were sculpted possibly to re-create a miniature model of the main features in the Choquequirau group (Niles 1987:186–188). This interpretation is intriguing, and it would represent another case of a small-scale model or animated double of the natural and constructed landscape as formed by other carved rocks and discussed below. However, I could not find the carved boulders described by Niles in the summer of 2001, and they may have been destroyed.

The fifth zeq'e of Antisuyu was maintained by a nonroyal ayllu. There are ten wak'as, two of which can be identified with or are related to carved rocks. Cobo describes the third wak'a as follows:

> [An-5:3] The third was called Subaraura. It was a round stone which was in the town of Yaconora. They offered it only shells, some whole and others cut in pieces (Cobo 1990:66 [1653:Bk.13, Ch.13]).

Niles proposes that An-5:3 corresponded with an outcrop that has the form of a ball, adjacent to the Yacanora road and within an area that still retains the name Yacanora (1987:193–194).

Bauer accepts her identification (1998:86). This rock supports coursed masonry constructions and a water channel runs along its southern base. Niles writes she could not find evidence of carvings, but I saw rows of vertical and horizontal cuts in 2001. The carvings were not refined and left the basic stepped outline typical of Inka rock sculpture.

Niles (1987:194) notices similarities in the style of masonry and of the accompanying terrace system between Subaraura and the T-shaped plaza complex further up the hill. She has documented the T-shaped plaza group as special-purpose architecture within the estate of Callachaca (1987:22, 106–114) and places An-5:7 in its vicinity (1987:197–198). Cobo describes An-5:7:

> The seventh was called Aquarsayba. It was a greatly venerated guaca, and they had the opinion that whatever they offered to it all the guacas received (Cobo 1990:67 [1653:Bk.13, Ch.13]).

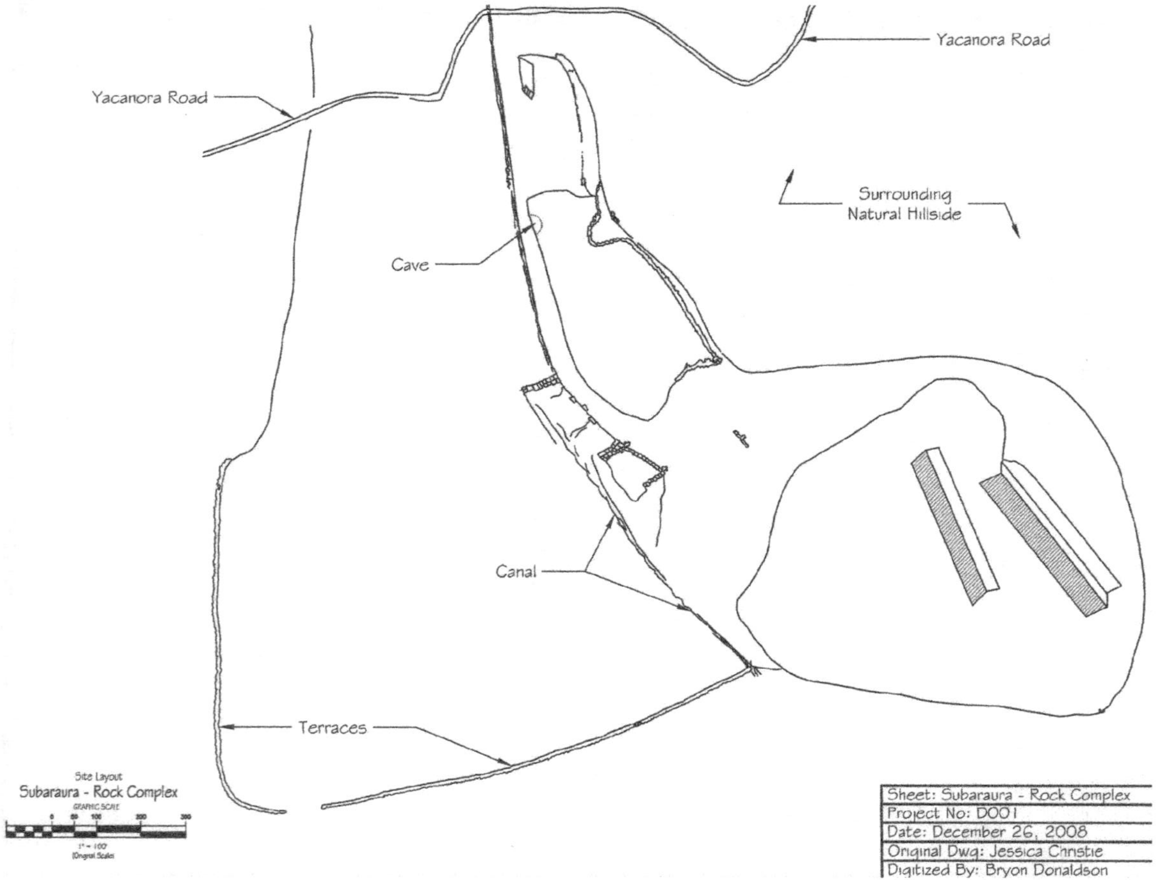

Figure 2.7. Callachaca, Subaraura, Carved Rock Wak'a. Drawing by the author.

Bauer follows Niles's identification (1998:86). The T-shaped plaza complex is formed by two long halls constructed in such a way that their rear wall is built into a hill and their front, freestanding wall containing multiple doorways opens onto the T-shaped plaza. This plaza has a stepped plan view or that of a T, the wide end of which is occupied by the long building facades while the small side of the T is bordered by a cliff (Niles 1987:107, fig. 3.40). The cliff contains a number of natural crevices modified by man which Niles interprets as tombs (1987:107–108). The best-preserved carving is a seat-like cutout facing downhill. Since Cobo does not tell us what Aquarsayba looked like, only that it was "a greatly venerated" wak'a, it is not possible to associate it with a specific feature, and thus Niles locates it somewhere in "the vicinity" of this plaza complex. The T-shaped plaza group will be mentioned again later in the discussion of sculpted boulders at royal estates.

On the sixth zeq'e line of Antisuyu, several wak'as may be connected with carved rocks located on royal estates. According to Cobo,

> [An-6:2] The second guaca was a curved stone named Comovilca which was near Callachaca; they offered it only shells.
>
> [An-6:3] The third was named Chuquicancha; it is a well-known hill which they held to be a house of the Sun. On it they made very solemn sacrifice to gladden the Sun (Cobo 1990:67 [1653:Bk.13, Ch.13]).

Niles suggests that the archaeological area of Rumi Huasi located uphill from San Sebastian, a community just outside of Cusco, held An-6:2 and An-6:3 (1987:198–203). She documents and analyzes Rumi Huasi as special, purpose architecture on Amaru Thupa Inka's estate Callachaca (1987:59–92). The site consists of two architectural complexes which integrate natural outcrops and are referred to as Lower and Upper Rumi Huasi. Lower Rumi Huasi includes four well-preserved structures, a worked boulder face, and an underground passage arranged on three terraces which sculpt an approximate U-shape on the face of the hill. The main buildings of Upper Rumi Huasi are composed of two large structures and a small walled space erected upon a high construction terrace which forms a broad U. The largest building may have been an Inka great hall or kallanka (Niles 1987:89). Niles identifies An-6:3 or Chuquicancha with the great hall of Upper Rumi Huasi (1987:199–202). She and Bauer offer several candidates for An-6:2 or Comovilca: an outcrop elaborated with a curved, niched wall at its base situated between Upper and Lower Rumi Huasi, the curving part of the cliff modified with masonry that separates the building levels at Lower Rumi Huasi, or a cave to the east (Niles 1987:202–203); the semicircular outcrop of Lower Rumi Huasi (Bauer 1998:88). Sherbondy (1982:47,183–184 n. 50) and Zuidema (1990:631, 635) place An-6:2 and An-6:3 in an area called Cusicallanca, about a hundred meters east of Rumi Huasi across a small ravine. Cusicallanca is heavily terraced and includes an isolated outcrop exhibiting a seat-like carving. This carved outcrop could represent Comovilca.

Bauer adds that the name Chuquic(h)ancha is mentioned by Cobo in the context of the Inti Raymi festival and by Molina in the context of the *Ccapacocha* festival (1998:87–88). In the above discussion of An-3:6, the sixth wak'a on the third zeq'e line called Mantocallas by Cobo, multiple locations were identified at and to which the Inti Raymi was staged and moved. Two major stations appear to have been Laqo and Kusilluchayoq, and another stopping point was Chuquicancha, where more animal sacrifices were conducted. Cobo's description of Chuquicancha (An-6:3) as a house and place of sacrifice to the Sun reinforces the Inti Raymi interpretation because this was the most important Sun festival. Molina's account of the *capac cocha/ Qhapaq ucha* festival (1959:91–94 [1575]) emphasizes that many sacrifices were made and buried at Chuquicancha in honor of the Sun, the Thunder, and the Moon. Molina confirms that Chuquicancha was a small hill above San Sebastian (also see Bauer 1998:87–88). In summary, the evidence is strong that An-6:2 (Comovilca) and An-6:3 (Chuquicancha) were situated within the Rumi Huasi complexes and that Upper Rumi Huasi and its kallanka were ritual spaces where sacrifices to the Sun were performed and buried.

The fifth wak'a was a fountain or spring, and the sixth wak'a was a house associated with the wife of Pachakuti Inka Yupanki.

> [An-6:5] The fifth was called Viracochapuquiu; it was a fountain which is in a flat place on the way to Chita.
>
> [An-6:6] The sixth was a house called Pomamarca which was on the said flat place. In it was kept the body of the wife of Inca Yupanqui, and children were offered there along with all the other things (Cobo 1990:67 [1653:Bk.13, Ch.13]).

Bauer (1998:89–90), Zuidema (cited by Bauer), and Niles (1987:198–199) propose that a complex of Inka structures with an adjacent spring which are still called Pumamarca and located approximately four kilometers northeast of San Sebastian represent An-6:5 and An-6:6. Pumamarca displays foundations of Inka buildings in fine Cusco masonry of stones which were exactly fitted without mortar. These foundations were later integrated into a colonial hacienda. Today some of the rooms are inhabited by a local family. Next to the buildings extends a garden where the spring is situated. This garden includes numerous carved rocks one of which has multiple corners adding up to twelve sides (figure 2.8).[17] The spring emerges from a cave; its water is

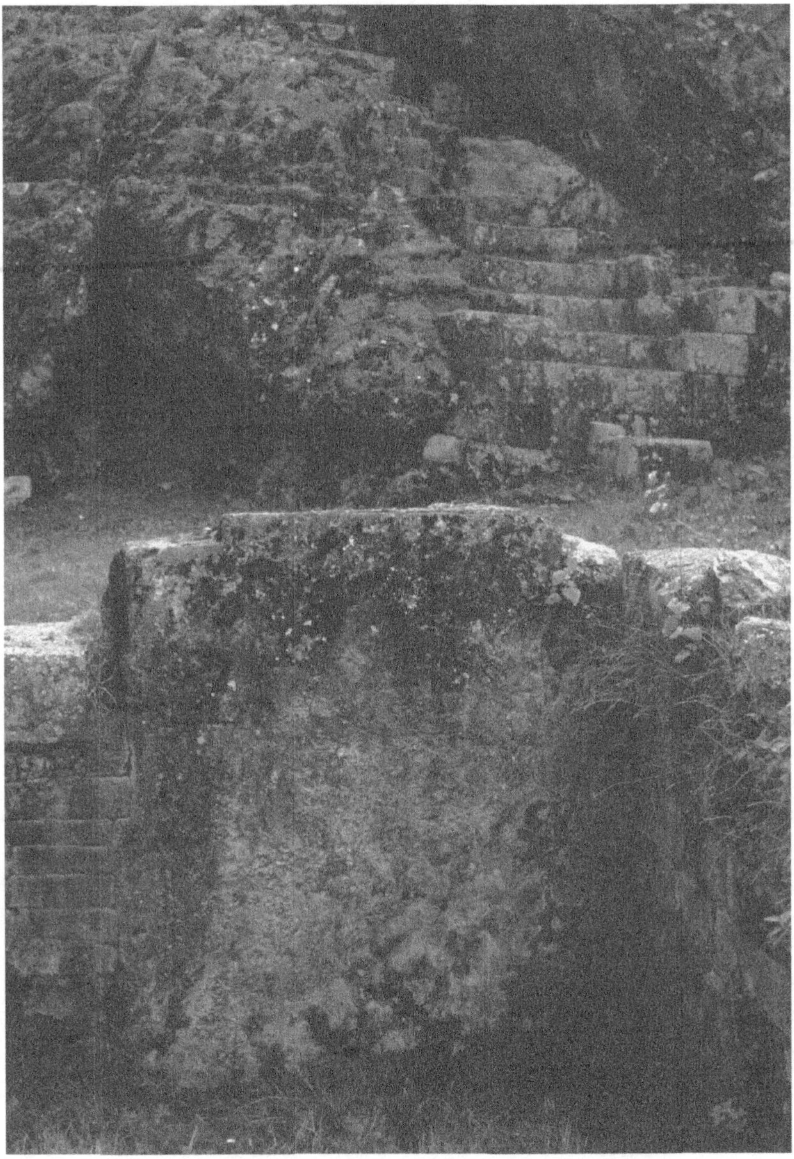

Figure 2.8. Pumamarca, Carved Rocks. Photograph by Brian Garrett.

channeled across the garden and eventually ends up in a constructed fountain basin below. Bauer cautions that the spring Viracochapuquiu, An-6:5, may have been another one with the name Virapuquiu located outside of Pumamarca and closer to Cusco (1998:89–90). His latter identification would be more consistent with the order of the shrines.

The historical context is contradictory. While Cobo says that the mummy of Pachakuti Inka Yupanki's wife was kept at Pumamarca, Santa Cruz Pachacuti Yamqui Salcamaygua notes that after the Chanka War, Wiraqocha Inka remained at Pumamarca until his death (1993:220 [1613]). Bauer refers to local landownership documents dating to 1570, which also suggest that Pumamarca was owned by Wiraqocha Inka (1998:89); and Rowe (1997:279) cites specific sources from 1548 and 1569 attributing Pumamarca to Wiraqocha. Other chroniclers state very clearly and describe in length and detail that Wiraqocha Inka withdrew from the Chanka War and remained at his estate, Juchuy Qosqo, the version I have accepted. Of course, Pumamarca could have been another one of his properties.

The Sculpted Rock Wak'as in Kollasuyu and Kuntisuyu

Although Cobo described many wak'as in Kollasuyu and Kuntisuyu as stones, Bauer could not identify any one of them as a sculpted rock (1998:96–134). His findings may imply that there were no carved rocks in Kollasuyu or Kuntisuyu in the immediate vicinity of Cusco. I have not encountered any myself. Van de Guchte has also commented on the selectivity of sites with carved rock complexes in Tawantinsuyu—a selectivity that is not solely the result of geological conditions (1990:54–57, illus. 2, 3). In other words, sculptures do not occur in just any outcrop that is composed of a stone type conducive to carving. He has since further developed his thoughts about why some rocks were carved and others were not, focusing on the principle of *alterity* or difference from other rocks (1999). In the Cusco Valley, sculpted boulders are clearly concentrated in the broad northern region of the city that is in Chinchaysuyu and Antisuyu.

INKA LANDSCAPE AND CARVED ROCKS—APPEARANCE, EXPERIENCE, AND PERCEPTION

The first section of the following analysis of the data presented above focuses upon changes in the physical appearances of the boulder sculptures through time and how such changes evoked varying bodily and cultural experiences between subjects and the carved rocks as objects of interaction. Central questions to be addressed are in which ways can we as citizens in a global twenty-first century world begin to excavate the meanings of carved rock complexes? It must be pointed out that the terms "experience" and "perception" take on diverging emphases among theorists: phenomenologists understand them from the personal perspective of the individual while Adam Smith defines them as rooted in culture (see chapter 1, introduction). The following will discuss layers of viewing carved rock wak'as embedded in time and space. Subheadings will structure three themes, and examples taken from the corpus discussed above will substantiate and illustrate such themes.

Shifting Appearances

Human experience and perception are grounded in physical appearances. In several cases, contemporary Cusquenos have altered physical aspects of stone wak'as and situated them in new social contexts, adding contemporary layers of experience and perception. The Mesa Redonda group (Ch-1:1) is an excellent example of this multilayering: the excavations by the Ministerio de Cultura in the early 2000s exposed Inka walls in their effort to restore the correct Inka appearance. Before their intervention, only the outcrop and the carved rocks had been visible and accessible, and local residents had adopted them as a social gathering place to soak up the afternoon sun. Now the area has been landscaped, is partly fenced off, has a guardian, and is most of the time devoid of people. Most residents I spoke with implied it was a positive change that "they cleaned it up" and access is more restricted. People who live in the neighborhood today acknowledge Mesa Redonda as a "wak'a." They did not mention the zeq'e system or a link with Pacariqtambo (see above)—they look at Mesa Redonda as it fringes their daily practices.

Sapantiana (Ch-2:4) constitutes another case in point. Today the site has become part of a small neighborhood park. It is surrounded by a playground, soccer field, and swimming pool and serves as a social gathering place. Although the carved rock complex appears rather open and accessible today for those who wish to climb it, in Inka times the containing walls on the east, south, and west sides would effectively have shielded the structures, carvings, and access to the water from public view. Even though we do not know the height of the walls, Sapantiana would clearly have been a more private space.

At Chacan (Ch-5:7), the most potent area of the entire site was without much doubt the cave through which the Chacan River continues to flow. In the perceptions of most precontact and traditional Andean peoples, caves are connecting links with the underworld. The Inka demonstrated their control over water by manipulating the waterflow through the canal crossing the river high above the cavern and the channels associated with the carved rock. New Age meditation groups continue to use the cave for their spiritual needs, which they attempt to align with traditional Quechua beliefs and practices. In the summer of 2001, I discerned several contemporary rock art paintings in bright colors altering the cave walls and marking the locale as liminal space in past as well as contemporary spirituality.

Field of Vision

Visibility constitutes an important phenomenological tool in the process of mapping the Inka landscape in its grandiose Andean setting. Phenomenologists treat visibility as a universal to be experienced by the human body in a physiological sense through time as long as the observation point is not obstructed by man-made intervention. Oftentimes ethnographic references are cited to reinforce the special relations indigenous people have maintained with a mountain visible from their community. We have to caution, of course, that we can reexperience visibility and sight lines but not necessarily re-perceive such spatial relations in the same way the Inka did.

I refer to Sabacurinca (Ch-5:6) as an excellent example to comment on the importance of visibility and physiological sight lines as a conceptual mapping device of the Andean landscape. Van de Guchte (1990:125–126) identifies six sacred mountains of Cusco as mentioned by Sarmiento de Gamboa, which can be discerned from Sabacurinca. Elsewhere van de Guchte (1984:546) has argued that one function of Piedra Cansada (Ch-4:6) was that of a point from which Saqsaywaman could be viewed and venerated; on the four roads departing from the Cusco plaza to the four suyus, the points from which the capital could be seen for the first or last times were commemorated by ritual gestures and physically marked by inconspicuous offerings, such as coca leaves and stone piles (*apachetas*); Johan Reinhard (2007) has convincingly shown that the function of Machu Picchu was to a large degree linked to the visibility of sacred mountains (*apus*) from specific locations within its center; Zuidema comments that in the zeq'e system, direct visual connections could be extended by the concept of straight lines (1982:431) and that sight lines have traditionally been used in the Andes for surveying, mapping, and long-distance travel (1982:445). As was the case with the imagined distances traveled by the Piedra Cansada (Ch-4:6), such sight lines mapped out a metaphorical socio-religious geography elements of which were pan-Andean, others were imposed by the Inka state, and still others were purely local.[18] In the second part of the analysis of Inka landscape, we will return to the theme visibility in the discussion of stone ideology and assess it as one tool in the state enterprise of organizing the empire into administrative and ideologically defined spatial units. Many sculpted rocks, such as Sabacurinca (Ch-5:6), were markers of such divisions.

Performance in Motion

Human experience and perception are framed and shaped by the names of a particular site location and by events that occurred there. Laqo and Kusilluchayoq (An-3:4, 3:5, 3:6, 3:7) provide insightful examples. Cobo's description of the Inti Raymi indicates that the hill of Manturcalla/Mantocalla had to encompass a larger area to accommodate numerous feasting and sacrifice events. Cobo further mentions other toponyms which imply that at least part of the festival could travel temporarily to different locations (Chuquic(h)ancha is one such toponym identified as An-6:3; see above). There was a procession with animals carried on a litter, which also suggests movement and changing locations. One conclusion is that complex festivals, such as the Inti Raymi, were locationally mobile and nomadic and cannot be fixed to one specific site. This continues to be played out by the present day Inti Raymi performance, which travels from the Qorikancha to the Awkaypata plaza and on to Saqsaywaman.

Tom Cummins (Houston and Cummins 2008:373–375) has explained the simple and frugal interior setting of Inka palaces by the documented practice of the Inka court in motion. He reasons that a palace was activated by the presence of the ruler when the court was in residence; and while it traveled, the remaining populace would have experienced and perceived the palace as a normal house. A similar multivocality may well have applied to rock art sites. They might have been activated for various rituals. Afterwards rock wak'as returned to a passive dormant state but always with the potential of becoming active again due to their embedded lifeforce/*enqa*, which will be assessed in the next section addressing essence.

INKA LANDSCAPE AND CARVED ROCKS—ESSENCE, IMAGINATION, AND STONE IDEOLOGY

The second part of the analysis moves beyond physical appearances to conceptual essences of stone which have been discussed by various writers as *enqa*/lifeforce (Bolin 2006) and changing states of being indicated by the Quechua verbs *cay/tiay* (Salomon 1998:8–9). An exploration of essences does not cancel out the phenomenological approach but complements it, since Inka carved rocks have been experienced by many individuals through time but at the same time held enduring essences. In the framework of Smith's (2003:74) political landscapes, spatial imagination emerges from the analysis of form and leads to spatial theory and philosophy. I will argue that Pachakuti formulated a specific stone ideology that was grounded in the sculpted rocks. Again, subheadings will structure six themes, and examples from the zeq'e line corpus will bring them to life.

Fields of Imagination

Mesa Redonda (Ch-1:1) functions as a great example of how Andean people have imagined sculpted rocks and their site settings through time. Cobo's consultants apparently connected the location with Manqo Qhapaq and his brothers and with Pacariqtambo, the Inka caves of origin (see chapter 3). Indeed there are caves on the slopes of Totocache/Tococache a short distance up from the site.[19]

From the Colonial Period to the present, Quechua descendants of the Inka have attempted to appropriate and re-create origin places and their narratives in their own environments. Gary Urton (1990, 1999) has convincingly shown that the Inka ancestors traveled to Cusco from Pacariqtambo in the south. They stopped at the mountain Wanakauri located to the southeast of Cusco while Mesa Redonda and the first zeq'e of Chichaysuyu are situated almost due north. Therefore these ancestors could not have passed by Mesa Redonda. While Cobo's intention was to collect data for his books in the spirit of nascent Western empirical sciences, his consultants did not respond with the historical facts Cobo was looking for. Their interest was to make Totocache a powerfully charged location and therefore brought it in connection with the origin myth. Urton (1990:41–70) discusses how certain natives of Pacariqtambo tried to prove descent from Manqo Qhapaq in order to claim noble status under the Spanish regime in the sixteenth century. Thus claiming links with the origin story may have been a way of foregrounding and elevating individuals as well as places by giving them a long history. Indeed Pacariqtambo as well as Wanakauri have been re-created at a number of other locations.[20]

When I inquired with today's residents in the neighborhood into the history and meaning of Mesa Redonda, the standard answer was "oh, es una wak'a." This indicates that Cusquenos remember the rock as a potent place but have forgotten its specific associations.

Doubling of Sites and Essences

The process of re-creating sites and objects is vital in Inka culture and seems to have to do with transferring essences and life forces. Some sites could be reproduced over and over—which was to a certain extent true for Cusco—

whereas others, including rock art sites, certain objects, and some powerful human beings could only exist in pairs.

I suggest that doubling in Inka visual and social culture happened on three tiers (see my full argument, Christie 2012). Tier number one presents a visual and literal formal resemblance between an object and its representation in another medium. In Inka rock art, including Q'enqo (Ch-1:2), the ubiquitous vertical and horizontal cuts creating platforms, steps, and seats duplicate and hence represent agricultural terraces. The zigzag channel on top of Q'enqo Grande resembles and thus stands for a meandering miniature river. In tier one, the emphasis is on a direct reproduction which closely resembles the original in visual form. Although Inka formal language leaned toward abstraction, Inka artisans were capable of sculpting realistic, lifelike animals, cornstalks, and man-made structures (see the stone of Sayhuite, Christie 2003).

Tier number two implies that the very process of reproducing transfers some of the spiritual essence of the original to its copy. Although the copy has to share some physical commonality with the original, the connection is primarily established by means of the observer's memory and social embeddedness. This particular level of representation was well documented by the Spanish writers with regard to the portrait statues which were made of Inka rulers—note that Pachakuti's statue may have been deposited in the cave of Q'enqo Grande—and has been analyzed by van de Guchte (1996). Huauques were not isolated images detached from social life; they were created to be active social participants. Although huauques and other ruler statues rested at times, they were brought to life in and through ritual and could take the place of the ruler himself. Thus some of the vital essence, life force, or spirit of the ruler would have entered his image. In Quechua, such a process is implied by the word *camac*, which Garcilaso de la Vega translates as "the power which animates." It means that each subject—living or inorganic—is related to its primordial double, which empowers it to fulfill its function in the cosmos. The emphasis here is that the being or "copy" to which the life force is transferred is perceived and treated as being alive (Taylor 1974–76:232–234; Salomon 1991:16).

I reason that a similar process of replication took place in rock art: the carved step patterns came to embody agricultural terraces and mountain slopes, and cut channels became miniature rivers. Like their counterparts, the human huauques, the rock art sites did not exist decontextualized, but they were activated by means of ritual events. If the spiritual context was related to agricultural terraces and rivers, ceremonial performances would have focused upon fertility-related offerings and the renewal of the cycle between life and death in which ancestors played an active role. Andean people perceived mummies (*mallquis*) as "seeds" which were instrumental in regenerating new and life-sustaining growth of crops. At Q'enqo, fertility libations into the zigzag channel would surely have been included in Pachakuti's funerary program.

Tier three may be explained through a facet from the semiotic sign theory developed by Charles Sanders Peirce and previously applied to the Inka by Gary Urton (2003:140–143) and by Lawrence Coben (2006:226–228).[21] According to Peirce, the "sinsign" is an actual existing thing and the "legisign" is not a single object but a general type (Peirce 1985:7–8). In the process of replication, legisigns are created and applied through replicas or tokens of their kind. Thus the replica becomes the sinsign or token of the legisign, signifying its general category. As an instantiation of the legisign, each replica or sinsign is contextualized in a particular cultural and historical setting. Meaning is generated in specific and personalized encounters, which is different from solely formal notions of copying. The generative process implied by the Quechua verb *camay* could apply here as well (Salomon 1991:16; Taylor 1974–76): in tier three, the focus is upon the multitude and variety of sinsigns an animating essence camac can produce, whereas in tier two, the process of transferring life forces is of primary interest.

Peirce's model may be applied to rock art as an analytic framework: on a direct visual level, agricultural andenes, mountain slopes, and rivers take on the role of the legisign and *camac*, while the rock carvings of vertical and horizontal cuts and channels become individual instantiations or replicas and *camasca*. Another good example is the word and concept of *wak'a* and the site of Q'enqo itself: wak'a is the legisign for most sacred places and objects while sculpted rock sites are some of its specific materializations or sinsigns.

The legisign-sinsign model might be taken one step further to include broader concepts of Inka visual culture as well as issues of hierarchy. I reason that the fundamental Andean complementarity of high versus low could be understood as a general legisign and the site sectors Q'enqo Grande and Q'enqo Chico (Ch-1:2) as one of its tokens. First and most importantly, the dramatic experience of high and low in the Andes was and is phenomenologically inspired, by which I mean that it can still be felt by living beings. Over time, it became socially anchored. In Quechua worldview, the high-low opposition was manifested by male mountains and fecund female valleys, both of which surely had *camac*. Mountains and valleys form visual and physical binary categories but have to unite in order to sustain life. This profound complementarity has been represented, socialized, and sanctified in innumerable instantiations or replicas over time. Most significant in our context are the divisions of settlements in upper and lower sectors, specific ayllus and community obligations assigned to these sectors, and the veneration of mountain deities, the *apus*, in contrast to the earth, *pachamama*.

One specific manifestation or token/sinsign of high versus low, which might be applicable to Q'enqo Grande and Q'enqo Chico and other carved rock complexes is the pan-Andean Hanan–Hurin division. Hanan and Hurin are Quechua words translated as "upper" and "lower," which, in the literature, usually referred to geographically higher and lower sectors of Andean settlements as well as to two moieties or halves, which are social divisions composed of specific ayllus. Researchers have documented Hanan and Hurin organization of Andean towns from Inka times to the present, even though we possess limited insights into how the Inka understood the Hanan and Hurin complementarity (for example, Urton 1981:40–43). The chronicler Garcilaso de la Vega describes in most detail the differentiations of Hanan and Hurin the way the Inka may have perceived them:

> In this manner, our imperial city began to be populated, which is divided into two halves called Hanan Cozco, which means high Cozco, and Hurin Cozco, which is low Cozco. The king requested that those who came with him would settle in Hanan Cozco and this is why they called it the high sector; and those gathered by the queen should inhabit Hurin Cozco and this is why they called it the low sector. This division of the city did not mean that these who lived in one half were superior to those in the other half but that all were equal like brothers, sons of one father and one mother. The only reason why the Inka wished that this division of the city and differentiation in naming high and low was made was that there would be a permanent memory of the fact that the king had gathered one group and the queen the other; and he [the king] instructed that between them there should only be one difference and recognition of superiority; these from high Cozco should be regarded and respected as first-born and older brothers; and those from low Cozco as second-born sons; and in the end, they were like the right and left arms in any important place and office because these of the high part had been brought by the male and those of the low part by the female. In replication of this, there was later the same division in all the large and small towns of our empire, which they divided by barrios and lineages, naming them Hanan ayllu and Hurin ayllu, which is the high lineage and the low one; Hanan suyu and Hurin suyu, which is the high district and the low one (Garcilaso de la Vega 1963:28 [1605:book I, chap. XVI] translation by the author).

Although Garcilaso stresses an essential equality of the halves of the pair, he highlights the following differences: Hanan has a higher geographic elevation, is associated with the firstborn and therefore oldest and original ancestors, and it is linked with masculinity. Hurin is lower in elevation, is the second-born, and connected with femininity. Most of the terms Garcilaso and other Colonial and contemporary authors use to explain the Hanan–Hurin categories describe immaterial aspects, except for the topographic layout of cities. I think we may be able to view certain rock art sites, such as Q'enqo Grande and Q'enqo Chico, as physical and more visual materializations of Hanan–Hurin. The following elements link them together as two complementary entities: their spatial proximity, their high versus low location, and a general resemblance of material features; that is, rock carvings.

I have documented additional rock art sites which contain possible Hanan–Hurin relations: these are the high and low Intiwatanas at Machu Picchu and the Yurak Rumi rock at Vitcos, which has a smaller version on the lower andenes (Christie 2012). All three sites share: first, the topographic differences of Hanan and Hurin; second, visual formal similarities, and in the Intiwatana and Yurak Rumi cases, the Hurin part constitutes a miniature version of Hanan; and third, a more open public space in the Hanan sector, whereas the Hurin part is smaller, with reduced and, therefore, more private framing space (see the formal analysis of Q'enqo Grande versus Q'enqo Chico above).

Thus certain rock art sites, such as Q'enqo Grande and Q'enqo Chico, may have constituted and may have been socially performed as micro-models of the Hanan–Hurin divisions (for performance aspects see Tom Cummins 2007:273–274 and also Mariusz Ziolkowski 1966:52 about two imperial representations of the Sun and their opposing ritual spaces). Ethnographic studies by Catherine Allen (1998:22–23) in the Quechua community of Sonqo document multiple levels of performing gender dualities in ritual as well as overlapping physical macro- and microcosms linked to a hierarchical order as conceptualized in the Quechua word *pacha*. Dual rock art sites as discussed here should be one instantiation or sinsign of these broader Andean notions (or Peircean legisigns).

Geography of Stone Ideology

Although Piedra Cansada (Ch-4:6) became a local wak'a on the fourth zeq'e line of Chinchaysuyu, it nevertheless reached into the larger abstract space of the Inka empire by constructing the network of spatial relations explained by van de Guchte (1984; see also Tilley 1994:8). Many rock wak'as were placed on trails and marked the zeq'e lines to map out the physical empire the Inka ruled, which was overlaid by a conceptual one (McEwan and van de Guchte 1992; Christie 2008). Rock wak'as and especially carved rocks formed Inka places where Inka identity was visualized and advertised by the rock sculptures. Based upon information provided by ethnohistory, archaeological excavations, and historical documents, it was most likely Pachakuti who reorganized the zeq'e system and the associated wak'a worship. It was his strategy of stone ideology to position carved rocks at selected sites in Cusco and throughout the empire to signal Inka-ness and Inka presence.

Another significant aspect pertaining to the geography of stone ideology was that it was Hanan centered at least in the immediate Cusco area: sculpted rock wak'as only appear on zeq'e lines of Chinchaysuyu and Antisuyu but not on those in Kollasuyu and Kuntisuyu. Chinchaysuyu and Antisuyu constituted the Hanan part of Cusco and rulers six, Inka Roq'a, through ten, Thupa Inka Yupanki, possibly belonged to the corresponding Hanan moiety. The earlier rulers one, Manqo Qhapac, through five, Qhapaq Yupanki, were part of the Hurin moiety or of the physically and hierarchically lower section of Cusco made up of Kollasuyu and Kuntisuyu (see, for example, Bauer 1998:39–42). This constitutes the widely accepted though simplified grouping of rulers into assigned suyus.

A consensus remains that Wiraqocha Inka, Pachakuti Inka, and Thupa Inka Yupanki maintained close ties with the Hanan moiety and particularly with Chinchaysuyu, which appears to have been the suyu charged with most ideological potency. It was the quadrant/suyu of the rulers and it was equated with Collana (the most important in groups of three zeq'es and, in the context of marriage classes, the one of ego) (Zuidema 1964:94, 122). Geographically situated in the northwest environs of Cusco, it encompassed the Urubamba Valley, where many royal estates displaying carved rocks were concentrated.

It thus appears that rulers six through ten may have been more closely linked with rock art than the earlier rulers. As will be discussed in the following chapters, Pachakuti Inka and Thupa Inka formulated Inka rock sculpture and its specific style into a political tool, which I classify as *stone ideology* and which became something like a signature of the state frozen in stone. Consistent with this argument is the observation that the zeq'e lines controlled by the panaqas of rulers six through ten are all located in Chinchaysuyu and Antisuyu, whereas the panaqas of the first five rulers were responsible for zeq'e lines in Kollasuyu and Kuntisuyu. However, this convention only applied to the immediate vicinity of Cusco. Chapters 3 and 4 will show that some of the most powerfully charged carved rock complexes functioning as origin sites were situated at a distance further south of Cusco in Kollasuyu and Kuntisuyu.

Dynamics and Fluidity of Stone Wak'as and the Zeq'e System

The presentation of data above leaves unanswered certain incongruities about the relations between social groups involved in the maintenance of zeq'e lines and wak'as and the potential multivocality of rock shrines. If Pachakuti was indeed the author of the geometric style of rock sculpting and instrumental in organizing the wak'as in the zeq'e system, we would expect that his panaqa controlled important zeq'e lines and many wak'as. Yet the ethnography does not outright support such an assumption; for example, according to Cobo, the panaqa of Inka Roq'a, the sixth ruler, was in charge of the second zeq'e in Chinchaysuyu, but four of its eight wak'as were established by Pachakuti (ninth ruler) and one by Wiraqocha Inka (eighth ruler); and wak'a number four, Viroypacha (Ch-2:4), continued to be used as a pleasure palace by the eleventh queen (see above).

I am speculating that Wiraqocha and Pachakuti may have overlaid new associations upon older shrines and that all three rulers continued to be remembered at wak'as on the second zeq'e line. In other words, a wak'a was not necessarily possessed by one particular ruler or ancestor but could have multiple histories weaving together disparate generations. It is fascinating to imagine how the wak'a itself would have asserted its position in this discourse over possession, but so far I have not located any specific documented episodes. If it was so, the zeq'e system would truly have been a dynamic one that could be changed, manipulated, and overlaid as individual rulers and—perhaps—wak'as saw fit. Zuidema (1964) analyzes in depth the multilayered and overlapping links between rulers, suyus, zeq'es, the categories of collana, payan, and cayao, and marriage classes. It was quite literally "not carved in stone." D'Altroy (2002:167) reinforces this point by noting that Wayna Qhapaq's and Waskhar's palaces were added as wak'as on zeq'e lines in the empire's last years. It was continuously reinvented to fit the ritual and social needs of the Inka capital and, within this ongoing process, the carved rock wak'as appear to have been specially linked to Pachakuti and Amaru Thupa Inka and their panaqas.

Further, the wak'as discussed on the fourth, fifth, and sixth zeq'e lines of Antisuyu serve as powerful additional case studies of the multivocality of carved rock shrines. Sculpted outcrops were not shaped or used for one single, particular purpose but were perceived and treated as living actors who could perform in multiple spatial and ceremonial, political contexts. From this perspective, it is possible that An-4:4 could be a quarry to obtain building stones and at the same time harbor personified rocks modeling other entities; the carved rock complex Subaraura was An-5:3 but also a trail marker on the Yacanora road—indeed Niles (1987:193) emphasizes that the road was intentionally built to make a bend toward this rock rather than continuing in a straight direction; An-5:7, or Aquarsayba, cannot easily be physically identified, but according to Cobo, it had the remarkable quality of sharing its offerings with all other wak'as; Chuquicancha (An-6:3) was the great hall and possible kallanka at Upper Rumi Huasi, a wak'a on the sixth zeq'e line, and an important ritual station for sacrifices to the Sun in the course of the Inti Raymi and Qhapaq ucha ceremonies. Frank Salomon (1991:16–19) brings to life this notion of wak'as from the rich linguistic and ethnographic perspective of the Huarochiri Manuscript. He demonstrates how new wak'as have constantly been created. Van de Guchte (1999:155) emphasizes that wak'as could lose their power and become *atisqa* or "conquered." Any interpretation and reconstruction of carved rocks has to take into consideration living, shifting circumstances of time and society.

The Context of Royal Estates

One of the contexts in which carved rocks performed appears to have been the setting of royal estates. Q'enqo/Patallacta (Ch-1:2) belonged to one of Pachakuti's royal estates, and Pumamarca has been linked with An-6:5 and An-6:6, but it was also a private estate, whether owned by Wiraqocha or Pachakuti. At the latter site, stone is presented in various degrees of modification and transformation and is connected with a cave and water. This ideational landscape (Knapp and Ashmore 1999:12–13) may be seen as a miniature model of the larger, better known, and more public constructed landscapes of royal estates, such as Machu Picchu, Ollantaytambo, and Chinchero (see chapters 5 and 6). I argue that Pumamarca should be viewed as a small private estate at the outskirts of Cusco, which did not require a public and political setting.

As the following chapters will discuss, royal estates were also building blocks in the larger state project to construct an imperial landscape intended to integrate the entire empire. Recent archaeological work has identified a number of potential private estates in outlying regions: for example, the site of Intihuatana nearby Vilcashuaman in Ayacucho. The two sites have sight contact (Christie 2011).

Summary

In the following, I will summarize major points of observation which have developed from the discussion of the carved rock wak'as on the Cusco zeq'e lines and which carry relevance for all sculpted boulders discussed in this book:

- Point number one: Out of the 19 wak'as presented above which were sculpted rocks, four or possibly five were Pachakuti Inka Yupanki's (Ch-1:2, Ch-1:3, Ch-2:4, Ch-5:6, possibly An-6:6), six were Amaru Thupa Inka's (An-1:7, An-4:4, An-5:3, An-5:7, An-6:2, An-6:3), and one was perhaps Wiraqocha Inka's (An-6:6). Such findings are significant because they link the majority of the carved rock shrines to two individuals, the ninth ruler and his eldest son. In the following chapters on royal estates and most importantly Machu Picchu, I argue that Pachakuti was the author of boulder sculpture: he commissioned the carving of rocks and promoted their use as shrines and integration with architecture on his estates, which set Pachakuti's properties apart from those of other rulers. It would be premature to speculate that Pachakuti newly commissioned his sculpted rock shrines on the Cusco zeq'e lines; he may well have reused and carved wak'as sanctioned by pre-Inka groups in the Cusco Valley. Machu Picchu and its carved rocks, on the other hand, most likely constitute his direct personal commission.

The position of Amaru Thupa Inka is interesting. As explained above, he was Pachakuti's eldest son and therefore the rightful heir to the throne, but instead his younger brother Thupa Inka Yupanki was nominated (Sarmiento de Gamboa 1942:129 [1572:chap. 42]). What is most significant in the context of the carved rock wak'as is that Pachakuti made Amaru Thupa Inka together with another son the supervisors of the Cusco wak'as and their worship. Thus Sarmiento assigns Amaru Thupa Inka a special role in the enforcement of wak'a worship mandated by Pachakuti (Martin de Murua makes a very similar statement 1962:48 [c. 1611–1615]). If it holds true that Pachakuti actively sponsored carved rock wak'as, then Amaru Thupa Inka would probably have supported the same. As a son respectful of his father's wishes—he was respectful to the point that he did not rebel when he was not named designated heir—Amaru Thupa Inka's shrines were made in the style of his father's.

The discussion further demonstrated that when a ruler commissioned a wak'a on a specific zeq'e line, his panaqa was rarely in control of the same zeq'e line. Affiliations of lineages with certain wak'as and lines changed over time. Ultimately, this line of reasoning implies that all reconstructions of the zeq'e system from Cobo to Bauer and me do not constitute timeless pictures of how the zeq'e system would have functioned. Rather, these reconstructions have to be seen as slices of history which existed at some point, yet have often been presented as frozen in time. Such realizations are reflected in the format of my book, which groups themes and contexts into "slices" as opposed to a conventional narrative, mirroring the tesserae of the mosaic introduced as a metaphor for the structure of the book at the outset.

- Point number two: It may be possible to take the correspondences between grid-like carved rocks, yupanas, and khipus explored for Ch-5:6, Sabacurinca, further from the material and functional level to a structural and more abstract level illustrating specific patterns of Andean thinking. This line of reasoning reconciles structuralism with the dynamics of lived experiences which coexisted among the Inka as well as in any society. Like the zeq'e system, carved rocks obeyed a structural order which was modified based upon need and circumstances.

Expanding upon Urton's (2003) theory of the binary code, I suggest that the Inka sculptors as well faced binary decisions when they produced carved rocks and yupanas. For example, before they began, they had to decide whether they wanted to create a portable (yupana) or a stationary (grid-like carved rock) counting device. When they were ready to carve, they confronted options of whether to cut in vertical or horizontal directions, a sunken or high relief, one platform or a grid of platforms; when they worked on a grid or row, they had to decide whether the grid/row would continue to the right or to the left. All these are decisions related to binary categories.

Urton (2003:42) cites the work of Mary Frame (2001), who, after analyzing the *tukapu* designs (squares filled with geometric motifs) on Inka textiles, concluded that one would have to seek the meaning of the tukapus in binary oppositions, which, according to her, are the fundamental principle in the structure of the textiles. I argue that the carved rocks and yupanas as well have a structure which can be explained by and is the result of binary oppositions. This is what the Aschers (1997 [1981]) call "Inka Insistence" (Urton 2003:44). The manifestation of this cultural insistence is the process by which a civilization repeats its practices of organizing daily life and thus defines its character and central values. It is related to structuralism in anthropology. Following Urton (2003:44–45), the Inka Insistence is the binary code. Very importantly, Urton links his binary code to certain aspects of Inka society, such as the social divisions of moieties and hanan, hurin. Urton argues that the khipu was based on a binary code because it served a sociopolitical system composed of binary and dualistic structures.

Elsewhere, I have examined (Christie 2006.:372–390) the geometric and symmetrical design of residential architecture at certain Inka sites and the manner to work stone in a geometric style. In a general sense, this geometric and symmetrical style also manifests binary decisions on the part of the artist, such as long or short, wide or narrow, right or left. Urton's binary code may well help to explain the structures of various forms of the visual arts in relation to Inka social organization.

As has been addressed above (under Essence, Imagination, and Stone Ideology: Q'enqo, Ch-1:2 and Ch-1:3), binary opposites, the acts of doubling and copying, presentation versus representation have been analyzed through many models and methodologies developed by scholars. Urton's binary code is rooted in structuralism and emphasizes organization, order, control in society and all cultural affairs, and to some extent geometry in the visual arts. Peirce's model (see above) of legisign and sinsigns, on the other hand, is agent-based and fluid. Sinsigns or individual replicas are never exactly the same; they change with their agent-authors, time, and space. Urton's structuralism and Peirce's legisign and sinsigns may be understood as the theoretical underpinnings of the approach outlined above, which seeks certain structural systems of social, political, and religious organization and visual arts styles while carefully balancing them with the acceptance of changing, personalized, and historical reinterpretations and replicas of those same structures.

- Point number three: The most fundamental insight to take from this chapter and carry over to the succeeding chapters is that carved rock wak'as on the zeq'e lines and elsewhere contain the potential to come alive: they function as living performing actors parallel to humans, a type of "stone people"; they talk, eat, and are dressed, they own properties and animals, they multiply similar to humans, they gain and lose powers and may be conquered and die. The older and recent ethnographic literature abounds with case studies testifying to all these remarkable qualities of Andean wak'as including carved rocks. Relevant sources have been cited above and will continue to be referred to in the following chapters. One of the richest and most insightful analyses of Andean wak'as is clearly Frank Salomon's introduction to the Huarochiri manuscript (1991). Among many significant things, he notes (1991:19) structures of correspondences between human social organizations, the societies of supernaturals and wak'as, and earth forms manifested in the natural settings of the Andean landscape. In the context of chapter 2, such correspondences correlate with the ayllus and panaqas charged with the maintenance of carved rock wak'as and lines in the zeq'e system and with the topography of the Cusco Valley. They also help us understand the many mechanisms of reproducing, copying, and doubling of sculpted rock complexes, wak'as, and visual art forms through time, which will further be explored in the succeeding chapters.

NOTES

1. It is meaningful that Molina (Molina in Bauer 1998:40–41) presents the names of ten panaqas and ten nonroyal ayllus according to their suyu: three panaqas and three nonroyal ayllus in Kollasuyu, three panaqas and three nonroyal ayllus in Chinchaysuyu, two panaqas and two nonroyal ayllus in Antisuyu, and two panaqas and two nonroyal ayllus in Kuntisuyu. Complementing Molina's account with other documents which mention the names of panaqas and nonroyal ayllus, one can reconstruct that at the end of the Citua festival, descendants of Manqo Qhapaq, the first Inka, from the Chima panaqa and descendants of Zinchi Roq'a, the second Inka, from the Raurau panaqa ran with members of the nonroyal ayllus Quesco and Masca toward Kuntisuyu. Warriors from the panaqas of Hahuanina (descendants of Lloq'e Yupanki, the third Inka), Usca Mayta (descendants of Mayta Qhapaq, the fourth

Inka), and Apu Mayta (descendants of Qhapaq Yupanki, the fifth Inka) joined with representatives from the nonroyal ayllus of Cuycusa, Sutic, and Maras and headed into Kollasuyu. Members of the panaqas Aucaylli (descendants of Yawar Waqaq, the seventh Inka) and Sucsu (descendants of Wiraqocha Inka, the eighth Inka) ran with those of the nonroyal ayllus Sanoc and Tarpuntay to Antisuyu. Descendants of Inka Roq'a, the sixth Inka, from the Vicaquirao panaqa, descendants of Pachakuti Inka Yupanki, the ninth Inka, of the Hatun panaqa, as well as those of Thupa Inka Yupanki, the tenth Inka, from the Capac panaqa left with runners of the nonroyal ayllus Uro, Arayraca, and Chavin Cuzco in the direction of Chinchaysuyu. Based upon the same number of panaqas and nonroyal ayllus in Molina's account and the order in which they are mentioned, researchers have suggested that each panaqa was paired with one nonroyal ayllu; for example, in Chinchaysuyu, the Vicaquirao panaqa would have been associated with the nonroyal ayllu Uro, Hatun panaqa might have been linked to Arayraca, and the panaqa Capac would somehow have been affiliated with the Chavin Cuzco ayllu (Zuidema 1964; Rowe 1985).

2. An invaluable aid in sorting through the immense corpus of ethnohistoric and ethnographic material and finding specific references is the three volume *Guide to Documentary Sources for Andean Studies, 1530–1900* edited by Joanne Pillsbury (2008). Another candidate for authorship of the original zeq'e system manuscript is Father Cristobal de Molina (?–1585) (see Urbano 2008). In his *Relacion de las fabulas I ritos de los Ingas*, written around 1575, Molina indicates that an account on the shrines of Cusco—possibly the original zeq'e system manuscript—existed and that he sent it to Don Sebastian de Lartaun, the bishop of Cusco (Bauer 1998:15). This implies that Molina might have been familiar with the original zeq'e system manuscript and could perhaps have been its author (Rowe 1946:300; Zuidema 1964:1 n. 1).

Another important list of shrines was compiled by Cristobal de Albornoz (c. 1530–1583) who was one of the principal leaders in the Spanish campaigns against indigenous religions in the Andes (Millones 2008). Most relevant to the present study is Albornoz's essay titled *Instrucción para descubrir todas las guacas del Piru y sus camayos y haziendas* dated 1582. In this treatise, Albornoz states that many Andean shrines were organized along zeq'es, and he provides a list of thirty-seven wak'as in the Chinchaysuyu region of Cusco. While most of these wak'as are also mentioned in Cobo's chronicle, their order, names, and spellings differ enough to conclude that Albornoz composed his list independently of the original zeq'e system manuscript and that Cobo seventy years later was unfamiliar with Albornoz's work (Bauer 1998:15).

3. The references to the other writers are given in Bauer 1998:50.

4. For a thorough discussion of *huauques*, see van de Guchte 1990:272–294 and Van de Guchte 1996:256–268.

5. Pedro Pizarro's account is dated to 1572, the same year in which Sarmiento de Gamboa completed his *History of the Incas*. Polo de Ondegardo was corregidor of Cusco from 1558 to 1560 and then again from 1571 to 1572 and likely found the mummy and golden image in Tococache in those same years, 1571–1572 (Sarmiento de Gamboa). According to Betanzos, the statues had been taken away by Manqo Inka in the 1530s (1996:139 [1557:part I, chap. XXXII]).

6. The association of Qochapata with Q'enqo remains, of course, speculative. Rodriguez Limache (2007:208) emphasizes that all the sectors of Saqsaywaman from Muyo Marka to Tambomachay were related, and INC is working to further define and clarify such associations. I group Qochapata with Q'enqo precisely because of their physical vicinity and similar material components which integrate sculpted rock outcrops, architecture, as well as cave openings. The issue of Inka predilection in sectoring and doubling or copying will be taken up under discussion.

7. Such an assessment may change once the Ministerio de Cultura begins work at Q'enqo Chico as is planned.

8. The historical continuity is worth noting: today's shamans go to Qochapata not only because Q'enqo has become too crowded. Knowingly or subconsciously, they appear to express a desire to reuse precontact wak'as.

9. He had a house and a courtyard full of birds, big and small monkeys, (g) *uacamayas*(os), big and small parrots, budgerigars, semicalos, turtle-doves *cucuri* and black birds, goldfinch, and many other birds from the sierra and from the yunga, biscachas and a fish pond, water fountains *Uiruy Paccha*, a stream of water (Guaman Poma de Ayala 1987:338[1615:334(336)] translation by the author).

10. And she had a thousand native entertainers: some performed, others danced, still others sang with drums and musical instruments, flutes, and pan flutes; and she had singers, *haraui*, in her house as well as in other places to listen to the above mentioned music that is *haraui* or love songs in Uaca Punco, and to the pan flute in Pincollona Pata, in Cantoc, and in Uiroy Pata, Cinga Urco (Guaman Poma de Ayala 1987:134 [1615:141] translation by the author).

11. Wank'as were usually unmodified or slightly shaped rocks which were treated as the petrified owners of villages and fields in rural communities; see the influential studies of wank'a by Pierre Duviols (1979b).

12. One well-preserved Inka example belongs to the Archaeological Museum in Lima. Games very similar to yupanas are still popular in the Cusco area. They consist of a grid of numbered wooden boxes. Players aim to toss coins in these boxes, and they score points depending on the higher or lower box numbers in which their coins land.

13. An interesting alternative identification is offered by Muriel. He associates Ch-5:7 with a small hill straight to the north of Ch-5:6. This hill is famous for a special grass called *Nihua*, which is still commonly used in toys for children and in house decorations (n.d.:14).

14. There is a dirt road leading from Laqo straight to the Pisaq road. It enters the Pisaq road near Llanllakuyoq and could be an alternate for the road mentioned in the 1636 document.

15. He was born in Pomacocha near Vilcashuaman as the first son of the Coya (principal wife) Mama Anarhuaque of Pachakuti Inka Yupanki, which made him the rightful heir. He did not receive his name until he was brought to Cusco and presented to his grandfather Wiraqocha Inka. Oral tradition has it that at the time of his birth, there appeared a supernatural serpent, or *amaru/o*, from a sacred mountain near Cusco and flew through the air. The child was named after this apparition *Amaru Thupa Inka*.

Later, Pachakuti Inka decided to designate his younger son from the same mother, Thupa Inka, as his successor. Amaru Thupa Inka did not challenge his father's request but accepted it and remained loyal to Thupa Inka. Pachakuti put him in charge of the city of Cusco, and he possibly became the head of Capac Ayllu (see Rostworowski 1953:238–239), Thupa Inka's panaqa, in violation of the Inka rule of inheritance, which prescribed that sons who did not take rulership had to belong to their father's panaqa. Possibly because of his collaboration, he was given a number of properties in the Cusco area and powerful positions in the political-religious hierarchy. Most significantly in the context of the zeq'e system, he was charged with the reorganization of the shrines of Cusco. Franklin Pease (2004:130–132) interprets the succession of Pachakuti as a discourse between Hanan personified by Thupa Inka and Hurin personified by Amaru Thupa Inka which may have culminated in a ritual battle which the Hanan party has traditionally won.

16. Other important aspects are considerations of changes over time and chronology. Was the Cusco zeq'e system devised in one complete and coherent design, or did it develop over time? While it seems more logical to understand the zeq'e system as one consistent design, it is very probable that individual zeq'es were assigned for attendance and upkeep to individual ayllus as time progressed.

17. Henry Conza Leon has pointed out that this sculpted rock presents an interesting parallel to the famous block with twelve sides in one of the palace walls in the center of Cusco (personal communication, 2002).

18. At the same time, it cannot be overlooked that some sight lines clearly had practical strategic and military functions. Vincent Lee (2008) has documented a network of station points on mountaintops and sight lines between Choqequirau and Machu Picchu by means of which smoke signals could have conveyed messages quickly.

19. Bauer adds a connection to still another origin place. He (1998:79) suggests that the carved rocks Mesa Redonda and Cinca represent An-2:2 described by Cobo in the following manner:

> [An-2:2] The second guaca of this ceque was named Pachatosa; it was a large stone which was next to [Diego] Cayo's house. The sacrifice was burned on top of it, and they said that the stone ate it (Cobo 1990:64 [1653:Bk.13, Ch.13]).

The identification of Ch-1:1 and An-2:2 focuses on the interpretation of place names and cannot easily be resolved. Bauer refers to a 1572 source which talks about

> the descendants and origins of the Hualla Indians of the town of Payatusan [Pachatusan], that is at the back of San Blas, in this city, to the east, near the terraces that leave San Blas towards Salinas (Loarte in Bauer 1998:79).

Bauer equates Pachatusan with Lucrepata, which is behind and to the east of the San Blas neighborhood. The upper areas of Lucrepata contain caves from which the Hualla people are thought to have emerged; and a short distance below, there is the group of carved rocks including Mesa Redonda and Cinca. Bauer's maps indicate that the first and second zeq'es of Chinchaysuyu and Antisuyu ran very close to each other (1998:maps 5.2, 6.1).

20. In one instance, the people of the contemporary town of Pacariqtambo have designated a cave as the origin place, which is physically closer than the historical one. In a second case, people in the communities of Mara and Pitic in the Departamento Apurimac identify three openings in a steep mountainside as the three caves of Pacariqtambo where "Atawallpa hid his gold."

21. I have discussed relevant aspects of Peirce's semiotics elsewhere (Christie 2012). Here semiotics is referenced as a secondary theory to help explain Inka concepts of doubling.

Chapter Three

The Birthplace of the Sun, Moon, and the Inka Ancestors on the Island of the Sun and the Southern Basin of Lake Titicaca

Chapters 3 and 4 address Andean and Inka origin places at Lake Titicaca and at Pumaurqu. I have selected these sites because they play two vital roles in the discussion of stone ideology: first, they link rock wak'as with the beginnings of the world and the emergence of the Inka dynasty; and second, they extend the spatial context of carved rocks from Cusco far into Kollasuyu and Kuntisuyu. The reader should remember that no carved rock wak'as were encountered on the zeq'e lines of Kollasuyu and Kuntisuyu in the immediate Cusco area. To position the origin wak'as in these same suyus was likely a strategic move in Pachakuti's stone ideology. As in chapter 2, the discussion will consist of a *Groundwork* section that summarizes the known ethnographic and archaeological data and attempts to reconstruct the uses of the carved rock sites, followed by an analysis of the rock wak'as in the broader context of memory landscapes (under the headings *Inka Landscape and Carved Rocks*).

Lake Titicaca mesmerizes visitors today with its deep blue waters, which form a striking color contrast with the yellow brown of the highland landscape of the puna; from the beach in Copacabana, one can observe unforgettable sunsets over the water, like bursting fireworks with wide spectrums of color. The lake is located between the fifteenth and seventeenth parallels of southern latitude and straddles the border between Peru and Bolivia (figure I.2).

Lake Titicaca constitutes the primordial site of creation in pan-Andean myth. Over time, many stories and legends have been connected with the lake, its islands, and with special places in the surrounding countryside, beginning possibly as early as 1100 B.C. and continuing into the present (Salles-Reese 1997:50–52),[1] thus constructing a complex memory landscape. The earliest archaeologically documented human occupations in the Lake Titicaca Basin go back to the Early Archaic Period before 6000 B.C. and possibly several millennia earlier (Bauer and Stanish 2001:33).[2] The indigenous people residing in the area today speak Aymara and are the inheritors of many cultural influences.

GROUNDWORK

The Origin Narratives: Recorded Data Contextualized as Essence, Imagination, and Stone Ideology

Stories and creation narratives that mention Lake Titicaca and take place in locales in its vicinity were recorded by Spanish writers in the sixteenth century and reflect the Inka version of such stories and myths. To complicate matters further, the individual accounts are not identical, but instead were shaped and influenced by the personal experiences and preferences of the writers. They construct a rich and layered memoryscape rather than historical accounts. It becomes our challenge—and we also are shaped by our cultures, academic training, and research expectations—to extract shared themes and concepts from the many differing versions and to collect them into a generalized Lake Titicaca origin mythohistory, as, for example, Gary Urton (1999:34–44) has done.

The following is a paraphrase of such a master narrative reconstructed by Salles-Reese (1997:53–54) from sixteen versions of the origin myth recorded between 1550 and 1653.[3] Wiraqocha, the pan-Andean creator god, appeared in the Lake Titicaca region when it still lay in darkness because neither the sun nor the moon had been created. He came to Tiwanaku, where he fabricated giants, man, animals, and all other things. Wiraqocha set certain rules of conduct which his creatures were to observe, but man disobeyed. In response, Wiraqocha sent a flood that destroyed everything except for a few privileged individuals who managed to escape punishment. When the flood receded, Wiraqocha traveled to Titicaca where he created the sun, the moon, and the stars. He formed the different lineages of humankind with the clay of Tiwanaku and gave each group its clothing, language, songs, agricultural systems, and religion. Wiraqocha sent some of the individuals he had just created to the mountaintops, others to the rivers and springs, and still others into caves. They were to emerge from those liminal locations, turning them into *paqarinas* of their ayllus and themselves into the founding ancestors. This is how the different lineages originated. After creation was completed and the earth populated, Wiraqocha journeyed throughout the different regions of the Andes. He revealed himself to the people as their creator, though not all recognized him, and those who did not were punished. When his mission on earth was finished, Wiraqocha disappeared into the sea.

The Wiraqocha story has to be understood as the cosmic pan-Andean creation myth which had been formulated in the Lake Titicaca Basin but became well known throughout the Andes (for the full argument see Salles-Reese 1997:45–88). It addresses the origin and creation of the the cosmos, the world, and humanity through the acts of Wiraqocha.

In the fifteenth century, as the Inka moved southward from Cusco, conquering and annexing the territory of Kollasuyu,[4] they overlaid their version of the creation myth onto the Wiraqocha story. For example, Betanzos mentions that Wiraqocha came to Cusco, where he called up out of the earth a number of important people. His account clearly links Lake Titicaca, Wiraqocha, and the Inka (Urton 1999:39).

Salles-Reese reconstructed another condensed master narrative from sixteen variants of the origin myth of the Inka dynasty compiled between 1552 and 1653 by Spanish chroniclers (1997:93).[5] On the Island of Titicaca, the Sun—personified as a man in a shining costume—summoned the Inka, represented by Manqo Qhapaq and his wife/sister Mama Oqllu, and adopted them as his children. Since the rest of humankind was living in a state of barbarism, the Sun conferred a civilizing mission upon the Inkas. These divine emissaries traveled to the north and emerged from a cave at Pacariqtambo near Cusco (the site of this origin narrative will be analyzed in chapter 4). These original Inkas, four brothers sharing the name Ayar and four sisters, wore clothing richly adorned in gold and carried corn seeds. One brother, Manqo Qhapaq, also carried a golden rod that would sink into the ground at the exact place where, according to the wishes of the Sun, they should settle. The rod was thrust into the ground at either Pacariqtambo or Cusco. Three of the original four brothers were turned into *wak'as*. The remaining brother, Manqo Qhapaq, had a son with one of his sisters, thus becoming the progenitor of the Inka dynasty.

The two master narratives summarized above are extremely important for our understanding of the politics and ideology of Inka expansion and of their strategies to justify and legitimize domination in Kollasuyu. In a first step, the Inka adopted the origin myth with Wiraqocha as its protagonist that was known to Andean peoples, but at the same time they manipulated it by introducing their own ancestors. According to the chronicles, Wiraqocha Inka, Pachakuti Inka, and Thupa Inka Yupanki conquered and integrated the altiplano (D'Altroy 2002:65–72); therefore co-opting the pan-Andean site of origin would have legitimized their military campaigns.

A story reported by Ramos Gavilan brings this point to life. An old priest at the sanctuary of the Sun in Lake Titicaca had heard that Thupa Inka Yupanki had publicly declared his devotion to the Sun. So this priest decided to travel to Cusco and gain the Inka ruler's support for his sanctuary. Thupa

Inka received him, and he described the shrine, its history, and its oracles in such a powerful way that the Inka ruler was fascinated and promised to come and see it. Thupa Inka came, took possession of the island, and made the Sanctuary of the Sun the most famous and principal shrine in the empire (Ramos Gavilan 1988:39–41). Indeed, the Island of the Sun became the final destination of a highly prestigious pilgrimage route which originated in Cusco. It may be viewed as a superb example of the construction of a political landscape as defined by Adam Smith (2003:10) in which the experience, perception, and imagination of Inka power structures were overlaid upon the altiplano in Kollasuyu.

One great challenge faced by Inka administration as their territories expanded was the salient ethnic diversity characterizing the subjects of the Inka and their own brief history and lack of a long reputable descent line. By forging the mythological link to the Lake Titicaca Basin, they gained the type of respectable history they needed to accredit themselves against their neighbors and subjects. Since their subjects were widely familiar with the myth and the Island of the Sun, they created a unifying factor; and by overlaying the birth of their own progenitors, they claimed that their dominion was foreordained.

Yet it was not only foreordained but was also sanctified by supernatural forces. By appropriating the Titicaca creation narrative, the Inka constructed three important ideological links with the Sun, the underworld, and the material and essence of stone. The Inka version explains the father-child relationship between the Sun and Manqo Qhapaq and Mama Oqllu, the first mythohistorical Inka king and queen. From then on, the Sun remained the main deity, backing up the supreme religious and political authority of all Inka rulers.

Secondly, Manqo Qhapaq and Mama Oqllu were called forth by the Sun from openings in a rock outcrop. This setting evokes symbolic links with stone, caves, and the underworld. The Inka version of the origin myth continues by telling us that the ancestors hiked underground to reemerge from the caves at Pacariqtambo. Thus, clearly, the ancestral Ayar brothers and sisters have ties with the underworld, which Andean people envision as streams of water circulating to the upperworld. They have the ability to transform from the realm of water and fluidity to that of substance and to become human beings. The ancestors demonstrate connections with the supernatural worlds below and above and, in this manner, the origin myths contain the foundation and core ideas of divine sanctioning of the Inka state.

Third, both emergence sites are rock outcrops. The outcrop in the Sanctuary of the Island of the Sun (see below) remains unmodified, but Copacabana, the mainland center which controlled the cult and pilgrimage on the island, is graced with a number of sculpted rock complexes (see below). Puma-urqu, which contains the caves of Pacariqtambo, is also carved. Stone is the material par excellence to visualize links with the vertical divisions of the cosmos. It is the essence of mountains, which are grounded in bedrock in the underworld and tower above the human world, seemingly reaching to the sky. Here the phenomenological perspective is useful, underscoring the timeless human experience of awe when confronting the dramatic heights and depths of high mountain chains, such as the Andes, which Western technology has not been capable of diminishing substantially. Inka rulers wanted to associate themselves with such a powerful substance, and this is one reason why they chose these origin sites and pursued the strategy of *stone ideology*.

The association between the ruler and stone went one step further. During their migration from Pacariqtambo to Cusco, several of the Ayar brothers were turned into stone *wak'as* (see chapter 4). In the Chanka wars, during the decisive battle, stones in the field metamorphosed into warriors rushing to Pachakuti's aid and making his victory possible. These stones were later venerated as shrines called *pururaucas* (see D'Altroy 2002:64). Humans could transform into stone and vice versa, illustrating the concept that in Inka and Andean understanding of the world, stone possessed a life force which could potentially ally itself with the ruler (see also the story of Piedra Cansada or Ch-4:6 in chapter 2).

From a linguistic perspective, Salomon (1998:9–11) has elaborated that this life force in stone is the most permanent and durable in the trajectory of all being through time from young, kinetic, and soft toward static and hard. Therefore it is reflective of the authority of the ruler.

The argument which surfaces throughout this study is that Pachakuti and his sons developed and practiced a stone ideology that was grounded in the origin myths and materialized in the design and layout, and in the architecture and sculpture of many Inka centers and royal estates. The Wiraqocha story further reinforces aspects of this stone ideology. Wiraqocha created the first humans at Tiwanaku, a settlement famous for its exquisite stonework found in building blocks and architecture as well as in stone sculpture. We know that Pachakuti was so impressed by Tiwanaku craftsmanship that he ordered his masons and stonecutters to copy their style (Cobo 1956:191–192 [1653:Bk.12, ch.13]). So, clearly, if the symbolic qualities and manipulation of stone were meant to assume a significant role in imperial ideology, Tiwanaku had to be integrated metaphorically and politically.

In a second step, as is outlined in the second master narrative, the Inka selected a second site of emergence and of the origin of Inka dynasties at Pacariqtambo (chapter 4). This site was purely Inka, the *paqarina* of Inka rulers, and was not shared by Andean people in general. In Inka myth and storytelling, it functioned as the place where the specific history of the Inka began.

Salles-Reese explores the continuities and changes in the two master narratives in depth; two points are of particular importance in the context of this study. First is the observation that Wiraqocha created the natural order represented by the hills, stars, plants, animals, and mankind as equal elements within a whole. In turn, mankind produced the social order, consisting of lineages with different customs, but who again relate to each other on an equal and horizontal level. Only the Inka myth of origin introduces hierarchies in both orders: within the natural order, the Sun is now privileged and reigns supreme above all other stars, which are also deified. In the social order, the Inka hold the highest status because of their very association with the Sun (Salles-Reese 1997:97–98). In this sense, the origin narratives become a political tool to justify Inka rule.

Second, the Inka origin myth establishes the sacred space of Lake Titicaca as Inka since ancient times. While the Inka claimed a special relationship with the Sun, archaeological, iconographic, and ethnographic investigations have shown that the solar cult in the Lake Titicaca region predates the Inka and has a much longer history (Salles-Reese 1997:95–100; Bauer and Stanish 2001:132–157, 196; Ramos Gavilan 1988:39–41 above). However, by expanding the temporal limits of Inka history and by situating the place of descent of the Inka dynasty at Titicaca, the myth justifies the conquest as a return to Inka patrimony and reclaiming of history, rather than as an appropriation of a neighbor's territory (Salles-Reese 1997:99–100).

In sum, there is a concentric progression toward specificity (Salles-Reese 1997:127–130). Although the protagonists in both narratives establish order where chaos reigned, the Inka modifications lead away from the general and the cosmic toward the particular and individual. Thus, the Sun assumes a special function, whereas in the Wiraqocha story all elements of nature were treated equally. Although the Sun's importance increases, his functions are progressively more restricted. The god Wiraqocha was the protagonist of the divine order in the pan-Andean creation story, but now the Inka as a specific historical group move into the position of leadership precisely because of their relationship with the Sun. In the social order, the Ayar brothers become the ancestors of the royal Inka panaqa while the lineages in the Wiraqocha narrative were not individualized. Ayar Manqo Qhapaq is simultaneously depicted as a mythical figure and as a historical hero, and thus becomes the character in which myth and history coalesce. This concentric progression toward specificity ultimately constructs the position of authority of the supreme Inka ruler and explains his divine sanction.

Ethnographic, Archaeological, and Phenomenological Data Pertaining to Copacabana and Related Sites with Rock Art on the Inka Pilgrimage Route from Yunguyu to the Sanctuary of the Sun and the Island of the Moon

Primary sources of information about these sites are ethnography and local consultants as well as archaeological surveys and excavations. The most pertinent ethnographic writings covering the mainland as well as the islands were composed by Alonso Ramos Gavilan (1621; see also MacCormack 2008:559–563) and Bernabe Cobo (1653; see also Hamilton 2008:152–155). Ramos Gavilan was born in what is now the Peruvian city of Ayacucho. He visited different regions on the coast and in the highlands of Peru and arrived at Copacabana in 1618. There he began writing his books, interviewing many elders of Copacabana and personally visiting the Island of the Sun. The resulting work, *Historia del célèbre Santuario de Nuestra Señora de Copacabana*, was completed and published in Lima in 1621.

Cobo lived near Lake Titicaca from 1616 to 1617. He arrived at the town of Juli, sailed around the lake, and visited the Island of the Sun. Unlike Ramos Gavilan, he did not begin to write immediately. Cobo continued to travel and collect information and did not complete his extensive project, *Historia del Nuevo Mundo*, which originally contained forty-three books divided into three parts, until 1653 (see chapter 2). In his chapters on the Islands of the Sun and Moon, he copied from Ramos Gavilan and added his own personal observations.

The most recent and most comprehensive research program on the Islands of the Sun and Moon was conducted by Brian Bauer and Charles Stanish from 1994 to 1996. In the 1994–1995 seasons, they surveyed both islands and identified more than 180 precontact sites. The survey was followed by test excavations and archaeoastronomy research during the second season. Their investigation of the Island of the Sun and its Sanctuary and of the Island of the Moon was thoroughly documented and contextualized by Bauer and Stanish (2001). Yet their research area was limited to the two islands, and they did not investigate Copacabana and its surroundings, which is the focus here. The islands and Copacabana were included in the broader study of the development of complex societies in the southern Titicaca basin by Stanish (2003). Ongoing surveys (as of 2010) on the mainland in the Copacabana area conducted by Jose Maria Lopez Bejarano (2007) are bringing to light the historical layers of occupation.

In this chapter, I will walk the pilgrimage route from Yunguyu to the Sanctuary of the Sun—as a "scribal pilgrim" —so to speak, stopping at major stations with rock art and carved rock wak'as and exploring the sites as Inka pilgrims may have experienced them, as well as contextualizing them within state-sponsored stone ideology, after Bauer and Stanish (2001:213–232),

Stanish and Bauer (2004), Cobo (1990:91–99 [1653:Bk.13, Ch.18]), and Ramos Gavilan (1988 [1621]). To my knowledge, Spanish writers never described the complete pilgrimage from Cusco to the Island of the Sun or its precise sociopolitical role. The accounts referenced become prolific only for the last section from the Copacabana area to the main Sanctuary.

Pilgrims entered sacred ground at the town of Yunguyu, which marks the access point to the peninsula on which Copacabana is located (figure 3.1). At Yunguyu, both sides of the lake come so close together that a land bridge or isthmus is created. Cobo (1990 [1653]:94, 96) states that the Inka had this entrance closed off with a wall extending from one beach to the other. There were gates in this wall staffed with watchmen and guards to control the people who came to visit the sanctuaries. Prospective visitors were sent to confessors and, according to the gravity of their admitted offenses, were punished with forms of physical penitence. It is intriguing that still today the Copacabana Peninsula is separated from the rest of the mainland by the national border

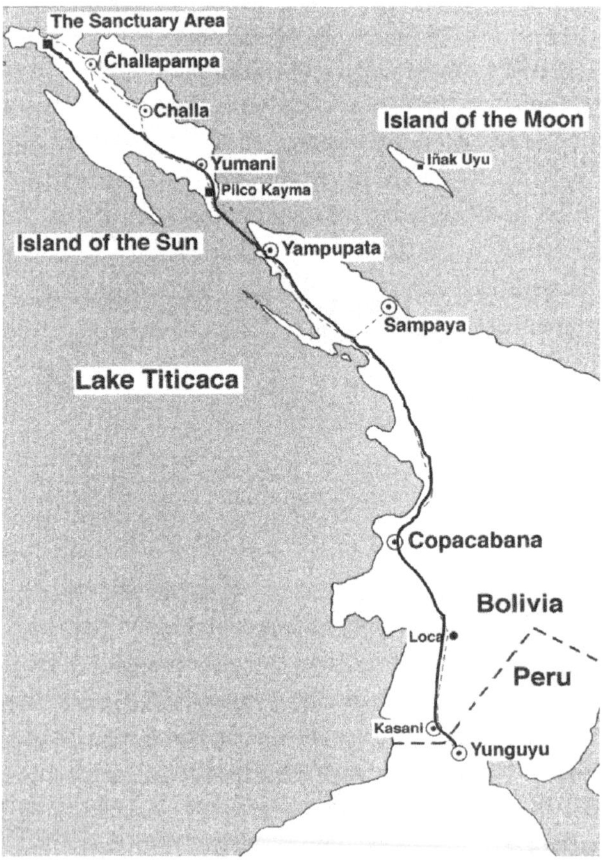

Figure 3.1. Map of the Pilgrimage Route from Yunguyu to the Island of the Sun. Courtesy of Brian Bauer; modified to retrace the pilgrimage route described here.

between Peru and Bolivia. Travelers must walk through a modern wall with an arch, pass customs and immigration, and obtain permission to continue their journey (see Bauer and Stanish 2001:214–215).

Ramos Gavilan (1988:127 [1621:Bk.1, Ch.20]) adds that pilgrims were well provided for by the Inka state. The emperor ordered that public hostels be built, which they called *tambos*, as well as special pilgrims' houses named *corpaguasi*, where the visitors were provided for as long as their journeys lasted. At Copacabana, they received food and drink, and those who were very poor were given clothing. All the supplies were collected in storehouses or *colcas* dotting the hills around Copacabana.

The physical appearance of Copacabana is poorly understood. The chroniclers tell us that it had at least one temple and that the pilgrims made another confession (Cobo 1990:96 [1653;Bk.13, Ch.18]). Ramos Gavilan is more specific in his description (Ramos Gavilan 1988:171–172 [1621:Bk.1, Ch.28]).

Since the Inka settlement is overlaid by the modern town, it will not be feasible to explore its full extent any time soon. Nevertheless, in 1983 Oswaldo Rivera Sundt conducted excavations in the main plaza of Copacabana. On the surface, he found ceramics and other objects that were Inka. Intermediate levels yielded ceramics from Tiwanaku IV and V, and in the lowest levels he identified ceramics and stone foundations of houses affiliated with Chiripa (Boero Rojo and Boero Rojo 1987:15). In the summer of 2008, locals stumbled upon an extensive archaeological site as they began to construct a new market area. Sergio Chavez, who is in charge of the salvage excavations, has stated that the earliest stratigraphic layers date to the 2,500-year-old *Yaya-Mama* religious tradition (see below) superseded by Tiwanaku and Inka layers which, in turn, are covered by Colonial and Republic deposits (www.canada.com/vancouversun/news/story.html, July 10, 2008). Chavez's preliminary data thus confirm the results obtained by Rivera Sundt in the 1980s. While there were apparently earlier settlements, this study is concerned with the Inka occupation.

Bauer and Stanish suggest that "the town of Copacabana served as the regional administrative center for the peninsula and was responsible for guarding and controlling access to the sacred islands" (2001:216). Ramos Gavilan reinforces such a classification by noting that Copacabana had a governor and that this governor "was of royal blood, second only to the person of the Inca; . . . This governor dressed like the Inca with the only difference that he wore the tassel on one side while the Inca wore it in front" (1887:27, translation by the author). Martin de Murua adds "it was known that the Inga Wayna Kapaj had his people and captains called and then he left for Quito, leaving one of his brothers who had previously been captain and governor to act in his name and on his behalf, after a certain time, this brother settled down in the village of Copacabana and married and had children, whose descendants continue to live in said village of Copacabana" (de Murua in Rivera Sundt 1978:73, translation by the author). Citing passages from Ramos Gavilan and Cristobal Vaca de Castro, Roberto Santos Escobar (1981) suggests that a daughter of Wayna Qhapaq who directed the *aqllahuaci* on the Island of the Moon married her brother Paullu Tupaj Inca in Copacabana, a union which established an important royal descent line. Such remarks speak to the continued influence Copacabana must have exerted on the royal family in Cusco. Its importance remained unchanged in the Colonial Period, when families from Copacabana attempted to regain their noble social status as well as their possessions by submitting *probanzas* to the Spanish Crown in which they tried to prove descent from the Inka royalty (Santos Escobar 1981).

Bauer and Stanish (2001:160) comment on the large number of small settlements on the Island of the Sun, more than sixty of which are less than one hectare in size. These numbers contrast with data from the mainland where several large sites have been located; for example, the town of Juli was at least 20 hectares in size during Inka reign, and Chucuito was at least 50 hectares. This type of evidence suggests that the regional center from which the pilgrimages were coordinated and supervised was not situated on the islands but on the mainland. Copacabana is the most likely candidate because it is past the wall of Yunguyu and, according to Ramos Gavilan and Cobo, it functioned as a major station on the pilgrimage route (Ramos Gavilan 1988:84–85 [1621:Bk.1, Ch.12], Ramos Gavilan 1887:13). Gavilan's writings demonstrate the extraordinary importance the Inka assigned to the Sanctuary on the Island of the Sun and to Copacabana. This unusually high position of Copacabana within the Inka state is reflected by parallels in the rock art. No other Inka settlement outside the Cusco region and Urubamba Valley exhibits such a large number of sculpted boulders in a style very similar to the Cusco examples (Hyslop 1990:121). It is most significant that four of the five known and visible precontact sites in and around Copacabana are carved rocks. Some scholars hesitate to attribute the sculpted boulders of Copacabana to the Inka (Trimborn 1967; Portugal and Ibarra 1957).[6] The main reason I believe it is justifiable to name the Inka as the artists of the carved rocks is the striking similarity of style with the rock art sites in the Cusco region (see also Arkush 1999, 2005).[7] I will now investigate five places (1–5) with empowered rocks in the Copacabana area.

The best known site within the city limits is Intinkala (1), situated near the cemetery on the road to La Paz. According to Portugal and Ibarra (1957:25), the Aymara spelling is *Intin Qala*, and, according to Aymara tradition, it was the place "where the sun sat." The Aymara continue to place offerings on the rocks on June 21, which marks the beginning of their new year.[8] The site consists of a group of finely carved boulders. The carvings are primarily

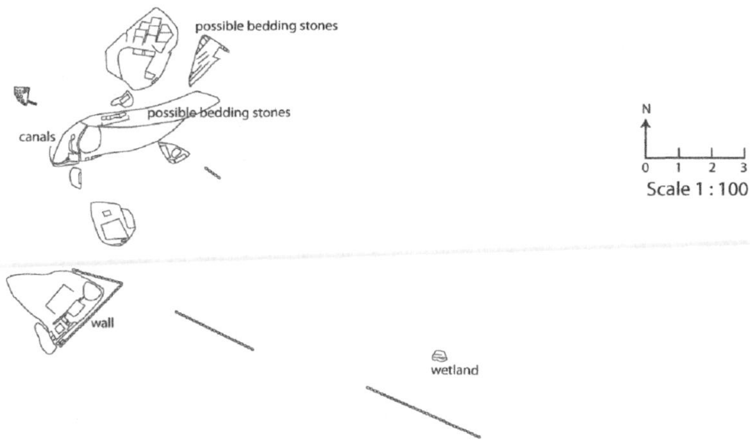

Figure 3.2. Copacabana, Intinkala, Site Plan. Drawing by the author and Brian Garrett.

vertical and horizontal cuts forming platforms and seats. The top of one large rock displays a conspicuous canal, the beginning of which resembles a snake head. The boulders are associated with several walls, one of which leads into the direction of a wetland (figure 3.2).

The drawing documents the spatial relations of all the features noted on the surface. In 1975, cleaning and excavation conducted by the Instituto Nacional de Arqueologia (INA) uncovered walls built directly on the rocks, traces of a paved floor, and a system of drainage canals (Rivera Sundt 1978). Roberto Mantilla (1972) describes some carved rocks at this site which he calls "Parque Arqueologico de Copacabana," which I could not identify. His site plan is at such a small scale that it is impossible to match it up with the present condition. A short distance further on the road to La Paz sits a second group of smaller boulders sculpted into seats and platforms. Local people call this second group Orcohawira, Rio Macho, or Rio Fuerte. The cuts are straight, forming full ninety-degree angles, and demonstrate high quality of craftsmanship (figure 3.3). In between the rocks is a wetland which appears to receive water from the higher slopes. Perhaps this water source is the reason why the site is referred to as "Rio." These carvings were first documented by Roberto Mantilla in 1968 (Mantilla 1972). INA and Elizabeth Arkush (2005) report scatters of fine Inka ceramics at and around both rock art sites.

Hermann Trimborn (1967:19–23) argues that the boulder with the largest number and best executed carvings at Intinkala functioned as an observatory of the sun. He reconstructs that a person seated on the lower platforms and facing east could have observed the sun on the equinoxes rising from behind a visually outstanding mountain formation on the horizon line. He interprets

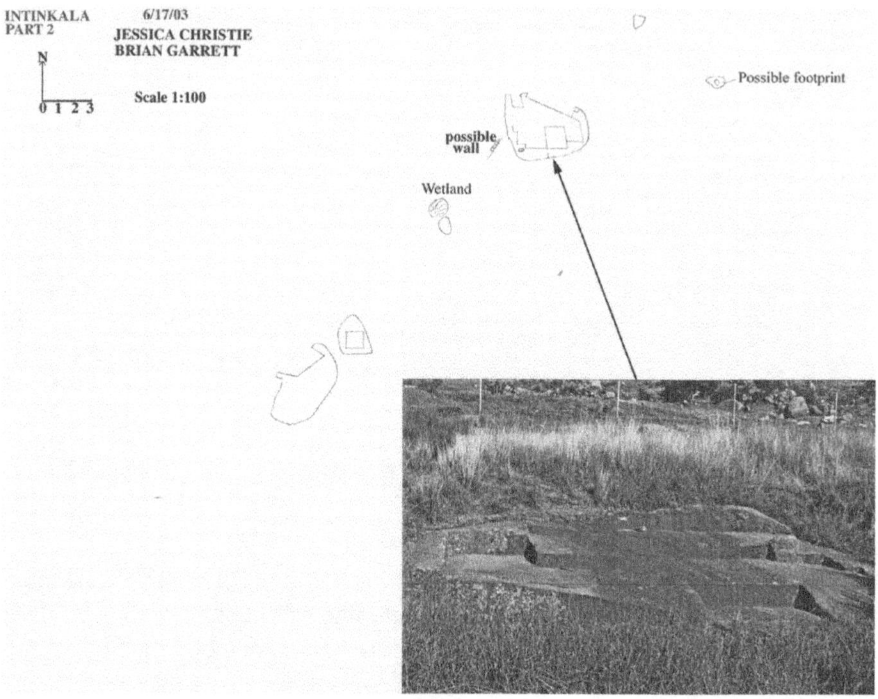

Figure 3.3. Copacabana, Orcohawira/Intinkala Part 2, Carved Rock. Drawing and Photograph by the author and Brian Garrett.

the upper platforms not as seats but as areas where a priest would have stood to greet the sun. In addition, Trimborn suggests that Intinkala as an observatory had a connection with the so-called Horca de l'Inka located further up in the hills (see below). From the higher position of the Horca, sunrise would have been observed earlier, and signals might have been passed on to the lower Intinkala station.

While the eastern orientation of the main carvings is clear, the function of carved rocks as observatories in general is challenging to prove. Trimborn's interpretation is a possibility, but it is based upon the present site condition and does not take into consideration original heights of the wall foundations which may have impeded visibility. Today observation of the movements of the rising and setting sun along the horizon line remains a vital phenomenological tool capable of linking the past with the present; yet we can squarely relive the bodily processes of perception and reception through which the Inka generated meaning or the fusion of the physical form of the site complex, its placement in the landscape, and the sensuous experiences and ideas and histories the Inka may have connected with it.

Phenomenology can be only one methodological tessera piece in the mosaic work of contextualizing sculpted rocks and has to be complemented by empiricist data gathering and ethnography. It is in this sense that I suggest the carvings of the main boulder might also have been used as counting devices. I have argued in chapter 2 that rock complexes with grid-like carvings which constituted wak'as on the Cusco zeq'e lines functioned not only as places where offerings were presented and burned, but also as locations where such offerings were accounted for and perhaps transcribed and recorded on khipus.

The main boulder at Intinkala could have registered three types of objects in quantities up to three decimal units (singles, tens, and hundreds).

Such a context could have been the primary function of the whole site of Intinkala; a very large number of pilgrims passed through Copacabana, and they all brought some type of offering for the local shrines. Pilgrims and offerings could have been counted on the carved platforms. As has already been pointed out, there were numerous storehouses in the surroundings of Copacabana. According to Ramos Gavilan (1988:127 [1621:Bk.1, Ch.20]), stored supplies were used to provide pilgrims as well as warriors and priests with food, drink, and clothing based on need. The Inka must have kept track in some form of the supplies being handed out and of new supplies coming in, most likely as tribute from subjected villages. These counting activities as well may have been conducted at Intinkala. The presence of walls—although their full extent remains unknown—indicates the creation of secluded spaces and restricted access. Although very speculative, one can envision that Intinkala was an architectural complex with carved counting systems which served administrative and partly ritual purposes.

It is further possible that Intinkala was the location where the idol of Copacabana was on display. Ramos Gavilan describes it and reports on its setting (Ramos Gavilan 1887:45). This passage is intriguing because Gavilan suggests that the idol was placed near carved rocks on the road to Tiquina. Today Intinkala and Orcohawira are indeed next to the main road to La Paz which passes through Tiquina and more than likely represent the location Gavilan is referring to.

However, it does not appear that Gavilan saw the idol there himself. The description of this idol does not match the style of Inka sculpture, and one has to conclude that it was probably a pre-Inka work made perhaps by the Yunguyus. Arkush (2005) proposes that it was most likely a Formative or Tiwanaku monolith which continued to be worshipped. Gavilan explains (see above) that the Yunguyus were the original inhabitants of Copacabana and had been relocated by the Inka to the present town of Yunguyu. Since idols are movable, it is easily conceivable that the Yunguyu idol was displayed at a later Inka rock art site. Just as the idols of conquered tribes were taken to

Cusco and exhibited in the niches of a room in the Qorikancha which was reserved for this purpose, the idols of subdued and relocated local groups may have been set up in rock art shrines at Copacabana. In this manner, the idols were taken and held hostage by the Inka, but they remained physically close by their original owners, who likely had access to them. When the Yunguyus visited their Copacabana idol, they would have found it inside an Inka sanctuary. The figure of the idol and its sculptural and architectural setting reified a discourse between two distinct stylistic and cultural traditions in which the dominant Inka state negotiated loyalty from the subjected Aymara-speaking groups which included the Yunguyus. This type of stage setting is characteristic of Inka policy in many parts of the empire (Arkush 2005:238–241).

In summary, the most significant aspect of Intinkala and Orcohawira appears to be the style of their rock carvings and the possible use of rows of seats/platforms as counting devices as well as the placement of the idol. The Copacabana region is the only one outside Cusco and the Urubamba Valley where rock sculptures in this form and number are found (Arkush 2005:238), which could indicate that Copacabana was the second most important administrative center in the empire, below only Cusco.

On a hill above Intinkala/Orcohawira sits a site nicknamed "Horca del Inca" (Gallows of the Inca) (2). According to Portugal and Ibarra (1957:26), local Aymara people continue to call this hill *Seroqa*, which is derived from *siri oqha* and means "voice of the hung." Such nicknames came from the description by Ramos Gavilan (1988:175 [1621:Bk.1, Ch.28]).

Although there are no rock sculptures at Horca del Inca in the style of Intinkala/Orcohawira, the natural stone was clearly modified, and therefore this site must be included in any discussion of rock art at Copacabana. Modifications are visible in two natural stone pillars which have been cut to support a lintel between them. The bedrock at the base of the pillars has been carved to provide the footing for a stone wall of which only a few blocks remain. Looking through the Horca del Inca to the west, the viewer takes in the wide blue expanse of Lake Titicaca, while the view east is blocked by natural rock formations. In the 1960s, Trimborn (1967:11–16) was looking for links between the Horca del Inca and features in the rock which might have marked station points of the rising sun. He notes that sunrise observed from the Horca del Inca occurred earlier than at the lower Intinkala, and signals might have been exchanged between the two sites. The astronomical alignments of the Horca del Inca were not fully understood until 1978, when Oswaldo Rivera Sundt documented that a hole in a crag to the northeast casts a well-defined spot of light onto the center of the lintel on the morning of the June solstice. At sunrise on the September equinox, the crag casts a shadow on the lintel, while the vertical pillars are illuminated (Bauer and Stanish 2001:207–209;

Rivera Sundt 1978:79–80, 1984; Arkush 2005). Thus the Horca del Inca, together with this hole, was probably devised as a solstice marker.

One important question which remains to be answered is that of chronology. The literature generally assumes that the Horca del Inca was an Inka site, but was this really so? Trimborn (1967:16) reports that there were "none or very few Inca potsherds" (translation by the author)[9] in the areas of the Horca del Inca or Intinkala. Although Intinkala should be attributed to the Inka mainly because of the rock sculptures in the Cusco style, the situation with the Horca del Inca is different. Archaeological sites with natural stone pillars supporting a man-made lintel do not exist in the Cusco region, and two other Bolivian examples illustrated by Trimborn strongly suggest that this type of monument followed a local and probably pre-Inka tradition. He understands Horca monuments as predecessors of the Inka "intiwatana" stones and assigns them to the local Aymara culture, which was also the opinion of Henri Beuchat (cited in Trimborn 1967:16). Despite the possible pre-Inka origin of the Horca del Inca, I would argue that the Inka continued to use it to determine the summer solstice and perhaps as a place of punishment (see above).

Furthermore, the Inka use of the Horca del Inca highlights the importance of the sun cult in Inka ideology. Constructing cultural landscapes which linked the local as the point of human observation and interaction with the horizon line, as marked by the positions of the sun, was one calculated state strategy to integrate the empire into an ideational landscape system (this will be discussed below). Along the pilgrimage route at Lake Titicaca, there were thus several shrines dedicated to the observation and worship of the sun. The main Sanctuary on the Island of the Sun was one among many, though it certainly held the most innate power and marked the culmination of the pilgrimage route.

Ramos Gavilan speaks of another shrine of the sun in the town of Copacabana (Ramos Gavilan 1988:171–172 [1621:Bk.1, Ch.28]. It is very difficult to match this important shrine with a recorded archaeological site. Since Gavilan indicates that it was situated "where today the convent stands," it is likely that its foundations lie underneath the Church of the Virgin of Copacabana. This would be consistent with the general Christian Catholic practice of building churches on top of pagan sanctuaries.[10] A second possibility is that Gavilan may have used the term *convent* in reference to the church structures at the foot of Calvario hill. Some kind of religious building may have stood in this location ever since the Via Crucis was built in colonial times. However, there are no known pre-contact structures in this immediate location, only further up on the hill adjacent to Calvario.

Of interest is Gavilan's reference to the "handmade tank where the Inca relaxed and bathed." This description appears to match a precontact monument known as "Bano del Inca," and situated in the former Hacienda Kusijata (3) a couple of kilometers to the north of Copacabana. According to Javier Escalante Moscoso (1997:371), the Hacienda Kusijata was built over the foundations of an Inka palace to which the Bano del Inca belonged. Rivera Sundt (1978:82) concurs that the royal family lived there, for which the Department of Ethnohistory of INA supposedly has evidence. Only the Bano was documented by E. George Squier (1877) in the nineteenth century and by Trimborn in the 1960s. Trimborn (1967:29–31) reproduced Squier's drawing and added a plan and section view created by his team. This monument consists of a large and open cylindrical basin which was carved out of a single piece of rock and now exists freestanding in the garden of Kusijata. The inner surface is highly polished, and the edge is recessed. There are two semicircular gaps in its edge to accommodate canals which served to fill it with water. Today one canal is connected and maintains water in the Bano, which is then diverted to fields.

The problem is that the location of the Bano at Kusijata does not match the description of the Sun Temple with the tank by Ramos Gavilan. A drawing and observations concerning bath facilities by Charles Wiener who visited Copacabana in the late nineteenth century add more spice to this discussion.[11] Wiener includes a drawing of three water spouts, which are surrounded by walls and release the water into a basin which may have been cut into the bedrock. Obviously, Wiener describes a Bano del Inca that is different from the well-known one located at Kusijata. I have never seen the one he depicts and have not found references by other writers; it is possible that Wiener's Bano del Inca may not exist anymore. However, the location given by Wiener ("the top of the hill Llallagua (?) at the foot of which are the church and convent of Copacabana") raises a number of questions. Is this location the same as Gavilan's for the Temple of the Sun with the Inka bath ("where today the convent stands")? Of course, one has to keep in mind that Wiener's report was written more than two hundred years after Gavilan's and that both writings were likely colored by personal experiences and subjective preferences. Excavations at the precontact site on the hill of Llallagua (?) next to Calvario might shed more light on this issue.

This hill Llallagua is mentioned several times by Ramos Gavilan. His spelling varies slightly (Llallagua, Llallinaco), but I suspect he is referring to the same hill which corresponds today with the Calvario and the slightly lower peak with the pre-contact structure immediately adjacent to it (figure 3.4).[12] A visitor climbing up along the cross stations arrives at a platform where the path turns in the direction of the center of Copacabana and winds up to the top of the hill occupied by the large Christian sanctuary. From the opposite site of the platform, a small trail leads up a second and lower hill, the top of which is covered with stone foundations which appear to be precontact.

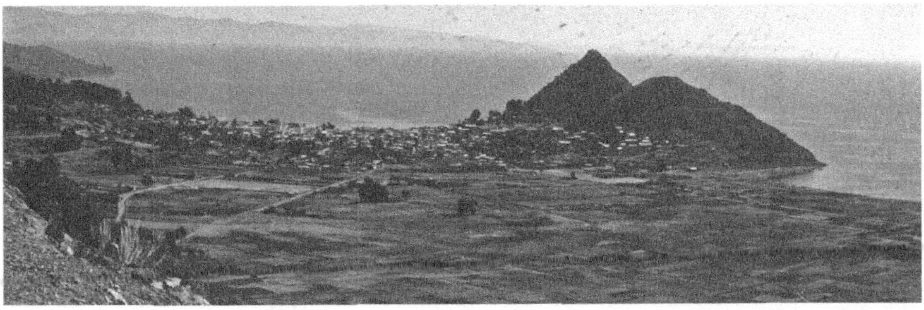

Figure 3.4. Copacabana, Llallagua or Two Peaks of Calvario Hill. Photograph by Brian Garrett.

Local people call this hill the Calvario of Santa Barbara. A surface investigation revealed the low walls of a rectangular platform with rounded corners facing toward the Christian sanctuary. Natural rock outcrops seem to be integrated into the rear façade of this platform. In June 2003, one prominent rock was covered with flowers and burnt offerings. One shaman showed me a crack on the rear side of this boulder and explained that it had been struck by lightning and is therefore a place for *brujeria* (black magic). A doll and burnt black candles in a niche constructed in front of this rock signal such a function (June 2012).

Like the first and lower platform on the main trail, this location is a popular place for shamans to perform their rites. The area is littered with burned remains and sherds from contemporary ceramic vessels and censers. However, the ceremonies on the lower platform are public and continue all day, as long as someone hires a shaman. Three to four shamans sit behind individual altar tables, and visitors pay them to conduct certain rituals for their well-being. I have never observed any ceremony in progress on the upper platform. This and the more isolated location would suggest that the precontact platform is a more secluded and private place today where rites and offerings of a nonpublic nature are being performed. This platform exists in a striking and meaningful opposition to the higher elevated Christian sanctuary. Very importantly, people continue to worship in both locations.

What the Inka did on the platform is far less clear. The existing literature, and this includes most ethnohistoric sources, seems to have skipped over this site. However, two references may be talking about the two Calvario peaks. In the mid-twentieth century, Max Portugal Ortiz (n.d.:21) describes the hill *Llallawa* where the principal Calvario of the town Copacabana is located. He cites another writer who saw a tapering stone pillar, about 3.50m tall, which was crowned by a flat and rounded stone which exhibited a cavity at its center. Portugal Ortiz further mentions a stone block called *Such-suna* which was used in folkloric rites. Both stones have since disappeared, but it is likely that they were originally associated with the precontact site on the lower Calvario hill.

In addition, there is one passage by Ramos Gavilan which may be talking about the hill with the platform.[13] Following Gavilan, I propose that the pre-contact platform with the rounded corners could have marked the endpoint of the races where the runners were rewarded or punished, and that its hill was Llallinaco. A similar event seems to have taken place in Cusco. Pedro de Cieza de Leon (1959:34–35 [1553:ii:VII]) states that Ayar Kachi appeared to his brothers on the hill of Huana-cauri and gave instructions of how puberty rites for young noble men and preparatory accession ceremonies for Inka rulers were to be conducted. In one episode of the accession procedures, the prospective ruler had to run up "a hill known as *Anaguar*" so that onlookers could see how fast he was and how brave he would be in war. The similarity and context of these state-sponsored competitions, the interest the Inka ruler took in them, and the high awards he gave out further reinforce the superior role and privileged position Copacabana held in Inka politics.

There are a number of complex rock art sites located in the environs of Copacabana. Most intriguing is Copacati (5), a carved rock complex situated on a steep, rocky hill near the road between Copacabana and Kasani, the Bolivian border station. Carvings are distributed at various levels of altitude. Approaching the site, the visitor passes a single, slightly rounded triangular seat below a pictograph portraying red geometric patterns. I have designated this portion of the site Lower Part, Carved Area I (figure 3.5). Triangular carvings are unknown in the Cusco region but abound at Samaipata in the southeast of Bolivia, where they are interpreted as seats. Thus triangular seats may constitute a stylistic feature that developed in Kollasuyu, in sites to the southeast of Cusco which are now in Bolivia. The arrangement of patterns in the pictograph consists of red squares forming the stepped design of a Southern Cross which is divided in two halves by two straight lines. It has been likened to the Kollasuyu flag, and hence the painting as well as the whole site of Copacati has been nicknamed *Banderani*. The pictograph is most likely not Inka, because it was not an Inka custom to paint on rocks (see also Escalante Moscoso 1997:371).[14]

Further uphill, there are two important features and a number of walls which I tried to identify as well as they can be observed on the surface without excavation. The first feature is a rectangular stone block with three carved steps descending from both upper corners (Carved Area II in Figure 3.5). The second feature is a large modified stone (Carved Area III in Figure 3.5). It is overgrown by a thick bush and sticks only partially out of the ground so that no further conclusions can be drawn without excavations.[15]

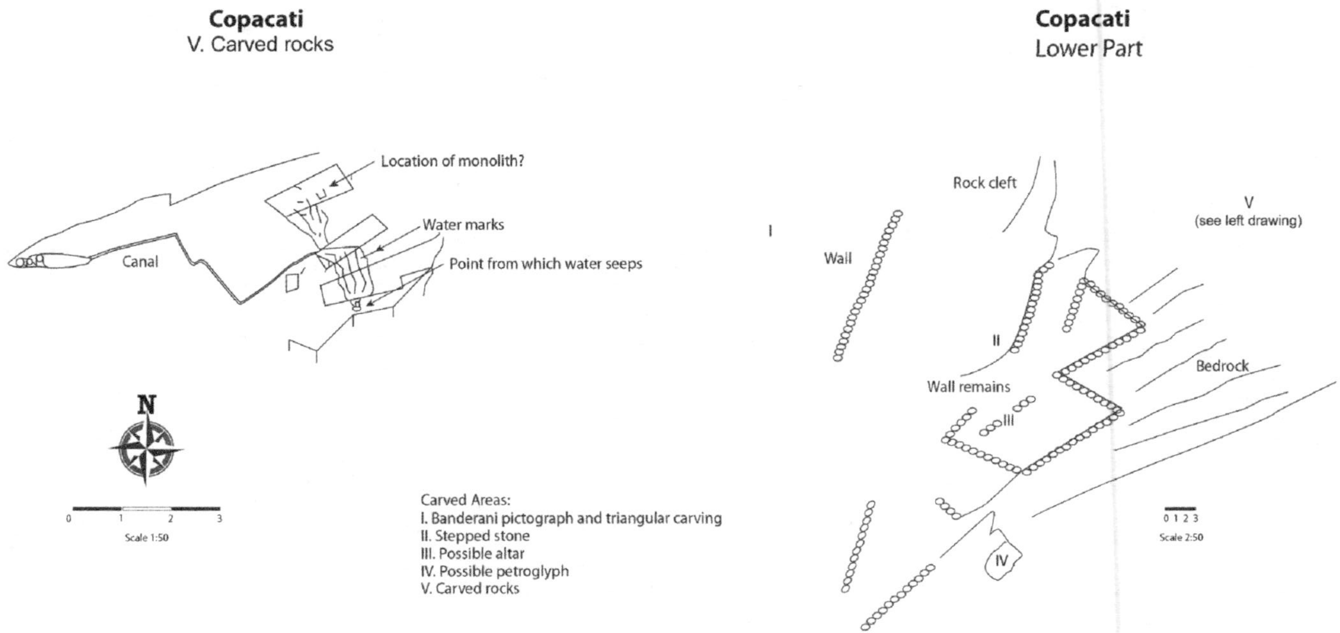

Figure 3.5. Copacati, Lower Sector. Drawing by the author and Brian Garrett.

A series of seat-like carvings is situated further up along the foot of the steep outcrop (Carved Area V in Figure 3.5). Porfirio Huanca Sanchez, a local guardian and guide at Copacati, pointed to a rounded depression in the wall above the upper seat and explained that during the rainy season in February and March water seeps out of this depression, trickles down the carvings, and collects in a crude canal which terminates in three deep cavities. While there was no water coming out in June 2003, water stains on the rocks confirmed Porfirio's statements. Most intriguingly, he referred to a small and shallow square carving on the lowest step, which is clearly defined by four ridges, and said that once another idol in the form of a monolith stood there, facing Lake Titicaca. Although there is only the formal evidence of the square depression to substantiate his interpretation, it identifies a possible location for the Copacati idol (see below).

Further along, but still at midlevel altitude, there are a number of rounded cavities in the rock, forming a crude design of the Southern Cross. These depressions look very smooth and are probably the result of water erosion, not man-made.

From here the trail leads straight up to the top, which forms two peaks. Between these peaks, a small stream trickles downhill. It irrigates a series of cultivated terraces, forms two canals on one level, and collects in a deep depression on a lower level before it runs down the cliff wall. The largest number, and technically most superb, of carvings can be found in the Upper Sector on top of the two peaks. On the higher peak, there is a group of seat-like carvings. These are examples which resemble the prototypes in the Cusco region. I believe the most important features of rock art are the two levels of finely cut platforms and shallow planes forming what the naked eye perceives as ninety degree angles and reason that all or some of the sculpted grids were not only places where offerings were deposited, but also where such offerings were counted (see Intinkala above).

On the lower peak, more seat-like carvings, steps, platforms, and shallow planes produced by vertical and horizontal cuts can be found. Some of them are composed in rough grid patterns and may have been used for counting purposes; others may have served as seats designated for supervisors and authorities. Along the edge of the outcrop toward the present road are two parallel rows of shallow rectangular carvings providing footing for two rings of walls. While we can only guess the height of these walls, they would have concealed the carvings and any kind of activities performed there from public view. The road is, of course, a modern construction, but it follows the ancient pilgrimage route closely.[16]

Copacati has been reported on by Portugal Zamorra (1977), Rivera Sundt (1978), Escalante Moscoso (1997), and Arkush (2005:220–222). Escalante

Moscoso (1997:371–373) connects Copacati with the site of Pasankallani about two hundred meters to the east. Pasankallani exhibits the remains of dwellings where the people who attended to Copacati may have lived. Furthermore, Ramos Gavilan mentions Copacati in the context of an idol which must have been similar to the Copacabana idol discussed above.[17] Gavilan's description of the location of the Copacati idol is so specific that he mentions the rock carvings. There can be little doubt that one of the functions of the Copacati rock sculptures was to serve as a setting for this idol, and the guide Porfirio Huanca Sanchez may have identified its location(s). Gavilan speaks of various stones and rocks in the context of the Copacabana and Tucumu idols so that it appears that rock art sites often accommodated idols.

As for the cultural affiliation of the idols, their description does not fit the style of Inka sculpture. Portugal Zamora (1977:301) and Arkush (2005:222)[18] suggest that the description rather corresponds to Formative or Tiwanaku period stone monoliths.[19] Karen Mohr Chavez (2001) and Sergio Chavez (2002) have defined a religious tradition which they name *Yaya-Mama*, which began to unify the populations on the shores of Lake Titicaca between 600 and 10 B.C. One identifier of Yaya-Mama is a particular iconography which is manifested on stelae found at Taraco and Chiripa. These sculpted stone slabs exhibit low-relief images of abstracted anthropomorphs surrounded by snakes and single heads with ray-like appendages. One of the Yaya-Mama sites, Ch'isi, is located on the Copacabana peninsula (see Mohr Chavez 2001). Thus, Yaya-Mama iconography fits well with Gavilan's description of the idols, and Yaya-Mama examples existed in the vicinity of Copacabana.[20]

Whether the Copacati idol dated to the Formative or Tiwanaku periods, the scenario seems to be that a pre-Inka idol was on display at an Inka rock art site. This was surely an act of Inka *wak'a*-taking. Polo de Ondegardo and others establish that it was one strategy of Inka conquest to confiscate the principal *wak'a* of a subjected group and display it in Cusco inside the Qorikancha and potentially at other sites (see Dean 2010:122). Here it is noteworthy that the Yunguyu idol was set up at a rock art site near its home and not taken to the more prestigious Sanctuary on the Island of the Sun or to Cusco.[21]

Another possible function of Copacati had to do with the counting of pilgrims and their offerings. As explained above, the sculpted rows of seats, platforms, or steps might have functioned as counting devices or *yupanas*, which tallied numbers to be translated into *khipus*. Given the fact that Copacati overlooks the pilgrimage route and is situated very near the first control station at Yunguyu, it is reasonable to argue that activities of accounting could have been carried out here. Inka administrators would have been interested in keeping records of how many pilgrims passed through and what kind and how many offerings they carried.

From Copacabana, the pilgrims walked approximately three hours to the port on a well-made road, sections of which can still be seen parallel to the present road. They had to pass two additional checkpoints or inns where they were given more corn, other vegetables, and dried meat, which were kept in storehouses (Ramos Gavilan 1988:127 [1621:Bk.1, Ch.20]). At Yampupata, there are still the remains of a small Inka site which may have served as a place for pilgrims to spend the night before crossing over to the Island of the Sun (Bauer and Stanish 2001:217).

Yampupata is situated at the end of a land formation that projects out from the Copacabana Peninsula toward the Island of the Sun (Figure 3.1; see Bauer and Stanish 2001:map 8.1). From here, the crossing to the island is shorter and straighter than from any other town. The pilgrims went in reed boats on this one to two-kilometer trip and landed at the southeastern tip of the island, which was called simply Puncu, meaning entrance. Ephraim George Squier (1877) crossed the strait in a reed boat in 1864 and described the landing place and the associated port. Most interesting is his plan of the small site of Puncu, which marked the beginning of the pilgrimage road on the Island of the Sun (reproduced in Bauer and Stanish 2001:219, figure 8.1). Today the walls of Puncu are barely visible, but the flattened plaza area in front stands out. The road itself remains well defined, while the rocky landing spot is abandoned and isolated.

From this landing place, the road climbs, and the pilgrims began their trip to the Sanctuary on the opposite end of the island. From Puncu to Yumani, the ancient road is easy to follow and leads through well-made terraces. The pilgrims soon came to the site of Pilco Kayma,[22] located below the road and nearer to the water. It is an imposing two-story, multichambered structure nested within the terrace system, which the pilgrims in all likelihood visited. At Yumani, they probably passed by the Fountain of the Inka to replenish their water supplies. It is a three-spout fountain that still provides abundant water, and today it overlooks the primary port of Yumani.

Near Yumani, the pilgrims climbed the slope, and the road continued along the ridge of the island. On this route, they passed the site of Apachinacapata, which was one of the largest Inka villages on this island. Today one can see the walls of several rooms standing in agricultural fields. The masonry is crude, and the door openings are surprisingly low. Past Apachinacapata, the road divided: travelers could continue along the ridge to the Sanctuary area, or they could take another road that led through several Inka villages along the northeast lakeshore. The ridge route is straight and very well defined, and still today one can walk from Yumani to the Sacred Rock leisurely in approximately three hours.

As the pilgrims approached the Sanctuary area, they had to pass several more checkpoints. I will paraphrase the reconstruction offered by Bauer and

Stanish (2001:224–231, 179–212) closely, because their work brings together current archaeological data and ethnographic sources about the Sacred Rock (see also Bauer et al. 2004). Although the Sacred Rock itself was not carved, it functioned nevertheless as the focal point of a rock sanctuary, and some of the activities conducted there may have been repeated at other carved rock shrines.

Cobo reports only one entrance to the Sanctuary area, which was called Intipuncu. Ordinary travelers were not allowed to approach the Sacred Rock itself but could view it from the Intipuncu gateway. There they had to hand over their offerings to the shrine attendants. Cobo describes the location of Intipuncu with enough detail that it can be identified on the ground today (1990:97–98 [1653:Bk.13, Ch.18]). Ramos Gavilan presents a colorful account of three gateways, and at each one the pilgrims had to meet with priests.[23]

Bauer and Stanish (2001:225–229) have identified the remains of the three gateways. The eastern outer limit of the Sanctuary area is defined by a low wall through which the lakeside route passes. From the point where this wall and the road intersect, the traveler gains a first view of the Sacred Rock. This entrance may be Pumapuncu, which Ramos Gavilan describes as the first of three gateways. Nonelite pilgrims may not have been allowed to advance any further, even though from this door they could see only the back side of the Sacred Rock, but they may have been permitted to gather and observe certain ceremonies performed at the Sacred Rock from a platform which Bauer and Stanish documented along the wall and near the ridge of the island.

Those who were allowed to go on descended a series of large stone steps. At the end of this descent lies the site of Mama Ojlia. This site contains a single rectangular building measuring approximately 16.80 x 6.75m below the road and a series of smaller structures above it, as well as a seasonal spring. Bauer and Stanish mapped and partially excavated this site and did not encounter any pre-Inka artifacts. They conclude that the Inka erected these structures as living quarters for shrine personnel, and in particular for the priests who guarded the gateways. Their findings are consistent with Cobo's report; he specifies that the single large building housed the *mamaconas* or Chosen Women (cited above); and with Ramos Gavilan, who also mentions the "Virgins dedicated to the Sun" (quoted above).

As the road rises again toward the Sacred Rock, it crosses an expanse of bedrock on which there are two large, natural marks resembling giant supernatural footprints. Both Ramos Gavilan and Cobo refer to them, and Cobo points out that they should be attributed to natural causes (cited above). Bauer and Stanish (2001:227) observed the remains of a wall running perpendicular to the road just before the footprints. A gap in this wall, where the road passed

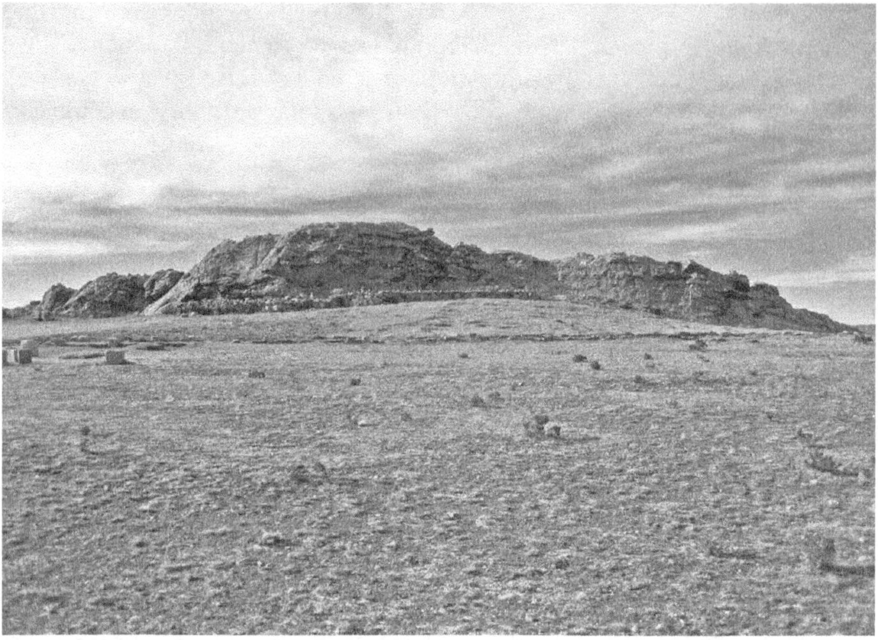

Figure 3.6. Island of the Sun, View of Sanctuary. Photograph by Brian Garrett.

through it, and three stone steps clearly indicate the location of a former doorway. The proximity of this door to the footprints provides strong evidence that this is the entrance named Intipuncu by Cobo and Kentipuncu by Ramos Gavilan (cited and quoted above). About fifty meters beyond the footprints, the road levels off and enters the plaza of the Sacred Rock. Although there are no visible foundations of another gateway, Bauer and Stanish interpret the entrance zone to the plaza as representing the third gateway which Ramos Gavilan calls Pillco-puncu (see above).

The immediate area of the Sacred Rock consists of the rock itself and a plaza adjoining it to the southwest (figure 3.6). Bandelier (1910:214–225) has provided us with the earliest academic investigation and interpretation of this site. The Sacred Rock does not contain any carvings. One side of the rock descends toward the lake while the other side forms a vertical front facing the plaza. This vertical rock front rises nearly 5.5 meters and lines the north side of the plaza for about eighty meters. The west side of the plaza is defined by an Inka wall which runs from the Sacred Rock to a smaller, isolated outcrop. Bauer and Stanish (2001:196–199) reconstruct this wall with a central trapezoidal doorway and eleven small trapezoidal windows. The east edge of the plaza was marked by the third gateway, while the south side remains open and is crossed by the Inka road that leads from the main entryway of the Sanctuary in the east to the Chincana in the west. The plaza area might have continued south of the road until the terrain drops off to the lake.

The descriptions of the Sacred Rock provided by Ramos Gavilan and Cobo (1990:96–97 [1653:Bk.13, Ch.18]) are truly impressive and may have implications for the worship at other carved rock shrines.[24] The two references show that Cobo copied from Ramos Gavilan. They are remarkable for their detail and specificity and raise the question of whether the Inka were accustomed to covering sacred outcrops with fine textiles and to pouring liquids. The answer is probably yes, although there is no direct documentation of other carved rocks covered with cumbi cloth, ethnohistoric accounts include many instances of the dressing of sculptures, most frequently of the small gold and silver figurines accompanying *qhapaq ucha* burials.

Although the cumbi cloth and the gold and silver are long gone,[25] Bauer and Stanish (2001:196–206) were able to verify some of Ramos Gavilan and Cobo's descriptions archaeologically. The cut andesite blocks in the plaza may be the remains of an elaborate construction that once stood in or near it. There is no source of andesite on the island, and therefore they must have been brought by boat from the mainland. The present arrangement of andesite blocks and a large slab of white sandstone forming a table surrounded by seats is recent, but the original construction could have been the temples of the Sun, Thunder, and Lightning mentioned by Ramos Gavilan (1988:93 [1621:Bk.1, Ch.13]) and Cobo (1990:97 [1653:Bk.13, Ch.18]).

The test units excavated by Bauer and Stanish in the plaza provided offering materials, such as fragments of miniature Inka bowls (2001:203). Most significant were the remains of a small stone canal which ran from the concave section of the Sacred Rock past the west wall of the plaza. Since there is no source of natural water in the plaza area, the canal must have been built by the Inka to drain liquid offerings from the plaza. Both, Ramos Gavilan (quoted above) and Cobo (cited above) speak of a stone basin in front of the Sacred Rock into which local people poured chicha for the Sun to drink. Furthermore, Bauer and Stanish (2001:206) identified at least three floors which once covered the plaza area. Most remarkable was the lowest one, which was made of crushed greenstone and set directly above bedrock.

Bauer and Stanish (2001:208–212) complemented their archaeological work with astronomical research. As seen from the Sanctuary area, the sun sets over a ridge, called Tikani, to the northwest. Together with David Dearborn, they identified two structures on this ridge that flank the position at which the Sun sets on the June solstice, as viewed from the plaza. Their findings indicate that part of the June solstice celebrations involved watching the sun set between these structures which so conspicuously marked the event. The elite stood in the plaza in front of the Sacred Rock, while the non-elite gathered on the platform outside the Sanctuary wall at the first gateway described above.

These data constitute a significant case study in the emerging pattern of special foregrounded rocks serving as observation points of the Sun in relation to sets of horizon pillars. The prototype of this scenario was the usnu stone in what is now the Plaza de Armas of Cusco.[26] On the Island of the Sun, a question arises as to the physical form of the two structures on the Tikani ridge. They are two rock-walled and rubble-filled buildings which hardly match the description of the Cusco horizon pillars given by the chroniclers. In August of 1995, Matthew Seddon and Brian Bauer specifically investigated these two structures (Seddon and Bauer 2004). They reconstructed two dissimilar and irregular buildings which were filled largely with rock rubble, gravel, and sand and did not contain any cultural remains. Intriguingly, certain formal aspects and higher quality masonry sections were oriented toward the Sacred Rock, leading Seddon and Bauer (2004:90–91) to confirm their previous interpretation of the two structures as markers of the June solstice sunset when observed from the Sacred Rock. This emerging pattern of sacred carved and unmodified rocks used by the Inka as observation stations of the movements of the Sun along the horizon marked by masonry pillars adds a significant functional category to the study of rock wak'as. On an ideological level, it binds together the metaphorical concepts of stone, the Sun, and political power.

The most elaborate architecture in the Sanctuary area stands roughly two hundred meters to the west of the Sacred Rock and has been called Chincana or labyrinth. It includes at least one carved rock. Ramos Gavilan and Cobo (1990:97 [1653:Bk.13, Ch.18]) describe the structures as a storehouse of the Sun and a house of the mamaconas in similar accounts.[27] Somewhat later in his book, Cobo refers to "a large building" on the other side of the entrance as "the retreat of the mamaconas, women dedicated to the Sun. These women served in making the beverages and finely woven cloth that were used in the services for the shrine. This house for the mamaconas was in the best location of the island" (Cobo 1990:98 [1653:Bk.13, Ch.18]). Cobo implies that the Chincana functioned as a "storehouse of the Sun" as well as a residence of mamaconas.

Today the Chincana is relatively well preserved, and its walls stand two to three meters high. The whole building complex may be divided into two architecturally different sectors. The east sector consists of small rooms, narrow passageways, and two small central patios which vaguely define a central axis of this sector.

This type of architectural layout clearly suggests private use and does not lend itself to public authoritative statements.

The western sector has larger rooms and more open spaces than the eastern one. The plaza, in particular, is significantly larger. A number of walls display large niches. The spring Ramos Gavilan and Cobo talk about still exists in a room on the west side of the plaza, but there are no more trees. The trickle of the spring water can be followed to the south end of this sector, where it runs down a slope next to a crudely carved rock. At the base of the slope, it forms a wetland and slowly filters into the ground.

This carved rock is one of two on the Island of the Sun. It exhibits several horizontal cuts and was likely part of the baths mentioned by Ramos Gavilan and Cobo. Other well-preserved Inka baths, for example at Phuyupatamarca, Wina Wayna, and Machu Picchu, consist of a descending series of walled basins and canals through which the water cascades. The construction techniques used are a combination of carved bedrock and masonry. If at some point there had been baths of this type in the Chincana, they would have had no more than two basins.Water supply is scarce in the altiplano environment of the Island of the Sun.

Bauer and Stanish's surface collections and excavations recovered only Inka pottery; there was no evidence of pre-Inka materials. They conclude that the Chincana was built as a state installation by the Inka. Following Ramos Gavilan and Cobo, they propose that the eastern sector served as the storehouse of the Sun, while the mamaconas and other attendant personnel lived in the west sector with its spring and tree garden as well as another possible Sun Temple (Bauer and Stanish 2001:195–196; Bauer et al. 2004:51–63).

As a hypothetical pilgrim, I arrived by the ridge route and am returning on the northeast lakeshore road passing through Kasapata. This site lies approximately halfway between the community of Challapampa and the Sanctuary area between two large hills. Kasapata covers an area of more than five hectares (Bauer and Stanish 2001:173; Bauer et al. 2004:74–81) which the lakeshore road from Challapampa to the Sacred Rock divides in half. I include Kasapata in this discussion because it contains the second carved rock on the Island of the Sun. In its relation to the standing architecture, it forms another example of the plaza-kallanka-foregrounded rock pattern discussed in chapters 5 and 6 (also Christie 2007a). The Kasapata example is of particular interest because early excavations by Bandelier (1910:203–214) and recent excavations by Bauer and Stanish (2001:173–179; Bauer et al. 2004:74–81) have brought to light material evidence of the kinds of activities that took place in these architectural spaces.

In the southern half stands a large Inka structure measuring 40 meters by 10 meters. Its north wall displays five trapezoidal doorways facing the road.

In between and high above these doorways were originally pairs of windows. The interior walls were lined with niches. The back or south wall is now largely destroyed. To the north of the lakeshore road, there are several foundations, isolated building blocks, and the carved rock. It is an upright stone slab exhibiting vertical and horizontal cuts. Some of the cuts continue below ground level, raising questions of whether its present position in the backyard of a private house is original. If this stone were turned ninety degrees and placed horizontally on the ground, it might show carved platforms similar to the ones on the Intinkala rocks at Copacabana. Squier, who visited the Island of the Sun in the late nineteenth century, produced a plan of Kasapata which shows rocks in the general area where the carved stone stands today (in Bauer and Stanish 2001:175, figure 7.5). Furthermore, Kasapata had well-made terraces that served as domestic house platforms.

Early investigators found the site in a more complete condition. Pentland describes adobe gables at the ends of the building with five doorways, and Squier notes that this same building was once stuccoed and painted (in Bauer and Stanish 2001:176; Bauer et al. 2004:74–77). Bandelier began excavations at Kasapata which he thoroughly documented and illustrated (see Bauer and Stanish 2001:176; Bauer, Futrell, et al. 2004:76–77). He excavated in the northern half of the site, as well as on the slopes of an adjacent hill. Among his finds were burials, a five-meter-long canal that ended in a tank, bones, pottery, grinding stones, stone beads, spindle whorls, and small metal objects. He also found intact vessels, such as domestic cooking pots, elite plates, and large jars more than one meter high, which had been used to ferment and store chicha.

Bauer and Stanish conducted surface collections and test excavations at Kasapata (2001:178; Bauer et al. 2004:77–80). The ceramics and other cultural material they recovered are solely Inka; they found no evidence of earlier occupations. One carbon sample yielded a date of A.D. 1680 ± 80, which would imply that the site continued to be in use after the fall of the Inka Empire.[28] The artifacts paralleled those recorded by Bandelier. Bauer and Stanish found pottery shards from large storage jars and areas of burned clay, which were possible cooking locations inside the large building. The outside units yielded domestic and finer ceramics, spindle whorls, copper pins, and animal bones.

The architectural remains, artifacts, and carved rock are evidence that Kasapata was a complex settlement, and visitors and scholars alike have offered various interpretations. Ramos Gavilan may be referring to Kasapata when he writes that "Topainga Yupangue founded a village of moderate size about half a mile before the Sacred Rock and sanctuary. In this village, he had his royal palace constructed" (1988:86 [1621:Bk.1, Ch.12], translation by the author). Cobo may also be talking about Kasapata when he describes "one quarter of a league before one reaches the temple, there was an impressive *tambo* or inn for the pilgrims to stay in" (1990:93 [1653:Bk.13, Ch.18]). Pentland interpreted the principal building with five doorways "as a place of worship. The Indians describe it however as having also been a residence of the Inga, and to have formed one of the many pleasure grounds which he possessed throughout the island" (Pentland cited by Bauer et al. 2004:74). Squier thought that the same structure at Kasapata was the Temple of the Sun, while Bandelier proposed that it was the palace of Thupa Inka Yupanki based on the information provided by Ramos Gavilan (see Bauer and Stanish 2001:178).

Bauer and Stanish emphasize public use of the large structure with five trapezoidal doorways and liken it to kallankas, which are long rectangular buildings with multiple doorways where festivals were held during rainy weather (Gasparini and Margolies 1980:196–219). They conclude that the principal building "served as a public gathering place in which ritual feasting and drinking took place" (2001:179). Their interpretation strengthens the plaza-kallanka-foregrounded rock pattern pointed out above (see chapters 5 and 6). I have documented the combination of plaza-kallanka-special rock in Cusco at the usnu, at a number of royal estates, and in several outlying administrative centers. Activities such as public festivals, feasting, and drinking have been securely reconstructed at the plaza in Cusco, which was surrounded by kallankas and royal palaces. The plaza at Kasapata probably staged a micro-version of such a setting where similar events were played out, although on a smaller scale. The specific reference by Pentland to a residence and "pleasure ground" of the Inka places Kasapata in the category of royal estates and underscores the inclusion of sculpted rocks in their layouts.

Furthermore, Bauer and Stanish note an account of the Capac Raymi festival in which Ramos Gavilan may be describing the site of Kasapata.[29] The "temple" with five doors referred to by Gavilan matches the principal structure of Kasapata closely. It is also relevant that Ramos Gavilan calls the plaza "Aucaypata." This is the name of the central plaza in Cusco, and thus establishes another link between the layout of Kasapata and Cusco. Ramos Gavilan's account reinforces that architectural spaces containing the plaza-kallanka-carved rock pattern were used for state ceremonies of the highest order, and I propose that similar activities took place at the other settlements which exhibit the plaza-kallanka-special rock layout.

Most pilgrims crossed over from the Island of the Sun to the Island of the Moon, where they worshipped at a temple staffed by women. This temple was dedicated to the Moon and Qoya, the queen and wife of the Inka, and its remains correspond with the archaeological site of Inak Uyu (see Bauer and

Stanish 2001:64–131). There are no known carved rocks on the Island of the Moon. From there, the pilgrims returned to Copacabana.

INKA LANDSCAPE AND CARVED ROCKS—APPEARANCE, EXPERIENCE, AND PERCEPTION

The foregoing reconstruction has demonstrated that the physical appearance of Copacabana and the Island of the Sun has undergone many changes through time. The shores and islands of Lake Titicaca were not a cultural no-man's-land when the Inka started to integrate them into their expanding empire. As noted above, occupation at Copacabana dates back to the Middle to Upper Formative transition or about 500 B.C. On the Island of the Sun, the archaeological work conducted by Matthew Seddon (1998, 2004) and Bauer and Stanish (2001:146–147) indicates that the northern part of the island began to turn into a ritual center, and a wak'a of island-wide importance was established during the Upper Formative (500 B.C.–A.D. 400). Seddon (1998; 2004) has documented that the site of Chucaripupata within the Sanctuary area was a major Tiwanaku center. The Inka usurped all existing settlements and turned them into a state installation which formed a core of their developing state ideology. Today Copacabana is a booming tourist town, and Aymara people residing on the islands are developing strategies to benefit economically from tourist visits.

Inka Centripetal and Centrifugal Landscapes

Since my focus is the Late Horizon period, I will now ask how would Inka infrastructure have shaped experience and perception of residents and of those people who traveled the pilgrimage road? I will contextualize Copacabana and the Island of the Sun as state-directed performance stages and mine potential parallels with the cultural landscape of Cusco.

Highlights of the reconstruction of Copacabana have shown that members of the ruling family resided there and that it was settled by trustworthy *mitmaq* or colonists from numerous other regions. In terms of physical features, it had a Sun Temple, a palace, storehouses, and carved rocks; with regard to ritual, elite initiation and puberty rites involving races were held probably at the double-peak hill which is the Calvario today. Passage along the entire pilgrimage route from the entrance into the Copacabana peninsula to the Sanctuary on the Island of the Sun was controlled by checkpoints established at architectural gateways. Most of these elements also apply to Cusco, yet the comparison has to remain broad since we do not know the full spatial layout of Inka Copacabana. One marked difference is that of access controlled by station points. In general terms, access to the capital was more open and welcoming. Nevertheless, there are the remains of one gate at Angostura which constitutes the southeastern entrance into the Cusco Valley.

Highlights of the reconstruction of the Island of the Sun target the Sanctuary of the Sacred Rock, which was covered with gold and silver plates and cumbi cloth. The plaza in front served as an observation point of sunset in relation to a pair of western horizon pillars; solstice celebrations were also held at the nearby royal estate of Kasapata. Restriction of access peaked in the progressive ritual exclusion enacted through station points at gates placed at short distances. In Cusco, similar experiences of controlled access and exclusion have been documented for those who wished to enter the Qorikancha and Inka palaces; for example, Garcilaso de la Vega (in Bauer and Stanish 2001:229) writes that prospective visitors to the temple had to remove their sandals; only the highest Inka royalty, accompanied by the priests of the Sun, could enter the temple area; visitors of lower status were not allowed to pass the entry gate and had to hand over their offerings to the priests. These are the same rules as described by Cobo for the Sanctuary on the Island of the Sun (Cobo 1990:96 [1653:Bk.13, Ch.18]). Inka palaces were enclosed by a perimeter wall, and visitors had to obtain permission to traverse two courtyards before arriving at the royal quarters.

The cultural landscape thus constructed would have been perceived as a dynamic interplay of centripetal and centrifugal forces: The state channeled pilgrims and encouraged them to take the trip, rewarding the needy with supplies. The man-made physical features of Copacabana answered these centripetal demands of the pilgrimage system. At the same time, this pilgrimage was not a steady continous movement forward and toward the shrine, but was interrupted, examined, and inspected by priestly officers at calculated points. Offerings were collected at many of these station points, possibly at Copacati and clearly on the Island of the Sun. We have heard that various people brought litters with their lineage idols and deposited them at Kasapata. This centripetal drive toward the Sanctuary overlaid the centrifugal forces implied by the origin narrative (see above), which emphasizes that Wiraqocha sent the ancestors out and away to found their ayllus and panaqas.

I conclude that Inka subjects would have perceived this landscape as a highly authoritative one. Control points at gateways structured the political landscape into a series of spatial units decreasing in size as pilgrims approached the Sacred Rock: first was the gate at Yunguyo, which marked the official entrance to the Copacabana peninsula and the islands conceptualized as the total unit of sacred space; station points at Copacabana and possibly at Copacati followed; and finally the staggered gates at the Sanctuary on the

Island of the Sun screened access. Each stop would have been perceived as an intervention of state power in the physical progress of the pilgrim. With regard to lived human bodily perceptions, cultural parameters and architectural closures would have dramatized the approach to the Sacred Rock and heightened expectations. Today, with all the Inka cultural overlays gone, we experience the bare natural outcrop set little apart from its altiplano surroundings, which can strike visitors as disappointing. There is a marked contrast with Cusco when the setting and cultural landscape of the capital are viewed in terms of a performance stage. The four main roads leading to the suyus facilitated a more open coming and going, which is manifested, for example, by the *qhapaq ucha* parties (see below) who brought offerings from all regions of the empire. These offerings were blessed in the center of Cusco, and then the parties returned to their home villages to present their offerings and sacrifices to the local deities. This kind of setting constitutes a centrifugal and redistributive landscape.

INKA LANDSCAPE AND CARVED ROCKS—ESSENCE, IMAGINATION, AND STONE IDEOLOGY

Carved Rocks as Building Blocks of an Inka Imperial Landscape

One building block forming the core essence in the construction of the Inka political landscape was the sculpted rocks and boulders. The style of rock art at Copacabana is most similar to rock art in the Cusco area. The vertical and horizontal cuts creating geometric volumes and masses identify the characteristic Inka style, and many authors have seen it as an indicator of the presence of the Inka state (van de Guchte 1990:228, 346; Arkush 2005:217; Hyslop 1990:102–128; Dean 2010).

Van de Guchte (1990:406, illus. 2) included a map illustrating the spatial distribution of carved rocks in Tawantinsuyu. This map makes clear how selective Inka rock art sites were, with Cusco in the center exhibiting the largest number of stone sculptures, the Lake Titicaca region and Samaipata marking the southeastern boundary, and Ingapirca and Pambamarca denoting the northwestern frontier. Very few sites are situated in between, or they have simply not been published. The geographical layout of the carved rocks and sculpted boulders appears to emphasize places situated far away from Cusco in the north and south. Since there are no obvious geological factors which would have influenced the selection of rock art sites, their layout was likely determined by ideological reasons and the ambition to map out an Inka political landscape. Ingapirca, situated in the south-central Andes of Ecuador, was commissioned by Wayna Qhapaq, whose strategy was to turn nearby Tumipampa into his state capital. As outlined above and in the next chapter, the two major sites in the south, Pumaurqu and Copacabana/Island of the Sun, were the Inka origin places. In both cases, the Inka integrated the sites by conquest and by overlaying their creation mythology and sculpting the local rocks.

It is conceivable that the Inka perceived the physical and ideological links between Cusco and these outlying rock art sites as an expanded zeq'e system that took its origin in Cusco. As Inka territory grew through conquest, they expanded the political and sacred landscape as defined in and around Cusco throughout the empire. The critical link in the Inka definition of sacred landscape was the rocks carved in a Cusco-centered style. Andean people had worshipped their stone wak'as long before the Inka. Concurring with Arkush (2005:239–241), I suggest that Inka administrators took into consideration local belief structures and understandings of the natural world, even regional and pre-Inka idols (see the Copacabana and Copacati idols above), but overlaid them with a specific Inka language written in stone that signaled the origin of Inka dynasties and their power sources.

There is preliminary evidence that extended zeq'e lines marked by foregrounded rocks and other wak'as existed in many parts of the empire. Van de Guchte (1990:228) and Ken Heffernan (1996:26–27) trace such a line going from Cusco straight to the west. Inka sites with carved rocks which lie either on or near this line are Inka Moqomallinan above Zurite, Quilla Rumi, Choquechurqo near Tilka mountain, Saihuite, Vilcashuaman, and Huaytara. If extended further westward, it would reach the coast near the Inka-modified site of La Centinela at Chincha. Heffernan (1996:29–30) presents evidence from the ethnographic literature that the expanded western zeq'e line played a role in the citua and qhapaq ucha ceremonies as described by Molina.

Chroniclers such as Polo de Ondegardo and Cristobal de Albornoz (both cited by Heffernan 1996:23) talk about zeq'e systems outside of Cusco, but no other shrine system similar to that in Cusco has ever been documented. Although speculative, the effort to expand sacred geography as defined in and around Cusco to the confines of the empire seems consistent with Inka state ideology. Zeq'e lines could be conceptual—while still marked by historically verifiable shrines—as many of the Cusco lines appear to have been; or they could be actual physical pathways to be walked on during pilgrimages and processions, as well as roads. Some of the completely straight and linear Nasca lines have been explained as ritual walkways (Reinhard 1992:295–297; Silverman 1990). During the qhapaq ucha ceremony, Inka officials from the provinces followed state roads to Cusco to bring their offerings and returned walking in straight lines (McEwan and van de Guchte 1992:361). In traditional highland villages, people continue to walk along straight lines in ritual processions (Reinhard 1992:295, fig.9).

Zuidema (cited by Heffernan 1996:27 and van de Guchte 1990:228) have outlined another long distance zeq'e departing from Huanacauri/Wanakauri mountain near Cusco to the southeast, following the Vilcanota River to its source at Vilcanota (today known as La Raya) Pass, and continuing on to Tiwanaku and Copacabana. This ritual line must have closely paralleled the aforementioned pilgrimage route from Cusco to the Island of the Sun in Lake Titicaca. Arkush (2005) and Stanish (2003:273–274) have documented numerous sculpted rocks and small outcrops situated on or near this main road. The final destination of this southeastern extended zeq'e line was Copacabana and the rock Sanctuary on the Island of the Sun.

I reason that the Inka materially and ideologically constructed similar long distance zeq'e lines between Cusco and the origin place at Pumaurqu/ Pacariqtambo; Cusco and Samaipata in Bolivia, a gigantic rock art site at the outer border of Antisuyu; as well as between Cusco and important northern sites at Cajamarca and in Ecuador. I have discussed elsewhere (Christie 2008) the multiple relations between carved rocks on roads, physical and conceptual zeq'e lines, and movement, and these issues will be explored further and contextualized in chapter 7.

Finally, we may link the relations between the centripetal and centrifugal forces of Copacabana/Island of the Sun and Cusco seen as performance stages and an extended zeq'e system under the Aymara space-time concept of *taypi*. Therese Bouysee-Cassagne and Olivia Harris (1987:19–28) reconstruct the first age in Aymara mythology as *taypi*. The Aymara word *taypi* signifies a center or central space, and the Spanish writers surely describe the Island of the Sun and Tiwanaku as such a center space. Its Aymara toponym *Taypi Qala*, or central stone, adds the dimension of stony essence. On the temporal level, taypi experiences a centripetal phase during the creation process of the Sun and human beings, which is followed by a centrifugal phase during which the ancestral humans are sent out to reemerge at their paqarinas or origin wak'as to found their ayllus. In spatial terms, this second phase leads from the center to the periphery and from the cultural order set up at the origin place into the uncivilized realms of outlying regions, when viewed in a horizontal two-dimensional direction. If viewed vertically, the centrifugal movement links the earthly terrain of the human world with the underworld sea below manifested by Lake Titicaca and the sky above linked with the nevados, through the various paqarinas. The concept of mountain and its stone essence again links the central and the peripheral and liminal spaces, and these ideas are compacted in the rock Sanctuary on the Island of the Sun. In Aymara mythology, this second age is called *puruma*.

I argue that the Inka reproduced a second taypi in Cusco: the founding events of the capital marked the centripetal phase, and the establishment of the local and extended zeq'e system and the road network corresponded with the period of expansion and diffusion. Carved and unmodified rocks and boulders were integrated as potent markers at selected locales—for example, the Cusco usnu became the taypi qala in the capital—and transformed pan-Andean ideas as outlined by Bouysse-Cassagne and Harris into Inka stone ideology.

How Some Contemporary Aymara Consultants Imagine Stony Places and Perform Rituals

To the present day, stone formations stimulate memory and imagination of local indigenous people. Although their explanations do not point to any overt political context, such as Inka stone ideology, their narratives delve into the past and memories of family and the local village. They weave rich memory landscapes we can compare with the archaeological record and with colonial source material.

It is insightful to closely paraphrase the tour of Copacati Porfirio Huanca Sanchez, the local vigilante, gave me in 2003. Porfirio says he grew up at Copacati and has learned a lot about Aymara history and traditions from his grandfather. At the ascent to the site, he offered an interesting interpretation of the triangular seat-like carving. He suggested that an idol was placed upon it, and people would have knelt in four depressions in the rock below the triangular carving to worship it. He may have pointed out one possible location of the Copacati idol (see above). At midlevel altitude, the first feature is a rectangular stone block with three carved steps descending from both upper corners (II in figure 3.5). According to Porfirio, this carved rock marked one of the gates at which visitors had to confess and repent before they were allowed to enter Copacati. His suggestion recalls the three gateways visitors had to pass on the Island of the Sun before reaching the Sanctuary of the Sacred Rock.

Further along at midlevel altitude, Porfirio explained that his ancestors would fill the round depressions forming an approximate Southern Cross with water from the upper source (see below) and observe the reflection of the night sky to make prognostications about the agricultural seasons. Looking up to the top with its two peaks and the terraces and stream in between, Porfirio said that in the rainy season the water will roar inside before it comes out of the ground. Porfirio's grandfather told him that above the terraces, on the ridge between the two peaks, there were at one time twelve structures. The stones of such buildings were reused to create and delineate the agricultural fields. If this is proved true, then it would probably have been the only settlement at Copacati where a number of people may have lived. On the other hand, Escalante Moscoso (1997:371–373) has suggested that the settlement associated with Copacati was Pasankallani, situated to the east. According to Porfirio, a number of man-made cave-like openings above Copacati are called Pasankallani.

In the Upper Sector on the higher peak, there is a group of seat-like carvings to which Porfirio referred as "Seven Astros" (Stars). Supposedly, this was a place where seven council members convened. Somewhat below these possible seats, he identified four areas of special natural and sculpted rocks which he connected with multiple stories.

In 2005, Ramon Calle (see below) explained the multiple features of Copacati in a remarkably similar way. In the Upper Sector below the council seats, which he also identified, he pointed out a shallow platform defined by traces of a quatrefoil frame which was used as an observatory, since the shadow of a person standing in the middle defines the cardinal directions over the course of one day. The grid-like carvings, which I have interpreted as a possible counting device (above and Christie 2005c), marked the foundations of a wall which formed part of a palace in Calle's view.

It is interesting that Ramon Calle's overall interpretations were so similar to those given in 2003 by Porfirio Huanca Sanchez, the local vigilante of Copacati, yet differed in details. I was left with the impression that both would see a carving or a special rock formation and connect it with an explanation and/or a story derived from their cultural as well as personal experiences. This would explain the individual variations, but the consistent context might best be approached through the model of a "figured world." Figured world is a term used by Dorothy Holland and colleagues (1998:52) to designate enduring culturally defined realms in which people share memory and identity and act to achieve particular outcomes which are valued by all. As individuals develop in specific settings of these worlds, and as they learn about their logics and inner workings, they bring their own ideas and narratives to bear and ultimately reinvent and change their "figured world" while at the same time reproducing it.

Furthermore, imagination and memory are performed and reinvented at rock art sites. Following is my narrative of the New Year's celebration conducted by the local Aymara people at the Horca del Inca in 2005. I used the ethnographic methodologies of participant observation, fieldnotes, and interviews and am aware that the process of retelling my experiences is not neutral. By necessity, I will describe the events from my physical visual perspective on this morning; conversations I initiated with local participants situated me as an outsider. Nevertheless, my narrative helps to visualize the dynamic cultural and political landscape indigenous people continually recreate at sculpted rock outcrops.

When I asked about the workings of the Aymara calendar, one person explained to me that it is based upon the activities and needs of the earth. The three months from approximately the September equinox to the December solstice signal the rainy season, when the rains slowly begin and become heavier, when the earth drinks, and when it is time to sow. From the December solstice to the March equinox, it continues to rain; water is still abundant, and it is the time of agricultural growth. The earth produces its new offspring. Between the March equinox and the June solstice, the dry season sets in; the crops have matured and are being harvested. After that, beginning on June 21, the earth rests after giving birth to the products the Aymara have by now collected. During the three months until the September equinox, the earth remains dry and sterile and has to generate new strength for the next agricultural cycle. This is why the Aymara celebrate and begin their New Year on the June solstice. They present and burn offerings which are intended to feed and nourish the earth as well as the Sun, so that both will be strong and provide in abundance during the next agricultural year.

The following is the detailed account of the New Year's celebration in 2005. People began to gather at the site of the Horca del Inca in the dark, very early in the morning of June 21. I had been advised to arrive early since space is limited in the rock formations. Before 6 a.m., local people sat in groups on the rocks, laughing and chatting; ever more people arrived. A few individuals unpacked bundles and quickly assembled offerings on paper trays, which they advertised to last-minute buyers. Activities began to concentrate around an eastern ledge in the rocks which offers a wide view of the eastern horizon. A line of people was forming, who all deposited their offerings in a growing pile at the easternmost point. Such offerings or *misas* typically consisted of a paper tray filled with small fruits, coca leaves, colored candy molded into significant symbols because the earth likes sweet things, colorful paper, resinous incense, and pieces of wool. By 7 a.m., sunrise was approaching, and the pile of offerings was sprinkled with beer and moistened with alcohol.

Two ritual specialists or *yatiris* wearing Aymara ponchos and hats began to make alternating speeches in which they invoked the Sun as the powerful force which binds humankind together globally. One of them aimed a magnifying glass positioned in front of dried grass toward the rising sun, and the first rays indeed ignited the grass. This fire was transferred to a piece of wood which set the offerings on fire. As the sun climbed fully above the horizon line and the offering pile burned, all participants stood facing east with their eyes closed and their arms raised and hands open. In this way, they felt the Sun's warmth and also attempted to absorb the new energy provided by the Sun. Sunrise on June 21 marks the end of the coldest night in the year, and manmade fires have to regenerate heat. A small group of Aymara elders lit a second and seemingly more private offering pile.

After such moments of meditation, the celebration continued more informally. Many people lit cigarettes to increase the fires and heat and shook hands and hugged each other, wishing "feliz año nuevo." More large beer bottles were shaken and then opened, spewing and splashing the liquid onto the earth before people started to consume it. Other people unpacked food and gathered for a picnic.

By 8 a.m., most participants moved to the actual site of the Horca del Inca and eagerly awaited the well-known light and shadow formations (see above). On the June solstice, the first beams of sunlight pass through a hole in the rocks and illuminate the center of the lintel of the Horca del Inca, but in 2005 the first light fell on the left side of the lintel (when seen from the front), and many people commented that it had moved since the year before. Yatiris observe closely how the light pattern descends the southern support of the Horca, and from its movements, they predict the degree of productivity of the next agricultural year. Meanwhile the spraying and consumption of beer and speeches by one of the yatiris continued at the Horca del Inca monument. Some people began to leave while others stayed, ate, and socialized as the Sun now illuminated and heated up the entire hillside.

This constitutes the official New Year's celebration of the Aymara in the Copacabana area, which is organized by Copacabana authorities. In 2005, it was advertised and promoted by means of a color poster, and it attracted a number of tourists. Javier Escalante Moscoso (personal communication, 2005), director of the Instituto de Arqueologia in La Paz, made me aware that it is also an effort to create a local copy of the famous Inti Raymi in Cusco, mostly to promote tourism.[30] It is a recently invented tradition growing in popularity and should not be confused with a cultural continuity maintained from the past—a point to which we will return. To some extent, the New Year's observation summarizes the public image of local culture and beliefs Copacabana wants to portray.[31]

Yet not all Aymara stand united behind such official strategies. In 2005, I spent some time with Ramon Calle, who is the chief guardian of the archaeological sites in Copacabana. His assignment is through the Instituto de Arqueologia in La Paz, which pays him for his work. Calle is a native of Copacabana and describes himself as a guide and ritual specialist. He told me that he did not like the official ceremony because pachamama was fed too much at one time. In order to make an offering which she receives well, one has to go to a quiet and more remote place and allow her to consume it undisturbed, which we did the following day.

Calle said everybody visits the Horca del Inka, but the really sacred sites where the ritual specialists go lie further up. He gave me a different tour of the Horca del Inka hill, and the following description is based upon his account. Slightly above the site of the celebration, there is an arrow-shaped opening in the rock. He explained that the Inka had this arrow filled with copper and that it marked the path of the priests descending toward the Horca. A flat area near the summit of the hill is the main place for offerings to be presented.

At the side facing toward the town sits a large rock which Calle identified as a frog. "Frogs bring luck," he said, "when a frog comes into your house, it is a good sign." It may be significant that Bouysse-Cassagne and Harris (1987:26–28), in their analysis of Aymara worldview, situate the frog in the liminal realm of the underworld waters of the puruma era localized in Lake Titicaca and that according to Ramos Gavilan people would set out idols of frogs on rock outcrops in order to attract rain. This rock appears to have been partly modified. Its back exhibits a weathered seat-like carving where, again according to Ramon, "the Inka" sat and bathed his feet in a cradle-like basin below, where water collects. Between the frog stone and the drop of the hillside, there is a flat stone surface with a depression where the Copacabana idol may have stood. Hearths with burnt remains below the head of the frog and at an adjacent boulder attested to the fact that indeed it is a location used by local ritual specialists and/or yatiris. On the way back, we stopped at the so-called balcony overlooking a small flat area. This stone balcony faces toward the bay of Copacabana, and its edge quite literally retraces the shoreline, inducing Calle to call it a model of the landscape.

Ramon's, as well as Porfirio's, identification and interpretation of places are fundamentally different from an academic archaeological discussion of Copacati and the Horca del Inka as, for example, given by Bauer and Stanish (2001:207–209) and Oswaldo Rivera Sundt (1984). During the past decades, an ever-growing corpus of literature has been debating the complex relations between indigenous people and outsiders, researchers and consultants, and whose interpretation is correct, some of which will be cited below. Rather than being a right or wrong issue, it seems to be a question of differing methodologies and perspectives. Both sides actively construct cultural and political landscapes which do not always merge, based upon their being situated in an academic setting versus an ethnic rural village.

Archaeologists traditionally are interested in material data, seen as the foundation of Western empirical science, and believe that their research leads to absolute universal truth as long as it is not invalidated by newly excavated contradictory data. Ramon and Porfirio, on the other hand, reach far back into their memory spaces filled with stories and wisdoms given to them by their parents, grandparents, and great-grandparents. I don't think they told me stories conveniently invented to impress tourists. Rather, their accounts form part of "invented traditions" by means of which the Aymara forge and define their identities in today's world (see, for example, Janusek 2004:8–9; Hobsbawm 1983; Friedman 1994). However, this does not imply that such invented traditions are disparate from the past or from Aymara heritage. Though invented, they build upon existing social practices and cultural conventions. In this manner, invented traditions reinterpret and adjust established ways of doing and understanding things, which often have a long history, to meet current needs. Thus they perpetuate cultural continuities at the same time as they change and reinvent them. Therefore, the New Year's ceremony and the storied explanations of rock formations all mark the multiple and dynamic relations of the present Aymara cultural landscape.

NOTES

1. It must be stated clearly that the chronology of the Titicaca Basin is still being debated among scholars. In the recent literature at the time of this writing, Charles Stanish (2003:89–90) proposes a general chronology divided into eight periods: Late Archaic (circa 5000–2000 B.C.), Early Formative (circa 2000–1300 B.C.), Middle Formative (1300–500 B.C.), Upper Formative (500 B.C.–A.D. 400), Expansive Tiwanaku (A.D. 400–1100), Altiplano (A.D. 1100–1450), Expansive Inca (A.D. 1450–1532), and Early Spanish Colonial (A.D. 1532–1700). Alongside this general chronology, he places local historical ones, providing a dual system for each area of the Titicaca Basin. Stanish (2003:139–140) dates the full cultural development of Pucara between approximately 200 B.C. and A.D. 200 and understands it as a historical period in the northern part of the Basin (2003:90).

2. The first sedentary populations living in permanent villages developed in the Early Formative beginning about 2000 B.C. (Stanish 2003:99).

3. The primary sources Salles-Reese (1997:184–185, note 23) consulted are Bartolome de Las Casas (1550), Juan de Betanzos (1551), Pedro Cieza de Leon (1553), Pedro Sarmiento de Gamboa (1572) two versions, Cristobal de Molina, el Cusqueno (1575), Pedro Gutierrez de Santa Clara (end of sixteenth century), Jose de Acosta (1590) two versions, Joan de Santacruz Pachacuti Yamqui (1613), Antonio de la Calancha (1637), and Bernabe Cobo (1653) five versions.

4. According to Stanish (2003:237), the Inca took control of the Titicaca Basin between A.D. 1450 and 1475.

5. Salles-Reese used as sources Francisco Lopez de Gomara (1552), Agustin de Zarate (1555), Pedro Pizarro (1571), Cristobal de Molina, "El Cusqueno" (1575), Martin de Murua (1590), three different versions of Garcilaso de la Vega (1605), Juan de Santacruz Pachacuti Yamqui (1613), Felipe Guaman Poma de Ayala (1613), Buenaventura de Salinas y Cordova (1630), Antonio Vasquez de Espinosa (1630), and four different versions of Bernabe de Cobo (1653) (1997:186, note 46).

6. Trimborn (1967:16) notes that he found "none or few Inca potsherds" in association with the Horca del Inca or Intinkala. Portugal and Ibarra (1957:24) claim that their careful study of the precontact monuments in and around Copacabana shows that they originated in the Tiwanaku period. However, they never explain what kind of study they did or the basis for their attribution.

7. Of course, it is always conceivable that some of the sculptures may have been begun during Tiwanaku times, and the similarities between Tiwanaku and Cusco carving styles have been mentioned above. So far and to my knowledge, there has been no archaeological report that presents solid material evidence that the Copacabana rock art sites would not be Inka.

8. In June 2012, a concert was staged at the archaeological site as part of the winter solstice celebrations.

9. Apparently his observations contradict those made by INA less than ten years later (see above).

10. Mantilla (1972:68–69) includes a historical reference which states that on the site where the Church of the Virgen of Copacabana stands today had indeed been an Inka sanctuary where pagan sacrifices were conducted.

11. On top of the hill Llallagua (?), at the foot of which are the church and convent of Copacabana, one finds galleries similar to those we have encountered at Rodadero (outskirts of Cusco) with the difference that there where the granite blocks formed continuous sections, the ancient architects have brought other blocks which they have placed on top of the first ones after they had carefully sculpted and polished them. The bath facilities, Bano del Inca, are admirably preserved, and still today the water falls through three openings into a basin with gravel (Wiener 1880:438, translation by the author).

12. The following is the description of another event held at Llallagua. The fact that Gavilan places it close to the lake reinforces the identification with the Calvario hill.

In the seat of Copacabana, at the hill called Llallagua, where one sees today the house of the hermit Santa Barbara, at the foot of which we descend to the lake in front of Pomata, there was an enclosure, which they called Taguakouyo. There they collected the chosen Virgins destined to be sacrificed and for the time being, they put them in beautifully adorned boats. They took them where one had to do the blood sacrifice for the Sun or Moon (Gavilan 1988:125–126 [1621:Bk.1, Ch.19]).

13. At certain festivities, each ayllu and lineage gathered their thirteen- to fourteen-year-old boys in a public place and in full public view, they whipped their legs, arms, and hands with leather slings until blood appeared. After that, the chief reprimanded them about their pranks and advised them that they were no longer boys but young men who now had to occupy themselves with community affairs and service to the Inca. Then they cut their hair and placed them in groups of two or three in a flat area near the lake. At a certain signal, *they had to race up the road to the top of the hill, which has the name LLALLINACO because of this game (the present Calvario hill), where the judges were waiting to reward or punish*. The prize for the first one to arrive at the goal was a canipo (a type of silver patenita), which they put on their llautos (*llautos* are the tassels which hung from the headdresses of Inka royalty and

highest nobility); or a chuspa made of Cumbi cloth, which are very interesting bags in which they carry coca leaves. These bags are so valuable that only the upper elite can use them. And the participants in these state-sponsored games had to be of the noble class, because later the Inca gave them high government positions, making them captains and governors, and their ears were pierced as a sign of nobility. But the ones who wore themselves out in the race were scolded by their parents and neighbors. They were shamed with hurtful words and with more whipping; they were given low service jobs benefiting those who had punished them (Ramos Gavilan 1887:33–34, translation and italics by the author).

14. The only Inka pictograph I know of is situated on a high rock wall near the entrance to Ollantaytambo. It represents a figure and overlooks the main road coming from Cusco.

15. Sergio Chavez has been working at Copacati and will produce a more detailed plan than mine and provide excavation data (personal communication 2003). Thus new information is forthcoming, and our understanding of Copacati may continue to change.

16. Lopez Bejarano (2009) has identified two roads leading from Yunguyo to Copacabana: one was the ritual pilgrimage road whereas the other served administrative purposes.

17. Besides the idol Copacabana, the Yunguyos had another one which they called Copacati whose name maintains the hill on which it stood, and one can still see remains of steps in this place. The idol was also of stone and had a most evil appearance, all wrapped by snakes like the statue of Laokoon. They prayed to it for rain in periods of drought. Father Almeida had this idol hauled into the village, and as he had it in the plaza in front of many people, one saw a snake come out of it. This event served Father Cura to shame them that they would have worshipped such a vile creature. In addition to these idols which functioned like communal gods, there were innumerable others . . .

Somewhat similar to the one at Copacati was another idol which Father Diego Garcia Cuadrado found in 1619 between Juli and Llave. It was of stone, three and a half varas tall, it had two faces like a Janus (Jano), but one face was male and the other female, with two snakes which climbed its legs, and on the head was a very large frog which formed the headdress. The idol was situated on a hill called *Tucumu* facing the lake. They worshipped it on a large stone block like the god of food (Ramos Gavilan 1887:45–46, translation by the author).

18. Arkush (2005:222) favors a Formative date because of the snake iconography. I concur that images of serpents do not abound in Tiwanaku sculpture.

19. The comments and suggestions made in reference to the Copacabana idol apply here as well.

20. Another possibility has been outlined by Teresa Gisbert (1997:628). She implies that Gavilan's description of the idols may have been tinted by European conceptions of Lucifer as the eternal serpent. She includes a striking representation of the wind god Typhoon wrapped by serpents and his fingers turning into serpents (Gisbert 1997:figs. 5 and 6).

21. Arkush (2005:222) raises the question of who worshipped at Copacati. According to Gavilan, the Copacabana and Copacati idols belonged to and were worshipped by the Yunguyus who were the original inhabitants of the Copacabana peninsula. They are described as being so poor that they could not offer gold or silver to their local idols (Ramos Gavilan 1988:191 [1621:Bk.1, Ch.32]). This scenario of local nonelites worshipping at an Inka site appears to be strikingly different from the situation on the Island of the Sun, where all pilgrims were carefully screened and only the privileged were permitted to approach the main Sanctuary of the Sun (Arkush 2005:220–222).

22. Bauer and Stanish remapped Pilco Kayma in 1996 and provide a detailed description accompanied by illustrations (2001:163–173). The function of this impressive structure continues to remain obscure. It is of architectural interest because its rooms exhibit Inka stone vaulting.

23. One enters all this through the aforementioned door Kentipuncu which is located two hundred paces before the Sacred Rock, where the Inca removed his shoes for the first time so that he placed his bare feet there; . . . to the right of Kentipuncu, one can see houses which at that time were the residences of the ministers of the sanctuary and of the virgins dedicated to the Sun. A little further (past the door), there is a natural bedrock, over which the trail to the false sanctuary leads. On this bedrock, there are the human footprints which we discussed before.

Before reaching this shrine, one had to pass through three doors situated little more than twenty paces apart from each other. The first door was called Pumapuncu, which means Door of the Lion because there was a stone lion which they said guarded the entrance. Before passing through this door, pilgrims had to confess their sins to the priest who resided there.

The second door was called Kentipuncu because it was completely covered with feathers of hummingbirds, which they call Kenti. Here they had to confess again to another priest who guarded this door. He advised the pilgrims to worship with devotion if they wanted to be favored by the Sun.

The third door was called Pillco-puncu, which means Door of Hope. It was adorned with green feathers of a highly valued and shining bird which is called Pillco and brought from the Chunchos. At this last door, the custodian priest effectively persuaded the pilgrim to examine his conscience rigorously because he could not pass with a burdened one, and in this case had to reconcile himself with the priest who was experienced in these matters (Ramos Gavilan 1988:94 [1621:Bk.1, Ch.15], author's translation).

24. [I]n the Island Titicaca, where, as the Indians assured, lived the Sun because some sorcerers saw it rise from this rock. It was the one to which they offered gold, silver, shells, feathers, and cumbi cloth, which was the finest that was woven in the whole world. And so state these Indians that the Inca had the rock of the Sanctuary covered with a curtain of cumbi cloth, the thinnest and finest that was ever seen among the Indians. The whole concave area of the rock was covered with sheets of gold and silver; and in a number of holes which one can still see, they placed the appropriate offerings. Depending on the ceremonies, some of which were more solemn

than others, they decorated the Sanctuary with curtains of cumbi cloth in different colors. This matches with the account of Brother Prudencio de Sandoval, in Book 13 of the History of the Emperor, where he writes that there were great riches when the Spaniards entered Cusco and that the temples of the idols were covered with sheets of gold and silver. Further, one can see today in front of the rock a cross which was erected there and a round stone in the form of a bowl in which they poured the chicha they gave the Sun to drink. I don't know with what kind of instrument they carved it because the circular bowl is extremely well finished.

They also say that the Inca kept a large golden incense burner there, which had as pedestals four lions made of gold or, as some say, of silver. It was a very valuable piece, but we cannot be certain whether it ever existed. Some elders have told that when the islanders learned that the Spanish were looking for silver and gold so eagerly, they hid most of the treasures they had by throwing it in the lake (Ramos Gavilan 1988:116 [1621:Bk.1, Ch.17]).

In this month, the Indians gathered in Copacabana, near the lake, in a large plaza, all the sheep and lambs from their hamlets which they could spare to sacrifice on the island. And to the sounds of flutes and music, they decorated the animals with colorful bands and ribbons and with much joy and happiness, they took them to the island where very rich and beautiful pieces of cumbi cloth covered the rock in the Sanctuary of the Sun. In the first night, they lit a great bonfire which was repeated later on the other islands, because when their inhabitants saw the smoke and fire, they followed the example of the main island. On the next day, the sacrifices of the sheep and lambs were performed; they also sacrificed many innocent children and sprinkled the rock of the Sanctuary with their blood. This rock was covered with plates of gold and silver, and because the rock glittered so much in the sunlight, the Indians said that no bird would dare to land; only by means of the cunning and order of the devil would birds come to this place (Ramos Gavilan 1988:149–150 [1621:Bk.1, Ch.24]).

25. Visitors in the late nineteenth century observed a large irregular depression near the center of the plaza which was probably a looters' pit. Bandelier talks about Inka objects which came from the immediate surroundings of the Sacred Rock (see Bauer and Stanish 2001:197–202).

26. One of the many functions of the usnu was to serve as an observation point of the sun (Zuidema 1980:318–331). At least a dozen chroniclers mention the existence of sets of small towers or stone pillars on the eastern and western horizons of Cusco (Bauer and Dearborn 1995:67–100, Hyslop 1990:61–62). One point of observation was the usnu in the Awkaypata plaza. The Anonymous Chronicler (in Bauer and Dearborn 1995:35 [c. 1570]) describes the four western pillars and states that

> when the sun reached the first pillar they prepared for the general planting and began to plant vegetables in the heights, as slower [to mature], and when the sun reached the two pillars in the middle, was the point and the general time of the planting in Cuzco, and it was always in the month of August. It is in this way that, to take the point of the sun between the central two pillars they had another pillar in the middle of the plaza, [a] pillar of well worked stone one *estado* in height, in a suitable indicated place, that they called *usnu*, and from there they watched the sun between the two pillars, and when it was exactly there, it was the time for sowing in the Cuzco Valley and its region.

Although the Anonymous Chronicler states that sunset between the western and central pillars was observed from the usnu, he does not say who watched it.

We might assume that he implies the ruler seated on the Cusco usnu. Dearborn (2000) makes the interesting argument that the distances between the pillars which the Anonymous Chronicler provides are much too wide to specify a precise date. According to Dearborn's calculations, the Sun would appear to set between the central pillars for nearly a week as seen from the usnu in mid-August. The large separation of the pillars also allowed most of the people gathered in Awkaypata to watch the sunset, and not exclusively the ruling Inka on the usnu. Dearborn concludes that the primary objective of solar observation from the Cusco usnu was not astronomical in the sense of fixing a precise date but ritual, meaning that all the people in the plaza watched the Sun set.

Dearborn's argument gives important insights into the discourse between the ruler and common people as it was acted out at the usnu in the context of watching the Sun. In the Sanctuary on the Island of the Sun, the grouping of the audience into status-based locales as outlined above seems to stage a similar scenario. Although individuals of high and low social status both observed June solstice sunset, the Inka ruler and the nobility stood closest to the Sacred Rock, whereas commoners had to remain at the outer entrance wall (Bauer and Stanish 2001:212).

27. Farther ahead of them [the Temples of the Sun, Thunder, and Lightning], in the ravine that drops in front of the road (between Juli and Pomata), there is the storehouse of the Sun which had a visually appealing layout and buildings before time destroyed it. Its innumerable rooms formed something like a labyrinth which the Indians call Chingana, meaning a place where one gets lost. In the center of the rooms, there was a garden with a grove of alders; its continuous freshness was sustained by a spring of sweet water which emerges there. In the shade of these trees, the Inca had strange baths of stone carved for the Sun and its cult (Ramos Gavilan 1988:93 [1621:Bk.1, Ch.13], author's translation).

28. Furthermore, this date may indicate that Ramos Gavilan and Cobo did not visit abandoned ruins but populated settlements and shrines at which perhaps Inka customs continued to be performed.

29. When the Indians celebrated the solemn festivals of the Sun, particularly the Capacrayme [at the December solstice] and the Intip rayme [at the June solstice], the individuals of the lineage of the Incas placed all the idols on their portable platforms (which they call ramp or litter) and decorated these platforms with many flowers, sheets of gold and silver, and many feathers. They performed their dances and festivals, and all came to the island, and they deposited the decorated platforms at a location called Aucaypata which had a great plaza, and there they held the festivals. There was a large temple with five doors. No Colla Indian was allowed to participate or be at these celebrations nor approach the area until the festivities were over (Ramos Gavilan 1988:176 [1621:Bk.1, Ch.29], author's translation).

30. Inti Raymis copying the successful Cusco festival have become increasingly popular in a number of places in the Departamento Cusco, such as Ollantaytambo and Espiritu Pampa. A major solstice celebration with a large audience takes place every

year at Tiwanaku and smaller ones at Copacabana, as well as at other sites. The idea that Bolivian authorities and Aymara yatiris would imitate a Cusco-derived and thus Inka tradition is interesting in itself, since many Bolivians and Aymara like to argue that Inka domination in their country was limited and that, for instance, the carved rocks have to be pre-Inka and Aymara.

31. In 2012, I revisited Copacabana and again attended the solstice sunrise ceremony at the Horca del Inca. The events had very much stayed the same, and one of the yatiris from 2005 continues to perform. However, more solstice ceremonies are being advertised; color posters invited visitors to come to one held at the rock Sanctuary on the Island of the Sun.

Furthermore, the observation of the June solstice and the New Year have been embedded in three-day-long festivities which attract many visitors to Copacabana. The city organizes stage performances of music bands, people build fires in the streets and shoot fireworks, and the market and restaurants stay busy until late at night.

Chapter Four

Inka Pacariqtambo—A Landscape of Power Relations through Time

In this chapter, we move to Pacariqtambo, Maukallaqta, and Pumaurqu as the Inka-specific origin places of their first dynasty. The *Groundwork* portion will thoroughly discuss the chronicles which narrate the story of emergence of the Inka ancestors and its locations, followed by the archaeological data which might support an identification of the archaeological sites of Maukallaqta and Pumaurqu with Pacariqtambo. The section *Inka Landscape and Carved Rocks* will explore the multilayered relations between the natural and political landscapes of the origin sites and the capital in Cusco through time.

GROUNDWORK

Textual Data in the Chronicles

As addressed in chapter 3, the Inka honored two places which, in their belief, were connected with their origins: one was the Island of the Sun in Lake Titicaca and the other was Pacariqtambo/Maukallaqta/Pumaurqu situated in the Province of Paruro to the southwest of Cusco, which related specifically to the mythological origins of the Inka state (Urton 1999:34–37, 45–54). Both sites exhibit special rocks with cave-like openings from which Manqo Qhapaq and Mama Oqllu, two of the mythic-historical founders of the Inka, emerged. However, the Sanctuary on the Island of the Sun was a pan-Andean pilgrimage site, while Pacariqtambo/Maukallaqta/Pumaurqu appears to have been the *paqarina* or special personalized origin place of the Inka dynasties. Many Spanish chroniclers intended to write a complete history of the Inka and thus began with their origins and their first ruler Manqo Qhapaq, but since they struggled to reconstruct a remote and quite possibly mythological past, their accounts differ and are not consistent with each other. A detailed comparison of the forty or so versions of the Inka origin stories recorded between 1542 and 1653 would go well beyond the scope of this chapter, and I refer to other writers who have done just that (Urton 1990, 1999; Pardo 1946; Bauer 1991,1992; D'Altroy 2002).

What follows is a paraphrased summary leaning heavily on the version given by Sarmiento de Gamboa, whom Urton as well as Pardo select as the principal source. Sarmiento's history was written relatively early in 1572 for Viceroy Francisco de Toledo, and, according to his own testimony, much of his information was obtained from *khipu kamayuqs* who later verified his story. Furthermore, Sarmiento was the first writer who described Pacariqtambo in more precise geographical terms and added toponyms which can be linked to locations existing today. The Inka first appeared at a place called Pacariqtambo ("house of dawn"), located approximately 33 kilometers south of Cusco. At Pacariqtambo, there was a mountain named Tampu T'oqo ("the house of openings/windows") with three windows or caves. The two lateral windows were called Maras T'oqo and Sutiq T'oqo, and the Maras and Tampu people came out of them. Urton (1999:46) thinks that the head lineage of the latter bore the surname Sutiq. The ancestors of the Inka emerged from the central window, Qhapaq T'oqo, which means "rich window." These ancestors were four brothers and four sisters paired as husbands and wives: Ayar Manqo-Mama Oqllu; Ayar Awka-Mama Waqo; Ayar Kachi-Mama Ipakura/Kura; Ayar Uchu-Mama Rawa. While all accounts say that the ancestors originated from Tampu T'oqo at Pacariqtambo, Martin de Murua, Guaman Poma de Ayala, and Garcilaso de la Vega add that Manqo Qhapaq and Mama Oqllu first appeared on the Island of the Sun and then traveled underground from Lake Titicaca to the cave at Pacariqtambo, thus linking the two origin narratives. According to Sarmiento, the Maras, Tampu, and Inka were born from the caves of Tampu T'oqo by the will of the creator god T'iqsi Wiraqocha, who functions as the founder of lineages. The newly emerged or "newborn" peoples allied themselves with the local inhabitants, and the eight Inka ancestors quickly assumed leadership. They prepared to go in search of fertile land where they would settle down and to conquer any groups who got in their way. As a reward for loyalty, they promised conquered lands to their followers and divided them into ten ayllus. These ayllus are very important for a number of reasons: they are known by name (see Urton 1990:22–29; D'Altroy 2002:51, table 3.2; and Pardo 1946:24); five have been associated with Hanan Cusco and five with Hurin Cusco; and their existence in Cusco has been documented historically.

The Ayar brothers and sisters began to walk northward in the direction of the Cusco valley together with the ten ayllus they had founded. Along the way, they tested the hardness or softness of the earth—for example, at Pachecti and Pallata—by plunging into the ground a golden bar, which they had carried with them from Tampu T'oqo. The Inka ancestors made several stops during their journey, which ended in Cusco. At Waynakancha, which may be a small Inka site called by this very name today located in the Yaurisque Valley near the Hacienda Waynakancha, Manqo Qhapaq and Mama Oqllu conceived a child. The group moved on to Tamboquiro, where Mama Oqllu gave birth to a son whom they named Zinchi Roq'a and who would succeed his father as the second Inka ruler. The next stop of the ancestors and their accompanying ayllus was Haysquisrro, which may correspond to the present town of Yaurisque. There three brothers and the four sisters conspired to rid themselves of Ayar Kachi, who was cruel, made trouble in all the villages they passed through, and disturbed peace among the ancestors. They sent Ayar Kachi together with a companion back to retrieve some items from the cave of Tampu T'oqo, which they said they had forgotten. When they arrived at Tampu T'oqo, Ayar Kachi entered the cave, and immediately his companion shoved a large boulder across the entrance, leaving Ayar Kachi trapped inside for all time.

The remaining ancestors and their ayllus continued walking north in the direction of Cusco and came to a mountain with the name Wanakauri. Ascending Wanakauri, they looked down upon the valley of Cusco for the first time. A rainbow[1] stretching over the valley was the sign that this was supposed to become their home, and the ancestors prepared to descend and take possession of it. At this point, Ayar Uchu was transformed into stone and became one of the principal Inka wak'as. They took time approaching Cusco, testing the soil again and again with their golden staff until it remained stuck in dense fertile ground. When they finally arrived at the locale that would become the center of the capital city, Ayar Awka was transformed into a stone pillar.[2] Thus only Manqo Qhapaq, his four sisters, and the young Zinchi Roq'a were left to found and build the city of Cusco.

The Material and Archaeological Data

Scholars such as Pardo (1946), Urton (1990), and Bauer (1992) have identified Pacariqtambo with the archaeological site of Maukallaqta and Tampu T'oqo with the sculpted outcrop Pumaurqu (figure 4.1). The main reasoning underlying such an identification is the fact that the chroniclers who specify the location of Pacariqtambo place it seven or eight miles to the south or southwest of Cusco, and the only major site in that direction and at that distance is Maukallaqta/Pumaurqu. Maukallaqta is situated near the village of Mollebamba, which is politically annexed to the present town of

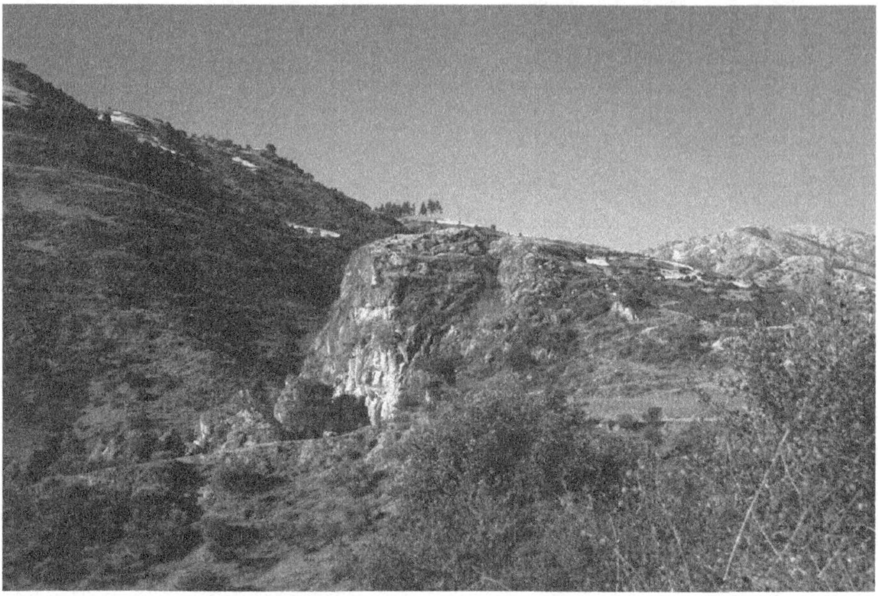

Figure 4.1. Sculpted outcrop Pumaurqu seen from the west. Photograph by Brian Garrett.

Pacariqtambo in the province of Paruro, Department of Cusco. The present-day town of Pacariqtambo was founded in 1571 as a *reduccion* (reduction) of eleven previously dispersed ayllus into a single, nucleated settlement. The policy of creating such nucleated settlements throughout the Andes was initiated by Viceroy Francisco de Toledo in 1570 and practiced for about five years. Urton (1990:32) notes the curious historical coincidence between the founding of the town of Pacariqtambo in 1571 and Sarmiento's completion of the *Historia de los Incas* in 1572. Surely, the khipu kamayuqs whom Sarmiento interviewed would have more likely responded with historical constructions based upon their most recent history rather than with a reconstruction of the remote history of the Inka Empire.

The name Maukallaqta means "old town," which implies that it was given to the archaeological site after the present-day Pacariqtambo was established. It is grouped into three sectors separated by dry *quebradas* (ravines): *Patanllaqta* to the east, *Chaupillaqta* in the center, and *Maukallaqta* in the west (Pardo 1946:5–9). Bauer (1992:111–114) further divided the sectors into seven zones, and his surveys registered more than two hundred structures. All the buildings of this large site are aligned along grid systems established within each zone. With regard to construction materials and techniques, finely cut stone masonry is found in the lower portions of walls and at principal doorways; adobe blocks were often placed on top of stone foundations.

In 1986, Bauer (1992:114–116) conducted excavations in Zone 1 of the Maukallaqta sector. Of special interest in Zone 1 is the central court constructed of well fitted stonework reminiscent of Cusco styles and framed by large triple-jamb niches, as well as the peculiar design that one of the niches on the north side constitutes the entrance into an inner chamber. Bauer (1992:116–123) argues that Maukallaqta was built by the Inka in affiliation with their origin myth and likely was the original Inka Pacariqtambo. The absence of extensive storage facilities speaks against its role as an administrative center; instead the central court may have denoted a ceremonial center. Bauer (1992:122) speculates, and I think with good reason, that this courtyard could have functioned as an oracle complex or temple for Manqo Qhapaq. The inner chamber might have housed an image or idol of the first mythical Inka, which would have appeared through the niche-entrance on ceremonial occasions, dramatically reenacting Manqo Qhapaq's emergence from Tampu T'oqo.

It is also important to note that Bauer (1992:116) found pre-Inka walls and ceramic fragments in Zone 1, which he tentatively dated to the Middle Horizon or Wari Period. The questions remain whether the people—Wari or other—who lived at Maukallaqta before the Inka regarded the area as sacred ritual space and especially whether they performed any ceremonial activities at Pumaurqu or perhaps even began to carve it.

The outcrop of Pumaurqu may have contained the cave, the actual Tampu T'oqo, from which the Inka believed their ancestors first appeared on earth. Most of the carvings are concentrated on the top (figure 4.2). As the visitor enters from the south, he or she finds two rounded felines sculpted in high relief and closely together (figure 4.3). Their heads are now missing, but they have given the site its popular name: Pumaurqu translates as "hill of the puma." From the location of the felines, the carvings extend to the northwest and northeast in gradually descending levels. There are large and small seats, most of them rectangular but some rounded or semicircular, wide and narrow platforms and planes, as well as steps. The stone is limestone, and weathering is responsible for sharpening the edges and for some grooves and small depressions amid the sculptures. Along the northwestern to the northeastern edge of the outcrop, there is a series of steps and shallow square depressions which likely constituted the foundations of a wall that once enclosed the sculpted top. The north flank is the tallest and steepest, and nobody from below was able to see the carvings or any activities performed on top. However, the wall would have been well visible, and it would have signaled to anyone that the top of Pumaurqu was sacred and secluded space to which access was strictly controlled.

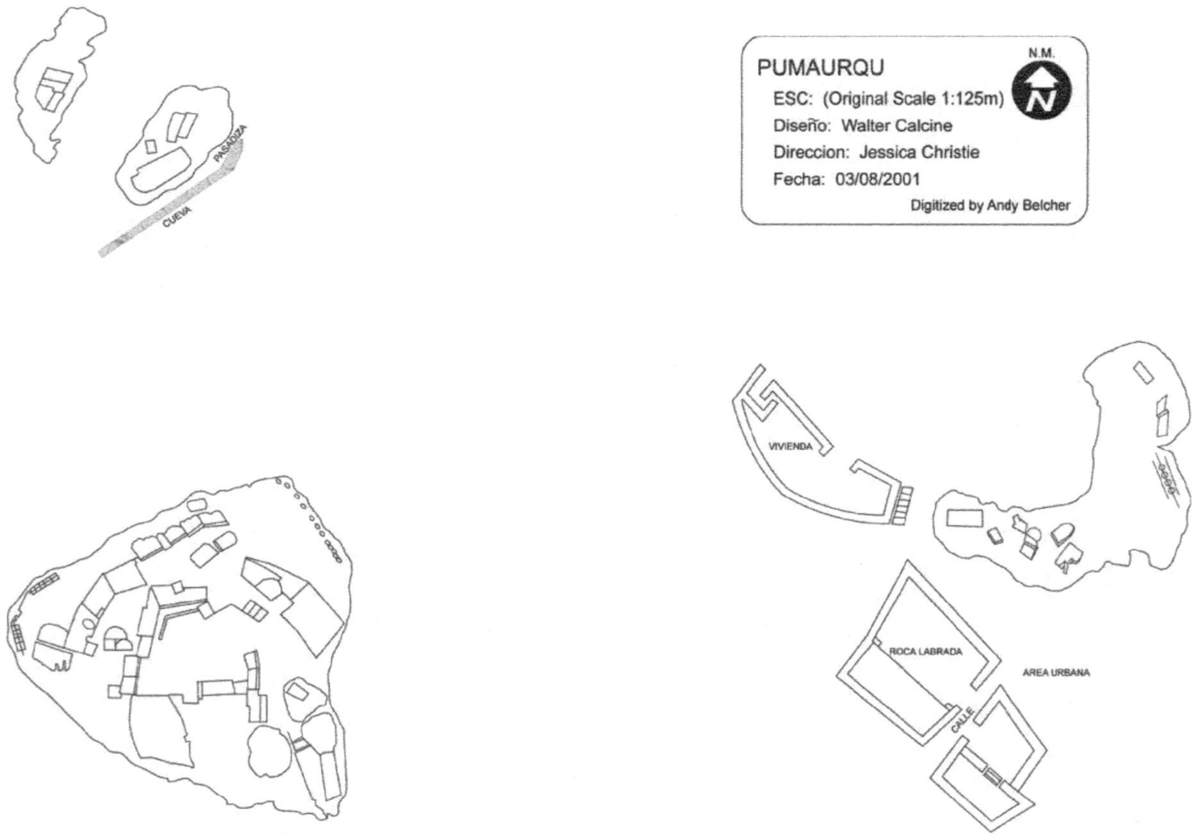

Figure 4.2. Pumaurqu, Plan Drawing. Drawing by Walter Calcine, directed by author, digitized by Andy Belcher.

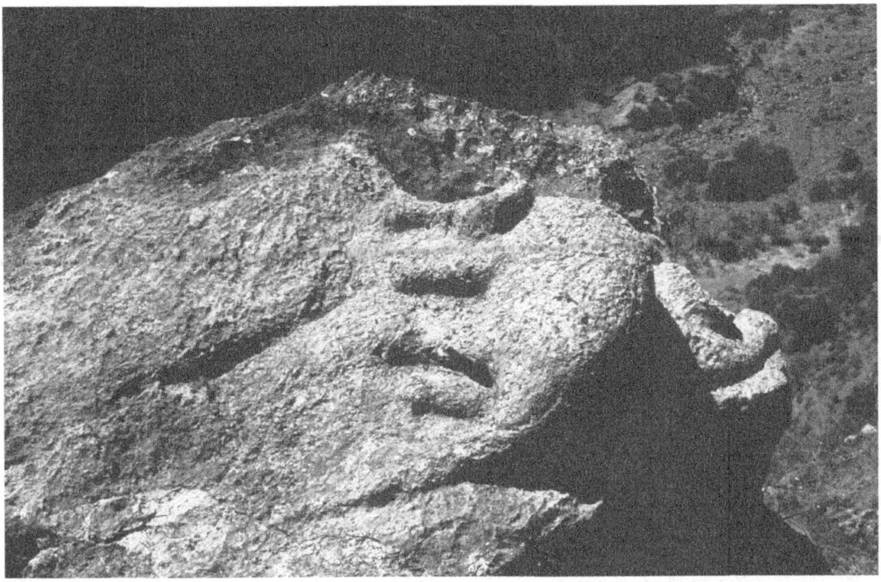

Figure 4.3. Pumaurqu, Relief Carvings of Felines. Photograph by Brian Garrett.

Below the almost vertical north and northeast side of Pumaurqu lies a rock fall area with large strewn boulders. Some of them are carved with precise vertical and horizontal cuts; others form cave-like spaces with the foot of the outcrop. The irregular manner in which these huge chunks of rock are distributed suggests that they may have split off the northern flank during an earthquake. How long ago such a specific earthquake took place is a tricky question even though earthquakes are common in this area. Pardo, who wrote in 1946, describes the rock fall area; yet his description of the south side and a *chincana* (cave-like passageway), as well as his drawing of the structures on the northeast side, do not correspond exactly with the present condition of the site (Pardo 1946:30–31). Thus, the evidence shows that there have been some modifications between the time Pardo initially reported on this site and today. Still, I suggest that the appearance of Pumaurqu has not changed substantially since Inka times, which is emphasized by the fact that the wall foundations along the edge of the top confirm that the drop-off was there in ancient times as well.

The visitor has to crawl through a cave-like space formed by fallen rocks to exit on the east side of Pumaurqu. After a short walk on a narrow trail at the base of the outcrop, the terrain widens somewhat, and the visitor stands between the foundations of three—likely Inka—houses and a carved boulder. The walls were constructed of fieldstones, and the masonry is coarse and unrefined. The two southern structures have two and one rooms and open onto a shared trail between them. The southwest wall of the one-room structure is formed by a sloping worked bedrock surface. The north structure opens to the northeast, where the ground drops off quickly and sharply. One opening is a regular doorway, while the other consists of two interlacing L-shaped walls with a passageway between them. The rear or southwest wall curves and displays three niches. This building may have served special ritual purposes because of the unusual L-shaped entrance/exit and the fact that the openings lead to a drop-off from where the eastern horizon and sunrise can be observed. The carved boulder sits at the drop-off and exhibits vertical cuts, horizontal planes, seat-like and semicircular sculptures. Below the houses and the carved rock descend a series of terraces and crude retaining walls, suggesting that Pumaurqu extended to the east and that the eastern side was a principal sector.

When I asked different people from Mollebamba where Tampu T'oqo was, they pointed to three irregular holes high on the rock wall on the northeastern side. Pardo (1946:10–11) adds two caves to the three openings on the north-eastern façade and thus accounts for five t'oqos. According to Garcilaso de la Vega (1963:30 [1605:B.1, Ch. 18]), the central window from which the Ayar siblings emerged was decorated with gold plates and precious stones; the side windows out of which the Maras and Tampu groups climbed were lined only with gold.

In an interesting turn of events, a new Tampu Toq'o was created in the late sixteenth century when the present-day town of Pacariqtambo was founded. Today these residents identify a small cave to the northeast of their town as Tampu Toq'o. In 2003, a young man from Pacariqtambo showed me this cave; I did not see any remains of offerings or burnt pagos (colorful offerings to the earth and the mountains which are assembled, wrapped, and burned by contemporary shamans). As Urton (1990:37) has indicated, the people of the new Pacariqtambo may have attempted "to take the origin place of the Inka and its inherent power away" from the small community of Mollebamba and to appropriate it for themselves by re-creating it close to their town.

INKA LANDSCAPE AND CARVED ROCKS—APPEARANCE, EXPERIENCE, AND PERCEPTION

Why Pumaurqu in Kuntisuyu?

At first appearance, Pumaurqu is an obscure outcrop in an isolated region that was relatively unimportant in state affairs when compared, for instance, to the Lake Titicaca Basin or the Urubamba Valley. Bauer's surveys in the Province of Paruro have shown that there are no significant natural resources and that agricultural land is of poor quality (see Bauer 1992:53–61), which makes the region economically unattractive. Why would the Inka have selected this site as the paqarina of their origin?

Pumaurqu/Maukallaqta are located in Kuntisuyu, which falls into the Hurin division of the empire and is of somewhat lower importance than the Hanan suyus. It is noteworthy that both Inka origin places, the pan-Andean pilgrimage site on the Island of the Sun in Lake Titicaca and Pacariqtambo, were situated in Kollasuyu and Kuntisuyu and thus in the lower Hurin half of the empire. Furthermore, it is significant that both origin places are sacred rocks: the Sanctuary on the Island of the Sun is a natural outcrop that was covered with gold sheets and finely worked cumbi textiles, while Pumaurqu was sculpted. On the other hand, in the immediate Cusco region there were no wak'as in the form of carved rocks on the zeq'e lines leading to Kollasuyu and Kuntisuyu (chapter 2). These observations might suggest that one reason for locating the origin sites there was to establish an Inka presence in these "underrepresented" suyus, and the markers of such a presence were highlighted sacred rocks.

Guaman Poma de Ayala's Pacariqtambo

Interestingly, in the seventeenth century, the physical features and even the different nature of the origin sites appear to have been forgotten or deemed unimportant, as is suggested by Guaman Poma de Ayala. It is instructive that he includes two visual representations of Pacariqtambo/Tampu Toq'o. The first is in the coat of arms of the first Inka(s), whom he describes as the "legitimate descendants of Adam, Eve, and Noah" (Poma deAyala 1987:73 [1615:79], 1956:59). The coat of arms contains four fields which display images of the Sun, Moon, one Star, and a mountain which is crowned by a human figure and has three cave openings at the bottom. This last field is labeled "Idol of Uanacauri," "Pacaritambo the inn where they appeared," "Tambotoco the inn with openings." A very similar image is repeated in the chapter on "Idolos y Huacas"(Poma de Ayala (1987:257 [1615:264], 1956:188). Guaman Poma (1956:187) further names "Uanacauri, Urco, Tanbo Toco y Titicaca" as specifically Inka gods. This interesting contraction of the migration narrative into one toponym signals that at the time of Guaman Poma's writing, in the early seventeenth centuty, the three locales had been conflated into a single origin place and ideological concept constructing Inka identity. The migration of the ancestors was no longer experienced spatially, but was perceived and imagined as one visual sign condensing three place names (if we extend the meaning of "Titicaca" from a god to the origin site). Both origin places took part in this colonial identity process.

Ranking of the Territorial and Social Landscape

The Spanish writers furnish rich material allowing us to explore how the relations between Cusco and Pacariqtambo/Maukallaqta/Pumaurqu were experienced, perceived, constructed, and performed. Sarmiento (1942:62–64 [1572:chap. 11]) tells us that the Ayar brothers and their sisters formed ten *parcialidades* or ayllus among the people who emerged with them from the lateral windows and among the people who already lived in the Pacariqtambo region. These groups moved to Cusco together with the eight siblings, and in 1572, at the time of Sarmiento's writing, descendants and members of these ayllus still lived in Cusco. Thus, there appears to have been a "living" connection between Cusco and Pumaurqu/Pacariqtambo at least up to the late 1500s. Sarmiento lists the ayllu names as follows:

> Chauin Cuzco ayllo, of the lineage of Ayar Cache; . . .
> Arayraca Ayllo Cuzco-callan; . . . of the lineage of Ayar Uchu—
> Tarpuntay Ayllo; . . .
> Guacaytaqui Ayllo;
> Sanoc Ayllo; . . .
> These five bands are Hanancuzcos, which means the group from the higher Cuzco.
> Sutic-toco Ayllo, which is the generation that emerged from the window called Sutic-toco . . .
> Maras Ayllo; these are the ones who say they came out of the window Maras-toco; . . .
> Cuycusa Ayllo; . . .
> Masca Ayllo; . . .
> Oro Ayllo; . . . (Sarmiento de Gamboa 1942:63–64 [1572:chap. 11]).

Urton (1990:25) stresses that Sarmiento's account is unique among the chronicles which talk about the origin myth in that he is one of two writers who address the creation of ten ayllus at Tampu T'oqo. The other text was written by Viceroy Francisco de Toledo, who mentions only three ayllus. I concur with Urton (1990:25) that the names of the ten ayllus and their organization into Hanan–Hurin may have been information directly obtained from Sarmiento's one hundred khipu kamayuq informants, who lived within the Cusco moiety system and may have known in person some members of the ayllu that Sarmiento lists by name.

Furthermore, it is very important that Sarmiento classifies the ayllus into Hanancuzcos and by implication Hurincuzcos, which reflects the moiety system of the capital and of the Andes in general. The eight Inka ancestors and ten ayllus in Sarmiento's version, as well as in other versions of the origin myth, are not treated as equals but are subjected to concepts of ranking. Certain aspects of these ranking categories seem to be reflected in the social organization of Inka Cusco and its relation to the periphery of the empire. Ranking begins with the emergence of the Ayar brothers and sisters. Sarmiento (1942:62–63 [1572:chap. 11]) presents them in two clearly defined groups: males and females. As he lists their names, the order of the brothers is based upon age and authority and the order of the sisters on age alone, with the oldest always coming first and receiving the highest authority. Sarmiento also divides the ten ayllus into two groups: those of Hanan Cusco and those of Hurin Cusco. One would suspect that the order of their names is not arbitrary but builds on those categories used for the Inka ancestors. Therefore, Chauin Cuzco ayllo and Sutic-toco Ayllo were likely the most influential while Sanoc Ayllo and Oro Ayllo were given the least authority. The latter formed a fifth rank in each moiety as opposed to the four members in each gender group of the ancestors, and thus would have placed low in the hierarchical system. Urton (1990:25–26) argues that "the fifth ayllus of the moieties represent the incorporation of low-prestige outsiders into the imperial sociopolitical and territorial organizations." He identifies Sanoc Ayllo or Sanuq Ayllu as people living in Sanu, which is now San Sebastian situated at the southeastern periphery of the city of Cusco. He interprets Oro Ayllo as a Spanish corruption of the Aymara word *uru*. Uru could designate pre-Inka populations of the valley of Cusco as well as lower-class people within the city in general. Further, Uru was the name of an ethnic group of fishermen and agriculturalists who lived around Lake Titicaca and whose descendants continue to occupy the floating islands which have become popular tourist spots. In the organization of ayllus in the town of Copacabana located on the shores of Lake Titicaca, there was also a group called Uru Ayllu. According to Zuidema (cited by Urton 1990:26), this ayllu was the fifth of the upper moiety in Copacabana and was referred to as *sullka ayllu*, meaning "youngest, least." Therefore, Uru seems to have designated low-status and junior groups of people in various spatial settings.

There thus seems to be compelling evidence that the ten ayllus of the Hanan and Hurin moieties were ranked according to an "age lending authority" classification similar to the ranking system of the four ancestral brothers and sisters. The critical distinction between the two cases is that the eight ayllus within Cusco are juxtaposed with two ayllus of outsiders, forming border groups who lived outside the city and could be non- or even pre-Inka peoples. Garcilaso reinforces such a dichotomy between the civilized center of Cusco and wild barbarian people outside when he describes how Manqo

Qhapaq instructed the residents of the towns and villages he had founded how to make their living:

> and that they [Manqo Qhapaq's subjects] owed worship and service in particular to the Sun and to the Moon because they [Sun and Moon] had sent them their two children [Manqo Qhapaq and Mama Oqllu] who had liberated them from the uncivilized life they had led up to that time and had brought them back to be fully human which they were now." (Garcilaso de la Vega 1963:34 [1605: B. I, Ch. 21]).

This discussion vividly demonstrates the living hierarchical connection that existed between Inka Cusco and Pacariqtambo/Maukallaqta/Pumaurqu as well as Inka tactics of incorporating outlying regions and people by elevating their rock wak'as. The issue of ranking leads to the less physical concepts of power, essence, and stone ideology.

INKA LANDSCAPE AND CARVED ROCKS—ESSENCE, IMAGINATION, AND STONE IDEOLOGY

Center—Periphery Dynamics

According to the origin story, the ranking system was established by the eight ancestors at Pacariqtambo and then transferred and adapted to the social and political organization of Cusco. This mythohistorical reconstruction raises some interesting issues about the relationship between Cusco in its role as capital and center of the empire and powerful sacred sites at the periphery. As outlined above, it is highly unlikely that the relations between Cusco and Pacariqtambo were economic in nature because the Province of Paruro does not have any significant natural resources; high-yield agricultural land is limited by the deep gorges cut by the Apurimac River and many of its tributaries, leaving only small areas of cultivable bottomland.

Instead, the interactions between Cusco and Pacariqtambo must have been nonmarket-based and related to ideology and cosmological principles. John Staller (n.d.) has discussed Andean interaction models between a core and the periphery based upon acquisition or trade and exchange and has suggested how geographic horizontal relations with certain sites or wak'as in the periphery can raise social status in the core/center in terms of a vertical hierarchy (for a broader discussion of core and periphery models, see Helms 1993; 1998:73–108, 164–173, who traces emergent hierarchies and social stratification to access to origin places and associated tangible and intangible wealth). Trade and exchange represent two-way interactions between two sociopolitical units which, in the Andes, are typically housed in different ecological zones. Both units are of compatible social status and exchange products they have in surplus for products they are lacking. Acquisition, on the other hand, constitutes a one-way interaction between a more powerful cultural center and a weaker peripheral sociopolitical unit. In the acquisition model, the actors hold contrasting status with regard to political hierarchy and spatial dimensions: high versus low, core versus periphery. Acquisition dynamics between core and periphery involve the appropriation of status objects and wak'as by the dominating party and are usually discontinuous. I concur with Staller (personal communication 2005) that the relationship between Cusco–Pacariqtambo and the Island of the Sun is a precise example of this type of acquisition model.

Sarmiento de Gamboa provides us with a specific reference of how Pumaurqu impacted Pachakuti's notion of identity as ruler and how he appropriated, altered, and integrated this site into the political landscape he was constructing.[3] We are informed that it was Pachakuti who instituted Pumaurqu as an official wak'a, and that he embellished the cave of origin with gold. This likely implies that he also commissioned the sculptures. The two puma reliefs could further tighten the connection with him, since it is said that he went to war with a puma skin over his head to claim ownership of the feline as his totemic ancestor (Salazar 2004:36). Thus Pachakuti appears to have been the one who turned a powerfully charged ancient site whose history and myths were passed along in oral traditions into a prestigious state-sponsored wak'a. Julien (2000:293–294) also concludes that he devised the Inka origin narrative and history marked by the rise of Manqo Qhapaq's descent line to imperial power. Her argument rests on the chronological observation that the demise of the Ayar siblings would have been correlated with military conquest, and thus origin mythohistory must have been constructed by Pachakuti in the process of imperial expansion.

I argue that one form of marking center–periphery acquisition dynamics in the Inka-specific political landscape was the use of rock art. Interaction between the core and these peripheral sites with rock sculptures would have played out in two dimensions: authority and prestige of the Inka state government in the core rose on a vertical scale as individual rulers established peripheral centers on a geographic horizontal level (see Staller n.d.). Distance to the peripheral sites may have affected the rise of authority in the center. Interaction also worked in the opposite direction. Many peripheral centers were founded by Inka rulers and the state government behind them, and thus owed their existence to the core and depended on continuous exchange. The origin site on the Island of the Sun and associated Copacabana is a well-documented example. Thus the acquisition of Maukallaqta/Pumaurqu grounded Pachakuti's claims and their material expressions in particular rock art to the origin site of the Inka state and of the Inka people, appropriated status-related intangible wealth for Cusco, and transferred part of it to those ayllus mentioned in the origin story. It is perhaps indicative of the high status of these ayllus that members of two of them, the Mascas and Tampus, appear to have lived near Ollantaytambo, which was one of Pachakuti's major private estates (see Bengtsson 1998:120–121).

Elements of Origin Narrative Turned into Imperial Stone Ideology

On a more subtle level, various aspects of state ideology and ruler legitimization may have been copied by Inka officials from the origin story or were reported as lived experiences by consultants of the chroniclers. In Juan de Betanzos's (1996:12–16 [1557:part I, chap. III, IV]) version of the origin myth, Ayar Kachi plays a special role. Betanzos informs us that Ayar Kachi was the first to emerge, together with his wife, Mama Waqo. After the four men and their wives had climbed out of the cave, all dressed in fine garments of wool woven with gold, they immediately went to the mountain *Guanacaure* (Wanakauri) overlooking the Cusco Valley. Here Betanzos adds an interesting feat performed by Ayar Kachi: as he reached the top of the hill, he took out his sling, placed a stone in it, and hurled it hard and far to a high hill where it struck with such force that it knocked down the hill and made a ravine in it. Then he shot three more stones, and with each one he created a ravine in the high hills. Watching Ayar Kachi's great strength, his three brothers became fearful that he would quickly overpower them, and they decided to get rid of him. The four men and their wives claimed they had to return to Pacariqtambo, since they had forgotten precious golden objects and fine garments inside the cave. They talked Ayar Kachi into entering the cave to retrieve these things, to which he agreed. As soon as his three brothers saw him disappear inside the cave, they sealed the entrance, leaving Ayar Kachi permanently trapped inside (see above).

Pedro de Cieza de Leon (1959:32–37 [1553:ii.VI, ii.VII]) also describes the great strength of Ayar Kachi; for example, after he was locked inside the cave, the earth shook so forcefully that many mountains collapsed (1959:32 [1553:ii.VI]). Further, Cieza de Leon's version of the origin story adds two other episodes supporting the line of reasoning that later rituals organized by the Inka state emulated acts first laid down in the creation stories. Episode one is that Ayar Kachi came back to his stunned brothers as a great bird with wings of colored feathers to give them instructions as to where and how they should found the capital Cusco; how Inka rulers should assume the fringe or crown of the empire, be dressed and attired and perform accession ceremonies; and how Inka youths should be initiated into the elite class (1959:33–37 [1553:ii.VII]). The second episode is that Ayar Kachi used *Huana-cauri* as a second name, and he and one of his brothers turned into figures of stone in human likeness on top of Huana-cauri (Wanakauri) hill. These could be the idols in human form Guaman Poma depicts on top of the conflated mountain toponym (Poma de Ayala 1987:73, 257 [1615:79, 264], 1956:59, 188).

Garcilaso de la Vega (1963:27–28[1609: B. I, Ch. 16]) relates multiple versions of the origin story, including one particularly interesting incident on the mountain Wanakauri. The four ancestral brothers and sisters rested on the summit of Wanakauri, and the golden staff sank easily into the ground. Therefore, Manqo Qhapaq told Mama Oqllu that the valley in sight was the place where their father, the Sun, wanted them to settle, and that they should call all the people in the region living a humble life—*como bestias*!—and teach and civilize them. Manqo Qhapaq went to the north and Mama Oqllu to the south, and all the people they met followed them to the site where they began to construct the new capital, Cusco, with communal labor. The people who had arrived with Manqo Qhapaq settled in Hanan (higher) Cusco, and the people who had come with Mama Oqllu settled in Hurin (lower) Cusco, establishing the moiety division which exists in traditional Andean towns to this day. This part of Garcilaso's account partly confirms Sarmiento de Gamboa (cited above), who names the ten ayllus grouped in Hanan–Hurin and the implied ranking system, one category of which is male-female. Garcilaso (1963:27[1609: B. I, Ch. 16]) mentions a sun temple on Wanakauri mountain which may correspond to an architectural complex that was recently restored by the Ministerio de Cultura.

The selected episodes from Inka origin mythohistory formulate three concepts which were acted out in Inka state ideology and ritual. First is the great importance of Ayar Kachi. Whereas, in most interpretations, Manqo Qhapaq is foregrounded as the leader of the Inka ancestors and as the founder of Cusco, Ayar Kachi and his sister emerged first from Tampu T'oqo, and Ayar Kachi was physically stronger than his brothers. In the episode on the mountain top of Wanakauri, Ayar Kachi defined the four quadrants of the world and the four-partite division of the Inka Empire by shooting four rocks with his sling. One might consider reading Ayar Kachi's acts as evidence that from the beginning the Inka state was not based upon absolute one-man rulership (here, Manqo Qhapaq) and that important responsibilities were shared by high officials (in this case, Ayar Kachi).

Second, the fundamental geographical and political divisions of the Inka Empire were laid down in the origin narratives; that is, the four-partite division into suyus by Ayar Kachi and the dual division into the Hanan and Hurin moieties by Manqo Qhapaq. The suyu division defined the identity of

Tawantinsuyu, the Inka Empire which claimed to be the land which incorporated the four quarters of the world. Franklin Pease (1994:38–39) reasons that Andean "creation" narratives should not be interpreted in the Judeo-Christian sense that God created the earth out of nothing but rather be viewed as acts of "ordering" a preexisting world which was in chaos. From his perspective, the four Ayar couples already foreshadow the four-partite division of Cusco and the four suyus upon their emergence.

Third, one may look at the origin mythohistory as a long narrative explaining the relationships between caves, mountains, and stone in the Inka construction of sacred political landscape and stone ideology (see also Christie 2008). The four brothers and sisters migrated underground through the underworld waters to emerge into the world of man through the cave of Tampu T'oqo. Therefore, caves become the vital portals between the two lower levels of the cosmos. From Tampu T'oqo, the ancestors climbed Wanakauri mountain, which functioned as the metaphorical access point to the world above. Here Ayar Kachi defined the four quadrants of the world, turning the summit of Wanakauri into a sacred center. Placing himself at this center, Ayar Kachi personified its vertical central axis as well as the living essence or *apu* (mountain deity) of Wanakauri. Later he turned to stone on top of Wanakauri (see Guaman Poma's Wanakauri idol), demonstrating the transformational qualities of stone as the material essence of mountains and the relationship of stone to human life. According to Maria Rostworowski (1999:14), transformed stones functioned as metaphorical landmarks for "the taking possession of a space." In a similar context, the Pumaurqu outcrop was carved.

I argue that these concepts were transferred to Cusco and built into the layout and design of the capital (Christie 2007a). The main plaza composed of Awkaypata and Kusipata took on the role of the sacred center, from which the four roads to the suyus departed. The usnu signaled the presence of stone and mountain and may even have been understood as a "miniature Wanakauri" by some Inka. When the ruler assumed his position on the usnu, he, like Ayar Kachi, personified the central axis and acted as the pivot of his empire, which sanctioned and oversaw the four-partite division. At the same time, one has to caution that this four-partite division was not altogether new. Rostworowski (1999:6–7) cites various Spanish writers who inform us that the pre-Inka village named Acamama existed on the future site of Cusco. Sarmiento de Gamboa (cited in Rostworowski 1999:7) states specifically that Acamama was comprised of four local districts. It appears that the four-partite division was redirected in Inka state ideology and political landscape to lay claims of control to the four quarters of the world. The paramount importance of Wanakauri in Inka stone ideology is evident by its copies in various parts of the empire, which are local mountains given the name Wanakauri. One such mountain is situated in the Culluctur Valley near the Inka settlement of Ingapirca in Ecuador (Heriberto Rojas 2006:36).

In another ritual, the Inka dramatized their ties with the other origin place on the Island of the Sun in Lake Titicaca. Beginning with the accession ceremonies of Inka Roq'a, they brought water from a spring in the bedrock of an island in Lake Titicaca to Cusco. Each new ruler was anointed with the Titicaca water as he took on the highest office (Joan de Santa Cruz Pachacuti Yamqui Salcamayhua cited in Sherbondy 1992:56–57). Such a ritual act implies that origin places could be symbolically displaced by moving water from the original source to a new site, just as ayllus were refounded by carrying water from the ayllus's previous water source into new territory, where it was poured into a spring on the new land. The new water source received the name of the old site of emergence, or *pac(q)arisca* (Albornoz cited in Sherbondy 1992:57). This conceptual moving of origin places was crucial in accounting for the fact that the sacred sites of emergence were not situated in the capital itself; it constituted probably the most significant way of tightening relations between the center and these so important but geographically peripheral places. Dean (2010:124–125) describes the process of replicating origin sites in a more far-reaching way: by carving into boulders and building walls as if they are coming out of outcrops, the Inka actualized and replicated their own emergence from cave openings. These Inka-style visual marks acted as new paqarina/paqarisca claiming land rights and ownership, especially of newly conquered territories.

A complete reversal in directions of power relations took place in the sixteenth century when the Callapina family from Pacariqtambo petitioned the Spanish for noble status by claiming descent from Manqo Qhapaq and thus deriving the authority of their request from their ties with Inka Cusco. In an impressive and very cogent study, Urton (1990) has documented how these individuals were successful in their petition. The main source Urton (1990:46–57) uses is the Callapina Document, the original 1569 version of which contains testimony by eight witnesses that Rodrigo Sutiq Callapina indeed was a direct patrilineal descendant of Manqo Qhapaq. His argument was so successful that his grandsons and great-grandsons copied it to support their own claims for noble status and thus exemption from taxes and *mit'a* labor in 1692 and 1718.

CONCLUSIONS

I have discussed a network of temporal relations which constructed the very dynamic political landscape between Cusco and the Inka origin places. First, there is the mythohistorical emergence event during which the four Ayar brothers and their sisters climbed out of Qhapaq T'oqo and subsequently migrated to the Cusco Valley to found the capital. Garcilaso and other chronicle writers connect Qhapaq T'oqo and Tampu T'oqo to the other origin sites at Lake Titicaca. In those versions, Wiraqocha created human beings, animals, and plants at the site of Tiwanaku. At his wish, the Sun and the Moon first rose from openings in the sacred rock outcrop on the Island of the Sun. The Sun, appearing as a resplendent personage with a human body, called forth Manqo Qhapaq and Mama Oqllu and sent them on a civilizing mission. They were ordered to travel to the northwest underground and reemerge at the cave of Pacariqtambo and to educate and elevate the standard of living of the people they encountered (Salles-Reese 1997:53–88, 93–108; Urton 1999). In this manner, origin narratives linked Inka Cusco not only with the emergence of the Inka dynasty, but with the creation of the world. The Inka explained themselves as special beings singled out and chosen by Wiraqocha and the Sun to populate the earth with humans—defined as the Inka—who would live more wisely and had greater knowledge than their predecessors. I suggest that this ingeniously constructed identity had its material reflections in the rock art (Pumaurqu!) and in Inka architectural stonework, both of which show visual parallels with Tiwanaku. These connections validated the Inka ruler and legitimized his campaigns to conquer the four quarters of the world, Tawantinsuyu. I argue that during the process of state formation, the oral traditions of origins were strategically spread among the local populations and among non-Inka societies to provide an ideological basis for the Inka right to rule. Perhaps the most ingenious aspect of this strategy was that Inka origin narratives were internally consistent with pan-Andean explanations of creation. While boldly weaving in the Inka paqarina, the stories would have been understood by all Andean peoples.

It has to be stated very clearly, however, that the ethnohistoric and ethnographic sources outlining the creation narratives find themselves at odds with recent archaeological evidence compiled by means of the painstaking surveys conducted by Bauer and colleagues (Bauer 1992, 2004; Bauer and Covey 2002; see the introduction). These archaeological data reconstruct a scenario in which the Cusco Basin was much more densely populated during the Late Intermediate Period than had previously been thought (Bauer 2004:76–78). Migrations of major groups have not been noted, and it appears more likely that the ancestral Inka were one local group who expanded their influence and rose to power. Therefore, the landscape constructed by archaeology is very different from that of the simple and uncivilized savages populating the Cusco Valley as described by Garcilaso. The chronicle writers collected and edited information from native consultants who shared their perceptions and imaginations of the past, whereas archaeologists reconstruct history applying

Western empirical methodologies, resulting in differing and competing versions of political landscapes.

On the second level, once the Inka dynasty had solidified its position in the Valley of Cusco, Inka rulers apparently felt the necessity to reestablish and formalize ties with their origin site. They built Pacariqtambo/Maukallaqta on top of an earlier settlement as a ceremonial center serving the actual place of emergence or Tampu T'oqo identified with Pumaurqu. The central court of the northern sector of Maukallaqta may have functioned as a temple complex associated with Manko Qhapaq, and Pachakuti was likely the ruler who had Pumaurqu remodeled into a state wak'a. One might say that in return for bestowing ideological legitimacy on the Inka dynasty ruling in Cusco, Pacariqtambo/Maukallaqta/Pumaurqu received material elaboration. Clearly, Inka rulers gave form to their place of origin at Pacariqtambo/Maukallaqta and Tampu T'oqo/Pumaurqu and assigned to these sites crucial roles in the cosmological justification of state government.

The third level of interaction leads into the Colonial Period. The Spanish established the new town of Pacariqtambo in 1571, and Urton (1990) has documented diligently how members of the Callapina family from this community petitioned the Spanish authorities in Cusco for noble status by tracing their descent line to Manqo Qhapaq. This represents a case in which individuals from what in the Colonial Period certainly was a peripheral center claimed high status and the privileges associated with it by citing their close ties with the Inka core during precontact times.

The fourth level is the willful move of the place of emergence from Pumaurqu to a small cave located to the northeast of the new town of Pacariqtambo. Obviously, this translocation had nothing to do with any central authority in Cusco and must have been a decision by the people of Pacariqtambo. I would interpret this fourth level not as another interaction with Cusco, but as an act of independence undertaken by the local residents. It constitutes an example of how origin places can be re-created and redefined by local people in more recent history, and it is part of the process of identity construction which is clearly rooted in an Inka past, yet at the same time partakes of the contemporary world.

Chapters 2, 3, and 4 lay out and discuss a multifaceted network of relations between Cusco, the capital, and the origin places at Lake Titicaca and Pumaurqu/Maukallaqta, which articulated the political landscape Inka rulers constructed. Paralleling very likely the process of redesigning Cusco, Pachakuti and his son Thupa Inka appropriated and incorporated the two origin sites according to the logic of the acquisition model (see above) into this Inka landscape. These acts of acquisition were early moves in the strategy of empire-building and established military and ideological centers in the Hurin suyus in an expansive horizontal direction. In Cusco, Pachakuti authored the mythohistory and ideology tied to these spatial connections into stone ideology, which I define as a sophisticated political rationale to consolidate human subjects and the land under Inka rule, personified by him as the vertical axis and visually expressed in the carved rocks. As demonstrated, the origin places on the Island of the Sun and Pumaurqu should be seen as the "drafting board" for stone ideology. Chapters 5 and 6 will show how Pachakuti and Thupa Inka articulated and spread stone ideology north into the Hanan suyus, and particularly on their royal estates. Machu Picchu and Chinchero have been selected as the primary case studies, and numerous other royal estates as well as administrative centers will be cited and materialize similar ideas.

NOTES

1. According to Sarmiento (1942:67 [1572:chap. 12]), the Quechua word for rainbow is *guanacauri*, which suggests that the mountain received its name later as a reference point in the origin story.

2. Sarmiento (1942:69–70 [1572:chap. 13]) says that this stone pillar stood near the Monastery of Santo Domingo, the Qorikancha of Cusco, while Urton (1990:38–39) locates the same pillar on the central plaza, Awkaypata. It seems reasonable to suggest that Ayar Awka turned stone pillar is one metaphorical layer of the Cusco usnu.

3. [S]ince he (Pachakuti) was curious to learn about the past and to memorialize his name, he went in person to the hill of Tambotoco or Pacaritambo, which is all the same thing, and he entered the cave from which they fully believe Mango Capac emerged together with the siblings who came with him to Cuzco for the first time, based on what has been said. And after he had seen everything thoroughly and given it some thought, he worshipped at this place with fiestas and sacrifices. He turned the window Capactoco into a doorway of gold and mandated that from here on, this site should be highly venerated and that all should comply with it. And to make this happen, he classified it as sanctuary and guaca, where they were supposed to ask oracles and to offer sacrifices (1942:107 [1572:chap.30], translation by the author).

Chapter Five

Machu Picchu

ROYAL ESTATES

The following chapters will discuss the role of sculpted rocks and outcrops on royal estates or country palaces of Inka rulers. At present, the identification of royal estates is still fragmentary, and our knowledge of what they looked like and what went on within their walls is in a state of flux. Sources of information are some of the chronicles, historical documents, and archaeological reports; and as new data become available, our understanding of royal estates will continue to change.

It was a primary duty of each Inka ruler to acquire property in order to construct his palace in Cusco and to commission country estates. These estates had to provide the economic support for his descent group. The Inka, like many Andean peoples, practiced a split inheritance system of succession. This means the oldest—or perhaps the most capable—son of the ruler succeeded his father while the other siblings inherited all of their father's properties and possessions. Their task was to maintain the deceased ruler's *panaqa* (descent group) and carry out its ritual obligations, fully supported by their inheritance. The next in line of rule, on the other hand, inherited a new title but started out with no material possessions. He had to accumulate his own wealth by whatever means. As a result of the split inheritance system, the fertile and desirable lands particularly in the Urubamba Valley were quickly taken by the early rulers, leaving later rulers with little to choose from. Pedro Pizarro reported of Waskhar, one of the last contenders for the imperial throne:

> Becoming angry one day with the dead, he said that he should order them buried and should take all their belongings from them. [He said that] the goods

> should belong to the living, rather than the dead, and that the dead had all the best property in the kingdom (cited in and translated by Susan Niles 1999:121).

It is important to emphasize that royal estates were privately owned by the monarch, and after his death, by his mummy and panaqa. Estates were kept up and its terraces cultivated by permanent resident servants to the Inka, called *yanakuna*. Other permanent residents were the *mamakuna*, women chosen to serve the Inka. They prepared food and drink, most importantly large quantities of *chicha* (Andean alcoholic drink made of fermented maize) for festivities, and they wove fine cloth. And there were workers with certain occupational specialties; for example, stone masons and those who built the agricultural terraces. All specialists were known as *kamayuq (Q.pl. kamayuqkuna)*. Royal estates thus functioned as private economically self-sufficient entities and were independent from the state apparatus. They contributed nothing to the state nor received anything from it (Rowe 1990:143; Murra 1980:38).

I think this independent political status of the country palaces is crucial to our understanding of visual culture in the context of estates. Thus their architecture was not so much designed to make a statement about government ideology but has to be first understood as the personal commissions of individual rulers. The same line of thought would apply to other visual media, such as the carved rocks. In addition, most country palaces were probably designed and constructed in one plan, since their patron was one specific emperor, and the property remained in the hands of this ruler's panaqa. When considering the carved rocks, this means that they should still exist in more or less original contexts and that they, as well, may have been the commissions of individual rulers. This is very different from the situation in Cusco where, according to Cobo, the wak'as on the zeq'e lines were connected with various Inka panaqas and ayllus. Because of the long occupational history, some rock shrines may have been carved long before the Inka, and links between wak'as and lineages may have shifted. I plan to show that Pachakuti Inka Yupanki in particular encouraged the integration of sculpted rocks into his estate architecture.

On the other hand, the fact that country palaces were personal commissions presents great challenges to scholars trying to understand how such palaces functioned. They display great diversity and among the ones I am familiar with, not two are alike. Indeed, even the palace complexes of the same ruler are different; for example, Pisaq, Ollantaytambo, and Machu Picchu were all possessions of Pachakuti, but their layout and design features are individualized and foremost respond to the land. It is more productive to look at Inka palaces and country estates as organizing structures for the actions of the ruler and not as artistic creations per se. They became animated when the ruler and his court arrived. Since the ruler (*Sapa Inka*) personified the center of Tawantinsuyu, this center became movable and traveled through him in the empire. Wherever he stopped and stayed, notions of centrality as originally defined in Cusco were activated and visually manifested in physical forms. After the court left, the architectural walls and sculpted outcrops in Inka styles were the permanent—though empty—reminders of his presence (Houston and Cummins 2004:384).

Lands for estates could be obtained in multiple ways: Rowe (1990a:142–143; 1997:280–281) suggests that Pachakuti founded Pisaq, Ollantaytambo, and Machu Picchu to commemorate certain military victories. His son, Thupa Inka Yupanki, as well, claimed five strongholds he had captured as personal estates (for details, see Rowe 1997:281–282). At another time, Thupa Inka won some properties in a game of chance played with the Sun himself (Albornoz and Cobo cited by D'Altroy 2002:127). In a later game, he deliberately lost some lands to a son to help him build his own wealth. Wayna Qhapaq was said to have seized the properties of his uncle Wallpaya and his associates after they staged a coup against the then youthful monarch (Cabello Valboa cited by D'Altroy 2002:130). Waskhar expanded his personal holdings at the expense of the state when he claimed a region east of Huanuco. Perhaps the most radical appropriation of properties took place at the close of the dynastic war between Atawallpa and Waskhar. Atawallpa tried to eradicate Thupa Inka Yupanki's panaqa and Waskhar's panaqa and took all of their families' holdings (see D'Altroy 2002:127–130).

The numbers of royal estates are clearly high and they vary in size, design, and architectural build out. Figure I.2 contains an inset map of principal Inka royal estates in the Vilcanota/Urubamba drainage; D'Altroy (2002:128–129) published an accompanying table listing royal landholdings in the same area as well as the sources identifying them. Alan Covey (2003) has conducted survey work in the Vilcanota area and has identified numerous small estates which are mostly agricultural terraces with single structures. These typically do not contain special rocks (personal communication 2003). Therefore this discussion will be concerned with some of the major well-known and documented country palaces, which are all located in the Urubamba Valley and its immediate surroundings.

This valley was and is known for its climate, which is much warmer than that in Cusco, and for its agricultural potential to generate crops which thrive in lower altitudes, such as maize and coca. Other authors have emphasized symbolic aspects pertaining to the Urubamba River: it forms a direct connection between the highlands and the jungle lowlands, the two worlds of the Andean cosmos; its ancient name *Willka Mayu* translates as River of the Sun while the mountain from which it originates was known as *Willkan Uta* or

House of the Sun (Valcarcel 1964:18) (today the nearby pass where the road from Puno to Cusco crosses this mountain is called La Raya). In his fieldwork at Misminay, Urton (1981:37–38) has shown that indigenous people conceptualize the earth like a ball floating in the underworld sea. The Vilcanota/Urubamba River flows through the center of the earth from the southeast to the northwest. It collects water from the smaller tributaries and transports it back to the sea, from where it is taken up into the sky and into the Milky Way to return to the earth in the form of rain. The Milky Way becomes the celestial reflection of the Urubamba River.

There are documentary sources which describe what went on in and around an Inka country palace (see Nair's investigation of Chinchero 2003; Niles 1999). Most specific are three sixteenth-century documents which refer to the properties Wayna Qhapaq owned in Urubamba and Yucay (Villanueva Urteaga 1971:1–148). The earliest is the report of an inquiry conducted in 1550 by the Corregidor Alonso de Alvarado to assess which properties belonged to whom in the Yucay Valley. The emperor's properties seem to be dispersed amid lands belonging to his family members, the Sun, and the native population of the valley.

But Wayna Qhapaq managed to convert the whole valley into his "own house" (Villanueva Urteaga 1971:2–3): his estate was a country palace with gardens and woods and a recreational facility. There was a meadow with a pond named Madre Cocha in which he bathed. He had another pond called Paucarchaca for his recreation. Nearby were small plots in which aji and camotes were grown (Villanueva Urteaga 1971:37). In Urubamba, he owned a hunting reserve; he had a water pond where he bred fish and totora reed and in another place he bred pigeons; he also possessed lands where they cultivated aji, coca, cotton, and mani, which are all exotic plants in this altitude (Villanueva Urteaga 1971).[1] *Quispeguanca* where the main palace of Wayna Qhapaq stood (see Niles 1999) is described as "unos tambos" (Villanueva Urteaga 1971:38) (a *tampu* is used for Inka state lodging on the road system). To the right of the palace, there was a ravine named Chicon with a wooded area, meadows, and a river, which during Inka times was left in its natural state, unbuilt and uncultivated (Villanueva Urteaga 1971:38). To build and maintain all his possessions, Wayna Qhapaq ordered one thousand mitimaes/*mitayuq* indians from Chinchaysuyo and one thousand from Kollasuyu to come and live in the valley. Together with the local inhabitants of Yucay, this resulted in a population of over 3,500 (Villanueva Urteaga 1971:2, 117, 124, 126). After Wayna Qhapaq's death, his mummy was transferred from Quito to his estate at Yucay and attended to by forty to fifty indians classified as *Apuyanaconas* (which means principal servants of the monarch) (Villanueva Urteaga 1971:3–4, 108–109).

Susan Niles has further pointed out that the design and concept of Wayna Qhapaq's country estate have to be understood as manifestations of the Inka view of a perfect landscape, "one that juxtaposes economic productivity with recreational gardens, mundane activity with ritual, and balances agricultural fields with parks and irrigation canals with ponds" (Niles 1999:152). With regard to ideology, the organization of the estate reflects the Inka view of a perfect social order: the monarch and the Sun are the central protagonists attended to by fleets of servants, guards, workers, weavers and other craft specialists (Niles 1999:152–153).

In the third document published by Villanueva Urteaga (1971:82–148), there is one reference which could be significant for our understanding of royal estates. Witnesses Martin Yupanqui, native of the Yucay Valley, and Lucas Chico, mitimae in the Yucay Valley, both testified that Wayna Qhapaq's lands in the Yucay Valley were never part of any of the four suyus into which the Inka Empire was divided (Villanueva Urteaga 1971:125–126, 129–130). This reference probably has to be understood in the historical context of the document which is a census for tax purposes, and yanaconas were exempt from taxes (see Villanueva Urteaga 1971:94). The reasoning of the witnesses would be that they should be treated like the yanaconas on royal estates and not pay taxes because they did not belong to any administrative province. But it is also conceivable that in addition to the special administrative status, the location of royal estates in something like an imperial "nowhere land" had metaphorical connotations hinting at a position of centrality comparable to Cusco.[2]

The remainder of this chapter and the following will address whether certain shared design and organizational principles underlie the main estates of Inka sovereigns with a particular emphasis on the role(s) of their sculpted rocks.

MACHU PICCHU

> Suddenly I found myself confronted with the walls of ruined houses built of the finest quality of Inca stone work. It was hard to see them for they were partly covered with trees and moss, the growth of centuries, but in the dense shadow, hiding in bamboo thickets and tangled vines, appeared here and there walls of white granite ashlars carefully cut and exquisitely fitted together. We scrambled along through the dense undergrowth, climbing over terrace walls and in bamboo thickets where our guide found it easier going than I did. Suddenly without any warning, under a huge overhanging ledge the boy showed me a cave beautifully lined with the finest cut stone. It had evidently been a Royal Mausoleum. On top of this particular ledge was a semi-circular building whose outer wall, gently sloping and slightly curved, bore a striking resemblance to

> the famous Temple of the Sun in Cuzco. This might also be a Temple of the Sun. It followed the natural curvature of the rock and was keyed to it by one of the finest examples of masonry I had ever seen. Furthermore it was tied into another beautiful wall, made of very carefully matched ashlars of pure white granite, especially selected for its fine grain. Clearly, it was the work of a master artist. . . . Dimly, I began to realize that this wall and its adjoining semicircular temple over the cave were as fine as the finest stonework in the world (Bingham 1997:173–174) (Figure f.3).

This is an excerpt of Hiram Bingham's report of his discovery of Machu Picchu on July 24, 1911. Ever since, his style of writing combining fact with adventure and the masterful architecture in a spectacular landscape setting have shaped a picture of Machu Picchu in the public mind as a place where one can still find mysteries and spiritual power. The aura of mystery was heightened by the realization that none of the Spanish chroniclers took note of Machu Picchu, and most authors concluded that the Spanish had known nothing about Machu Picchu. It quickly became a pilgrimage site for backpackers, naturalists, biologists, writers, and New Age spiritualists and a destination for tourists worldwide, especially since it was chosen as one of the New Seven Wonders of the World in 2007. Today Machu Picchu features as an icon of Peruvian national identity celebrated widely at the one hundredth anniversary of Bingham's discovery in 2011. The *Groundwork* section in this and the following chapter will be in the format of a guided tour. The succeeding analytical sections will explore changing perspectives of experiencing and seeing Machu Picchu as well as the mechanisms by which these perspectives were constructed into Pachakuti's stone ideology and later into a modern capitalist ideology.

GROUNDWORK

Prior Research on Site and in Archives

Although Bingham has widely been credited with its (re-)discovery, Charles Wiener (1880:545) had heard about ruins at *Huaina-Picchu* and *Matcho-Picchu* while he was in Ollantaytambo in the late 1870s. He set out for another excursion to the east but apparently never found the site. Hiram Bingham, who was professor of Latin American history at Yale University as well as the director of Yale's Peruvian expeditions, was clearly the first Euroamerican to enter Machu Picchu and develop scholarly opinions. He also excavated and collected large quantities of ceramics, human remains, and associated grave goods, which were taken to Yale and housed in the Peabody Museum at Yale.[3] Bingham (1979) analyzed these materials and had drawings made of a large number of them. At the same time, it must be noted from the outset that Bingham was not an archaeologist by training, that he taught history and geography courses at Yale, and that his initial interests in Peru were commercial as he was exploring ways to open South American markets for U.S. products (Lopez Lenci 2007:104–105; see below).

Over the past years, scholars under the leadership of Richard Burger and Lucy Salazar have reexamined the materials collected by Bingham; and only now, many of these objects as well as the results of their academic studies are being made public through the Machu Picchu exhibition and associated publications (Burger and Salazar 2003, 2004; Salazar and Burger 2004). The culmination of these efforts was the opening of the Machu Picchu Museum in Cusco in 2011. Archaeological work at Machu Picchu resumed in the 1930s under Luis Valcarcel, who directed a program of cleaning and restoration for the occasion of the Fourth Centenary of the founding of Spanish Cusco celebration planned for 1934. Reconstruction, mapping, and excavations have continued since then, but records were not always kept properly and therefore there exists some doubt whether everything we see at Machu Picchu today reflects exactly the design and forms the Inka built (see MacLean 1986:35).

Archival research has offered important insights from another angle. In 1990, John Rowe discussed two sixteenth-century documents which mention a village by the name of Picchu in the Urubamba Valley. One is the report written by Diego Rodriguez de Figueroa about his mission to Titu Cusi in 1565. He said that they slept the night from the 6th to the 7th of May 1565 at a bridge in an ancient style over the river *vidcos* (Vitcos) on the way to *tambo* (Ollantaytambo) and to *sapamarca* (Sayacmarca?) and to *picho* (Machu Picchu) (Rowe 1990a:140). The second document was written in its original form in 1568. Over the next 250 years, multiple copies were made, some of which were found in the Convent of San Agustin in Cusco. This text lists properties and their owners in the Urubamba Valley between Ollantaytambo and Chaullay. It is written that this territory was conquered by Pachakuti Inka Yupanki and that he claimed for himself the main part of these lands. All of the named properties are located at the bottom of the valley where coca was grown. Rowe (1990a:142) argues that if the agricultural land at the bottom of the valley belonged to Pachakuti, sites like Machu Picchu located at a higher altitude were likely part of the royal estate of the ruler. He (1990a:142–143) further suggests that Pachakuti founded his estates to commemorate military accomplishments. Pisaq and Ollantaytambo memorialized earlier conquests in the Vilcanota/Urubamba Valley, while the founding of Machu Picchu should be understood in connection with his campaign and taking of Vitcos. He reached Vitcos by the river route below Machu Picchu following the valley.

It is the consensus of scholars today that Machu Picchu formed part of the royal estate of Pachakuti Inka Yupanki.[4] Burger and Salazar's (2004:27) ceramic analysis confirms such an attribution since they did not identify pottery antecedent to the imperial Cusco style. Furthermore, Cusco masonry styles using precisely cut and fitted stone blocks in high-status buildings point to the reign of Pachakuti. Therefore it was a private commission and existed as an economically self-sufficient unit independent from the state apparatus. This is reason to argue that the sculpted rocks were also commissioned by and carved during the reign of Pachakuti. It is in this context that we will now look at their form, placement, and functions. I will re-present greater Machu Picchu as the petrification of a personal ideology between Pachakuti and the mountainous setting.

The Machu Picchu Carved Boulders

To begin with, we have to clarify that what we call Machu Picchu today was not the complete royal estate. Smaller sites, such as Chachabamba, Wina Wayna, Phuyupatamarca, and Sayacmarca, formed part of greater Machu Picchu, and they were at least partially connected to central Machu Picchu by Pachakuti's landholdings. Direct associations depend on to what extent we can identify Pachakuti's properties with certainty. Many of the small sites contain special carved rocks, and we will return to them later, but at this point we will begin in Machu Picchu proper.

Construction supposedly started in A.D. 1450, it was burned in A.D. 1562 and abandoned ten years later (see K. Wright et al. 2000:1; Salazar and burger 2004:27). The population of Machu Picchu is estimated at three hundred to five hundred permanent residents and up to one thousand people when the emperor was in residence (K. Wright et al. 2000:21; Salazar and Burger 2004:30). Analysis of the skeletal remains has shown that natives of the highlands as well as the coast were buried in the caves surrounding Machu Picchu (Verano 2003:115). This suggests that Pachakuti employed *mitmaqkuna* (these were people resettled by the state to provide their services) labor to construct and maintain his estate—the same practice Wayna Qhapaq used for his royal estate at Yucay and Urubamba, which is well documented there (Villanueva Urteaga 1971:2–3). Machu Picchu's population estimates are considerably lower than those for Wayna Qhapaq's properties (see above).

The chronology thus established paints a very homogenous picture of Machu Picchu, that it was built and used as one design. It supports the attribution of the architecture and the carved rocks to Pachakuti. Nevertheless, Bingham's initial observations (1979:232–233) of two distinctly different masonry styles and techniques are valid. There are many more variations from the precisely cut and fitted Cusco style, but they do not necessarily indicate different occupational phases as Bingham interprets them; instead, they could simply represent lower-status buildings or have been constructed by non-Inka mitmaqkuna workers. A thorough analysis of Machu Picchu masonry styles is still a promising topic for future investigations.

The archaeological maps of Machu Picchu produced by Bingham (1979:fig. 219) and by Kenneth R. Wright in 2000 (K. Wright, et al. 2000; R. Wright, and Valencia Zegarra 2001) clearly show that Machu Picchu proper consisted of Upper and Lower Agricultural Sectors in the south and an Eastern and Western Urban Sector toward the north separated by the Main Plaza. The discussion will proceed from south to north.

On the cascading agricultural terraces, the resident farmers of Machu Picchu cultivated the crops required to meet the subsistence needs of the royal estate. In the upper sector, one terrace has been enlarged to form a plaza on which sits an intricately carved rock (figure 5.1). Valencia Zegarra (R. Wright and Valencia Zegarra 2001:8–1) have determined that this rock was brought to the site and thus deliberately set into its present location. It is granite and has been cut to give it a large and flat, slightly slanted surface. On the northeast side, the remaining upper rock mass has been sculpted into a stepped triangle overlooking this surface and orients the rock to the southwest or toward the kallanka (see below). Steps and a ring solidly projecting out on one side were sculpted into the lower portion of this rock (figure 5.2). Wright and Valencia Zegarra (2001:10) further point out that most of the scattered rocks on the plaza with this so-called Ceremonial Rock are large river pebbles brought up from the Urubamba River. The plaza is delineated by walls to the west and southeast.

On the southwest side of the plaza, leaning against the next higher terrace, stands a small structure with an open façade. A kallanka with eight door openings is situated on a still-higher terrace directly overlooking this plaza. I will show in chapter 6 that kallankas were closely associated with plazas and special, oftentimes sculpted rocks on many estates. In general, this Machu Picchu plaza and Ceremonial Rock probably hosted a variety of public events. The multiple relations such events forged between outsiders and residents inside the estate as well as between Machu Picchu and the surrounding mountainscape and other settlements connected by roads will be discussed in the *Inka Landscape and Carved Rocks* sections below.

From this outer plaza with the Ceremonial Rock and kallanka, the Inka trail leads to the Main gate and enters the western sector of Machu Picchu which was likely considered Hanan.[5] The central and most extraordinary example conveying the messages of Inka stone ideology is found at the Temple of the Sun and its associated features. As referenced above, this was the first build-

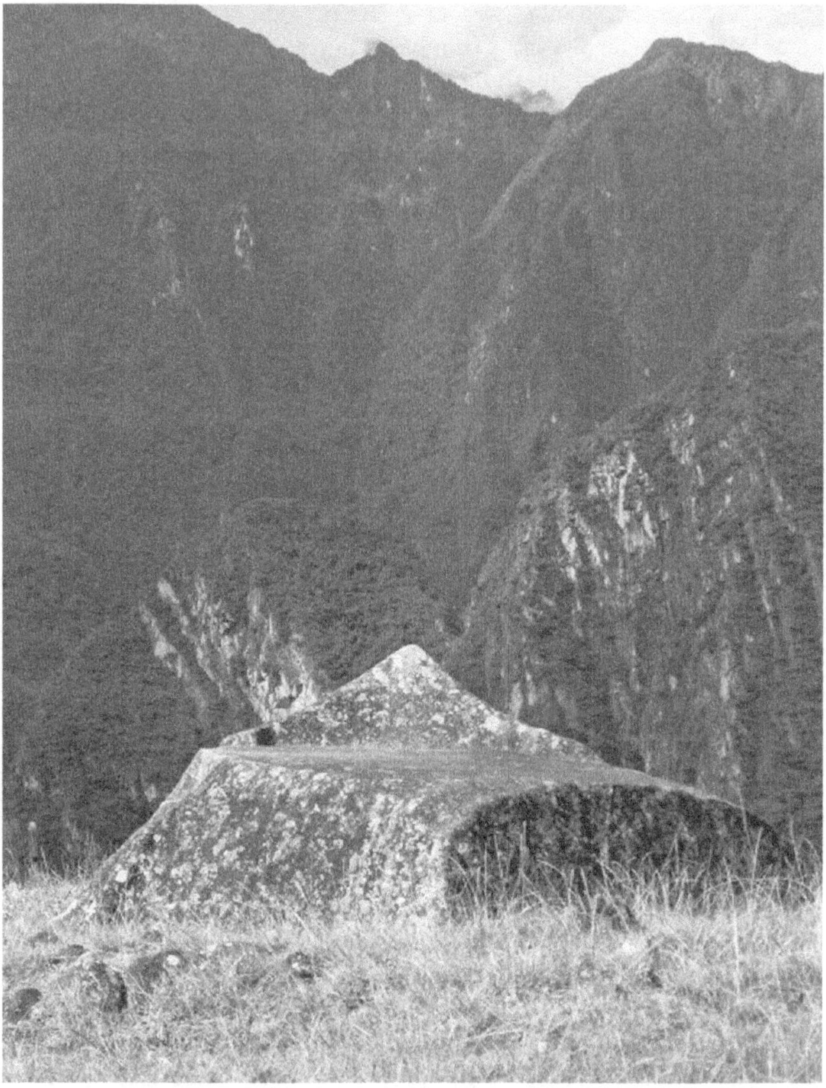

Figure 5.1. Machu Picchu, Ceremonial carved rock below Kallanka. Photograph by Brian Garrett.

ing group Hiram Bingham stumbled upon during his adventurous discovery and he already suspected that the building with the curved wall was a Sun Temple, which has been confirmed.[6] This Temple of the Sun is strategically located in Hanan Machu Picchu: to the south, it connects with what is be-

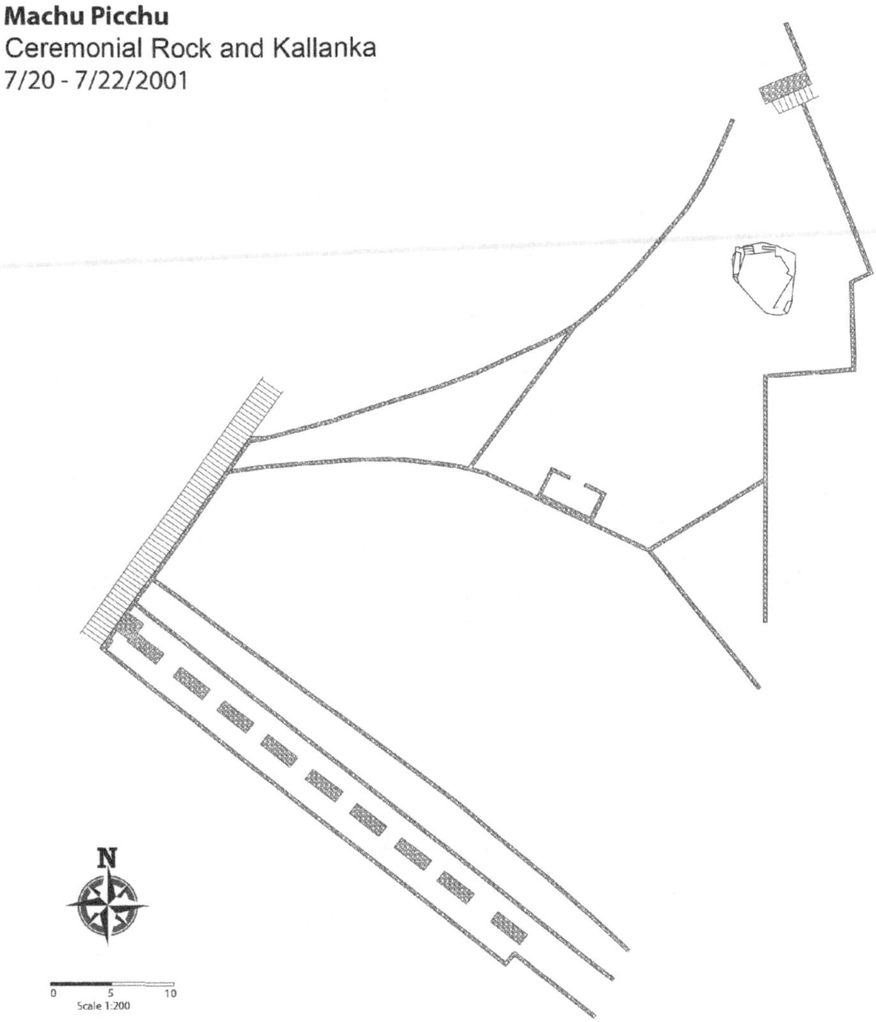

Figure 5.2. Machu Picchu, Drawing of Ceremonial Rock, Plaza, and Kallanka. Drawing by the author and Brian Garrett.

lieved to be a Priests' House; to the north, the temple complex is bordered by a principal stairway followed by the Royal Residence. Therefore, by location alone, this temple marks one of the most foregrounded places at the entire site.

The Temple of the Sun is part architecture, part sculpture, part bedrock, and part natural cave formation. A curved wall of tightly fitted stone blocks

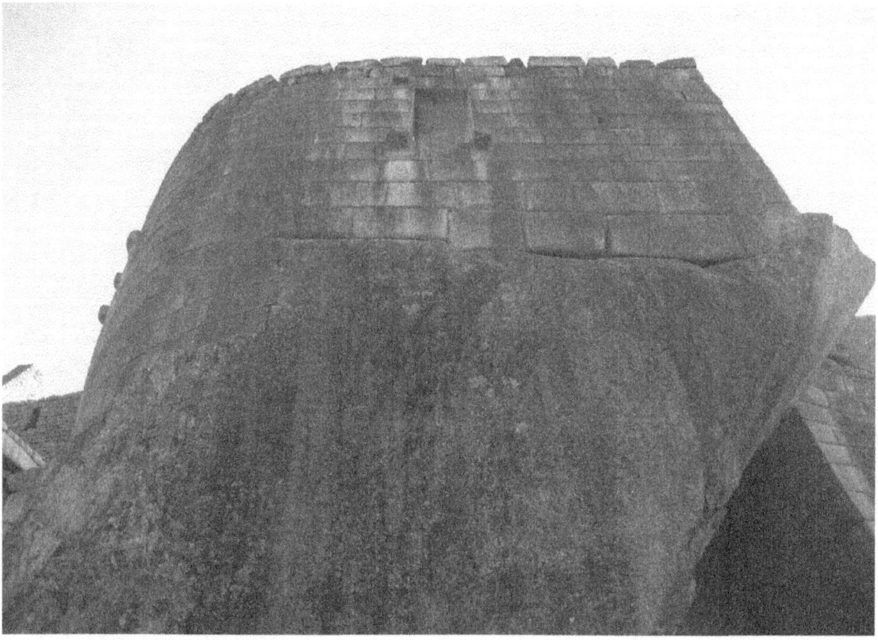

Figure 5.3. Machu Picchu, Temple of the Sun, curving wall and cave with stonework. Photograph by Brian Garrett.

was built on and around a rock outcrop, almost enclosing the top surface but not quite (figure 5.3). This wall leans slightly inward, and the stone layers diminish in size as they go up. It is considered one of the most illustrative examples of Inka stone masonry exhibiting the imperial style of Cusco which is typically associated with commissions by Pachakuti Inka Yupanki.

The interior wall is lined by niches and has three windows, one opening to the north, one to the east, and one pointing southeast. The top of the outcrop has been modified but is not exactly sculpted in an aesthetic sense.[7] The east window admits the light of the rising sun to illuminate this carved rock. It has been observed that the straight ledge cut into the rock surface bisects the sunlight passing through the eastern window at sunrise on June 21, which is winter solstice at Machu Picchu (Wright and Valencia Zegarra 2001:21–22). Or, according to the observations conducted by Dearborn and White (1982:250–252): the shadow of a plumb bob hung in the east window will touch the carved straight ledge beginning in May. As the sunrise point moves north, the angle between the plumb-bob shadow and the ledge decreases daily until they become parallel on the day of the solstice. Johan Reinhard (1991:55) has added that the position where the Pleiades rise is close to that of the sun at the June solstice, and that the Pleiades were and still are linked to crop fertility and the forecasting of weather among Andean peoples. Therefore the Temple of the Sun is one demonstrated example in which a (crudely) carved rock was used as an astronomical observatory in conjunction with architecture. Thus the element of sunlight and the techniques of sculpture and architecture, a natural outcrop versus stone worked into geometric building blocks, combine and are unified to construct a facet of the perfected and ordered Inka landscape.

Even more disparate elements are being brought together at the lower level of the Temple of the Sun. The outcrop has a cave formation in a cleft at its base which Bingham quickly identified as a Royal Mausoleum, but human remains were never found.[8] Instead, this cave displays a diversity of stonework: there is a beautifully sculpted stepped white stone at the entrance and a low stone on the ground carved, perhaps, to resemble the Intiwatana (see below); longer steps lead deeper into the cave; walls of exactly fitted stone masonry with tall niches and pegs line the natural contours of the cave; other stone masonry of the same quality is fitted between blocks of bedrock, while the outer bedrock surface remains unmodified. Most of this masonry is structurally unnecessary and makes an aesthetic and metaphoric statement. Wright and Valencia Zegarra (2001:29, 31) observe a stunning shadow play on the sculpted steps when looking out from the inside of the cave.

To the north of the Temple of the Sun stands a *wayrona*, which is a three-sided structure remaining open on one long side, which in this example faces east. It has a reconstructed thatched shed roof. The back wall of the wayrona is lined by a bench from which any occupant enjoys a spectacular view east beyond the Urubamba Valley to the Putukusi and San Gabriel peaks and the lower mountains of the Veronica Range (see book cover). Most certainly, this structure played a pivotal role in the ceremonies conducted around the Temple of the Sun.

The wayrona further held a strategic position in the fountain system as individual fountains circle around it. This system contains sixteen individual fountains which were fed by the main canal that originated at a spring in the flanks of Machu Picchu mountain. The first fountain sits at the end of the canal and to the west of the wayrona; Fountain 2 is located next to this building, and Fountain 3 in front of it. The design of the basin of Fountain 3 is highly unusual because it is formed by carved and smoothly polished rocks. It also marks the point which projects out farthest to the east in front of the wayrona structure. Fountain 4 is situated next to the principal stairway under which the water flows into Fountain 5. Fountains 6 through 16 are built in the more standard form: a small carved channel concentrates the jet of water so that it cascades into a shallow rectangular basin below. It has an outlet for water to flow to the next fountain following gravity flow. Each basin is surrounded by side walls with niches leaving an entrance opening. According to Wright and Valencia Zegarra (2001:23), large numbers of *aryballo* (large Inka carrying vessels with double handles and pointed bottom) potsherds found in and around the fountains attest to their use as domestic water supply. This practical function appears logical given the vicinity of the fountain system to the Royal Residence and expected water privileges of the elite.

At the same elevation as the complex of the Sun Temple and separated by the fountains and the stairway sits the Royal Residence. Bingham (1979:97) first called this building complex "King's Group because of the extremely solid character of the walls which enclose it, and also because it seems as though no one but a king could have insisted on having the lintels of the doorways to his houses made of great solid blocks of granite each weighing about three tons." There is no material evidence to support such an interpretation—indeed Bingham (1979:97) did not find any cultural deposits in this group. Fragments of numerous pots came from the stairway outside. Nevertheless, it is commonly understood as the private house of Pachakuti and his family (see Salazar and Burger 2004:343–347), because it is the neighbor of the Temple of the Sun and both groups share neatly fitted Cusco-style masonry which makes them visually stand out in the Hanan sector. The Royal Residence or Conjunto 3 is a small kancha with one peculiar carved feature in the courtyard. The surface of bedrock projects out of the floor, and adjacent to it is a small rounded depression sculpted into stone with a tongue-like carving attached to it. This carving is unique, and its meaning remains entirely mysterious. If any comparison can be postulated, it may be linked to the mortars in Conjunto 16 (see below).

West of the Royal Residence extends the quarry, and further northwest follows the Sacred Plaza. This plaza is surrounded by major buildings on three sides while the fourth, southwest side remains open and overlooks the precipitous slope to the Urubamba River. On the eastern side sits the Temple of the Three Windows, another wayrona-type building with three walls and an open façade facing the plaza. The eastern wall displays the famous three exceptionally large windows which look down upon the Main Plaza and the Veronica mountain range beyond. A closer look at the wall construction reveals that originally five windows were open and that the outer ones were converted into niches.[9]

Regarding carved stones, two examples are of interest: a stepped stone was placed next to the upright support in the front line of the Temple of the Three Windows. Bingham found this carved rock in the same position it is in today as photographs from his expedition prove. It will be discussed below.

Near the western wall of the Principal Temple which closes the north side of the Sacred Plaza sits a polished and sharply angled cut stone. It has no clear parallels in Inka rock art.

The northernmost compound in the Hanan sector is Conjunto 5 with the Intiwatana stone. The platform of the Intiwatana is accessed by a monumental stairway. At the entrance to this platform on the east side, there are a series of "image stones" facing the Main Plaza. "Image stones" constitute a category of carved rocks that is specific to Machu Picchu.[10] I define them as rocks modified in situ to resemble mountains visible behind them; it would appear that the artisans sculpted them facing their models. The carving style of the image stones is very different from the characteristic vertical and horizontal cuts of the Cusco style. Image stones are irregular and organic; the stone material is subtracted in any way feasible so that the resulting forms outline certain mountain silhouettes in the distance. Often times the man-made alterations are barely noticeable, and image stones may pass for natural rocks when not seen in relation to the surrounding natural setting. This example in Conjunto 5 mirrors Mount Yanantin and Putukusi mountain and may include a crude canal between the sculpted peaks. The purpose of image stones was likely to establish ritual connections with the mountains and their gods, which has been pointed out by several authors (see Wright and Valencia Zegarra 2001:49–52; Reinhard 1991).

Two sets of stairs carved out of bedrock lead to the final level of the platform, which constitutes the highest point in the urban area and the location of the Intiwatana stone. The Intiwatana was sculpted from the rock that formed the peak of the natural pyramid formation of Conjunto 5. It consists of a base exhibiting horizontal planes and vertical cuts in the typical Cusco style, which is topped by a four-sided protuberance tapering slightly from the bottom to the top. Bingham (1979:52, 1997:200) named this sculpture Intiwatana, meaning "the place to which the sun was tied" in Quechua, borrowing the name from a similar example in Pisaq. Early discussions interpreted it as some kind of sundial which cast significant shadows that would somehow tell the seasons. Bingham (1979:52) explains that the Inka would have been "in terror each year, as the shadows lengthened" and the sun went farther and farther north or south, fearing that he would never return. He postulates that Inka priests may have tied "a mystical rope" to the Intiwatana post "to prevent the sun (from) going too far away and getting lost." More recent researchers have been unable to confirm such a literal function (see Wright and Valencia Zegarra 2001:53–55; Reinhard 1991:48–53; 2007:63–70). Reinhard primarily elaborates on the relations between the Intiwatana and the surrounding mountains. When viewed from the Intiwatana stone, significant snowcapped mountains align with the cardinal directions. He proposes a visual relation between the Intiwatana protuberance and Wayna Picchu mountain, and that the Inka may have registered the similar shadow formations the sun casts on both (Reinhard 1991:51–53; 2007:67).

David Dearborn, Bernard Bell, and I have investigated an alternative scenario in which the Inka would have observed the sun setting behind the San Miguel ridge from the Intiwatana stone. There is evidence that there were pillars on the San Miguel ridge to mark sunset points on significant days in the year. In our scenario, the Intiwatana sculpture would have functioned as a type of usnu or throne where Inka astronomer priests sat to observe the movements of the sun (Bell and Christie, ms).[11]

There are more spatial networks surrounding the Intiwatana. In June of 2003, workers of the Ministerio de Cultura told me that there are two other similarly sculpted stones outside the ceremonial center, totaling three Intiwatanas. The second Intiwatana is situated below the first one in the valley, near the Hydro Electrical Plant where the railroad tracks end. The site consists of several small carved rock outcrops on a slope associated with wall segments. The rock outcrops and projections may be part of a continuous bedrock surface which is partially concealed by soil. The principal outcrop is beautifully sculpted into precisely cut platforms of diminishing sizes and a rounded projection on top, assuming an irregular pyramidal form. Downhill follow broad carved steps and walls and uphill, there are two small but deeply cut seat-like carvings and a cave formation. Bingham (1979:105) reports that he found a burial cave beneath the principal carved outcrop. Therefore it seems likely that this second Intiwatana was linked with ancestor veneration, an association of carved rocks which was first proposed by Uhle (1910).

None of the carvings form a gnomon or vertical projection that would exactly replicate the upper Intiwatana at Machu Picchu center. In addition, the low-lying location of this site limits the visual horizon and makes solar observations highly impractical. Thus the name Intiwatana appears to have been derived solely from a superficial stylistic resemblance between the two carved rock complexes.

At the upper Intiwatana, a third type of carved rock sits near the western steps leading to the main platform. It is a flat stone with a V-shaped depression resulting in an "arrow" form that points south.[12] At the top of Wayna Picchu, there is another similar "arrow" stone pointing south. From the higher elevation of the Wayna Picchu arrow stone, the nevado Salcantay is visible to the south (Reinhard 2007:67 and fig. 2.20). Mount Salcantay was one of the revered mountains of the Inka, and it is likely that the two arrow stones were intended to point toward it. Thus they created a sight line and conceptual bridge linking the settlement of Machu Picchu with Salcantay personified as an *apu* (mountain god/spirit) as well as between the carved rocks and the mountain, both of which share the material essence of stone.

Wright and Valencia Zegarra (2001:71–74) note two additional arrow stones also pointing south in Conjunto 16 in the Eastern Urban Sector. In my opinion, arrow stones constitute another group of carvings peculiar and unique to Machu Picchu since I did not find arrow stones at other sites. They fit well within the overall context of Machu Picchu rock art, which established a landscape of multilevel relations emphasizing mediation and dialogue between the ordered settlement and the untamed mountains.

The examples so far discussed constitute the most important carved rocks in the Western Urban Sector or Hanan division of Machu Picchu. Like most traditional Andean settlements, Machu Picchu was probably divided into an upper Hanan and lower Hurin sector (Eastern Urban Sector) which are separated by the Main Plaza. The Hanan area is clearly foregrounded by the fine Cusco-style masonry which announces the most important building groups of the Sun Temple, the Royal Residence, and the Temple of the Three Windows and by sculpted rocks positioned in significant spatial networks. The Hurin area appears to some extent subordinate, but it also contains some very intriguing examples of rock art.

Most enigmatic is the Temple of the Condor in Conjunto 17. This conjunto is an intricate place with entrances from different directions (figure 5.4). At its center opens a small and irregular courtyard area with a flat roughly triangular rock exhibiting what may be carved eyes and a beak framed by two white semicircular stone sections suggesting a feather ruff characteristic of condors. At the end of the beak, there is a small hole. This stylized condor head has to be seen in connection with the rock formations behind it, which form two outstretched wings. The whole composition leaves the impression that the condor is swooping down to catch prey. A prominent rounded stone between the wings may be a second representation of the head, and thus an implied movement could be staged with the condor taking off again and appearing in a state of flight. Around and below the condor sculpture are many levels, stairs, and caves forming an irregular maze. Most outstanding are the caves underneath the wings, which display a number of niches. Stylistically, the condor figure is very different from the geometric cuts since it builds semi-figurative relations between the sculpture and the bird and further suggests an entering of the untamed habitat of the condor into the domesticated center of Machu Picchu.

The condor figure is accompanied and surrounded by walls and other carved rocks. It shares the open plaza area with a boulder that contains a row of horizontal cuttings which were possibly bedding stones for a wall. Such a wall may have served to privatize the space with the condor since any visitor taking the north entrance would first have stood in front of the boulder with the wall and was forced to go around it to see the condor.

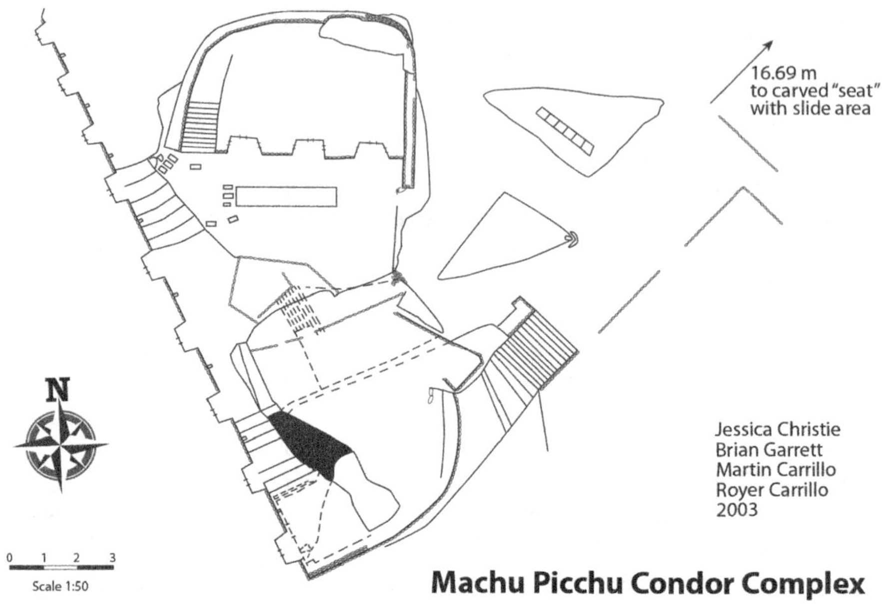

Figure 5.4. Machu Picchu, Temple of the Condor, Conjunto 17, plan. Drawing by the author.

Standing in the small plaza, a partial view opens of the sculptural and architectural construction which sits quite literally on top of the right wing of the condor. Inka builders cut two long horizontal seats or platforms into the bedrock and positioned a wall with three tall trapezoidal niches on top of the upper one. These niches display barholds or cordholders on each side and smaller niches in the rear wall. Bingham (1997:197) thought that the three niches received mummy bundles and that ropes were tied from barhold to barhold to hold them in place. Although there is no concrete evidence to confirm such a specific interpretation, the upper area likely functioned as some kind of shrine where sacred objects could be presented and displayed. Again we must note its secludedness, controlled access, and limited visibility, defining it as an elite space par excellence.

The western border of Conjunto 17 is a long wall with large trapezoidal niches. The eastern limit is less neatly defined; most conspicuous is a slanting natural rock with a platform and possible seat cut into its upper end.[13] Sitting on this upper flat plane, the view is oriented toward Hanan Machu Picchu, the Sun Temple, and the Royal Residence; perhaps a purpose of this carved rock was to provide a seat from which visual connectivity between the most powerfully charged sanctuaries of Hanan (Temple of the Sun) and

Hurin Machu Picchu (Temple of the Condor) could be established. A physical connection between these building groups existed in the fountain system: Fountain 16, the last and lowest of the series of fountains that begins above the Sun Temple, was a private fountain of Conjunto 17 (Wright and Valencia Zegarra 2001:89–90, 105–107).

The remaining part of the Eastern Urban Sector consists primarily of residential constructions in the form of kanchas, some of which follow a more regularized design than others. Interspersed are a number of special and/or carved rocks. Conjunto 12 is the smallest conjunto at Machu Picchu and situated so far east that the ground begins to slope. This location is responsible for the lack of a symmetrical kancha layout. Its dominant feature is a beautifully carved rock shrine connected with walls, rooms, and small caves on multiple levels. The carvings on the upper level consist of steps and a series of three platforms of diminishing sizes, carefully calculated and cut. At the outer edge, there are three small rounded protrusions which are commonly interpreted as image stones, likely referencing the conspicuous rounded Putukusi peak situated immediately behind them from a frontal viewpoint. This rock shrine overlooks and almost projects out over the Urubamba Valley, and thus the physical and conceptual connections with the landscape and the mountain on the other side of the narrow valley canyon are strongly noticeable.

Below the carvings, one encounters a maze of caves, small rooms, platforms, and walls created by natural overhangs and openings into the rock and by man-made constructions. Given the small space it occupies and its crammed design, there could not have been much practical use of Conjunto 12. The evidence points more to a ritualistic context in which people interacted with caves, stone, and mountains. This might be reinforced by the large number of niches in the walls, which probably stored objects needed in ceremonies. Some artifacts of such nature were uncovered in 1988, one of which was a small bronze figure with male sexual organs (Wright and Valencia Zegarra 2001:79–80).

A number of interesting carved rocks are integrated into Conjuntos 14, 15, and 16. Conjunto 16 is often referred to as Group of the Mortars because two mortars are conspicuously visible on the floor of building 1. They form deep depressions in two circular stones. Bingham found another stone close by, which could have served as a rocker, and he thought that they were grinding stones (Wright and Valencia Zegarra 2001:70–72). Another opinion is that the Inka filled the mortars with water and watched reflections of the sky, which is strengthened by the consensus that building 1 did not have a roof (see Wright and Valencia Zegarra 2001:71). It has been an Andean practice to observe reflections of the sky in small water pools as explained by Porfirio Huanca Sanchez at Copacati (see chapter 3). Further, the two mortars establish an interesting parallel with a similar carving in the courtyard of the Royal Residence (see above). A fourth but very small example can be seen in Conjunto 7, building 1.

Conjuntos 14, 15, and 16 are of special interest for this study because natural and carved stones have been remarkably blended with the architecture. Behind building 3, there is a shrine marked building 7 on Wright's map, which is a smaller version of the shrine on the wing of the Condor Temple. It consists of two long sculpted platforms surrounded by walls on three sides. Above the upper platform, there are three niches much smaller than those in the Temple of the Condor. The style of masonry changes within this small complex: the lower left wall displays carefully fitted Cusco-style masonry while the rest of the walls are of crude stonework. Wright and Valencia Zegarra (2001:71–74) call this structure a "stagelike shrine," yet no artifacts were directly associated with it.

The top of Conjunto 15 is formed by a large rock carved into several horizontal planes and partially surrounded by walls. Below this complex are caverns to which a hole at the foot of the carved rock connects. This is another setting in which rock art and caves repeat the theme of mediation, relation, and connection between the above and the below. Cut stone blocks and bedrock juxtapose the degrees to which the Inka altered stone material and thus entered into a dialogue with it. The geography of Machu Picchu greatly facilitated such interactions. Building areas were sloping and had to be leveled, and they were often interspersed with granite boulders of varying sizes. At times, such boulders were carved and shaped as substitutes for walls to integrate them more harmoniously into structures (Conjuntos 7 and 8). At other times, floors for higher structures were built on top of boulder groups, creating caves. Thus some of the stonework and caverns may simply be the result of practical considerations.

Conjunto 9 consists of three highly symmetrical units, each entered by a double-jamb doorway from the path overlooking the Main Plaza. The geometric regularity of these kanchas suggests that they were high-status households and by extension that noble families also resided in Hurin Machu Picchu.[14]

Conjuntos 7 and 8 are not symmetrical, but they attract attention because several walls have been substituted by natural rocks. This solution should be understood as another level of the interplay between the natural and man-built environments. This kind of solution is absent in the principal units in Conjunto 9, which articulate cultural order. Conjunto 7 is bordered on the north by a platform with huge jagged rocks understood as the remains of an unfinished temple (see Wright and Valencia Zegarra 2001:60–63).[15]

The northern end of the Eastern Urban or Hurin Sector is occupied by a building group oriented toward a large modified stone that is commonly

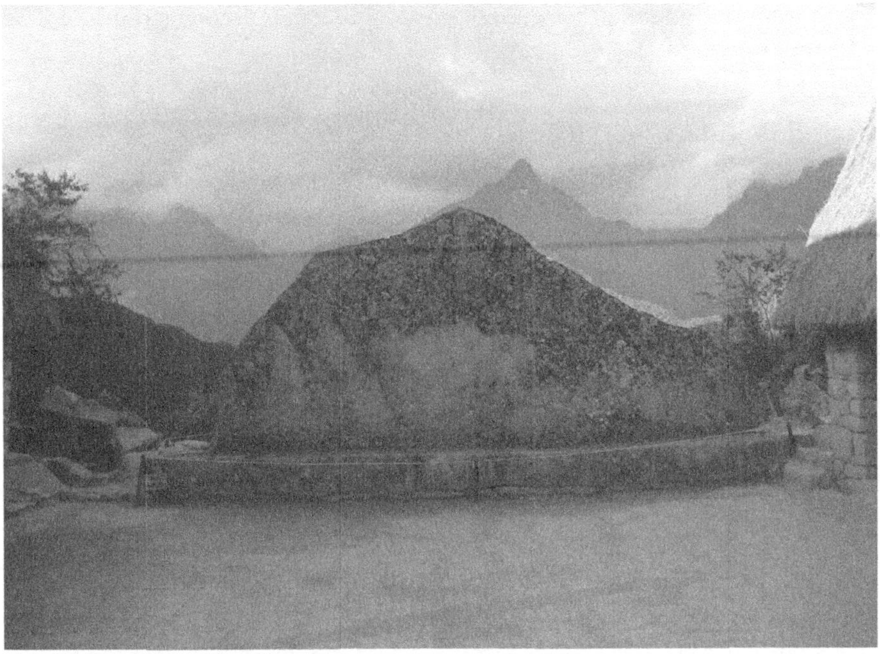

Figure 5.5. Machu Picchu, Eastern Urban Sector, Sacred Rock at the Northern End.

referred to as the "Sacred Rock." This Sacred Rock measures approximately 25 feet in length and sits on its own masonry platform (figure 5.5). It has been carved to resemble Mount Yanantin in the background. The design and spatial setting of this rock are quite dramatic and strategical. It is located next to the ancient and present entrance to Machu Picchu from Wayna Picchu and on a northeastern ledge overlooking the mountain slopes, which makes the visual connection with the surrounding mountain scene more immediate. To the west, the Sacred Rock faces a narrow rectangular plaza framed by two fully restored wayrona buildings. As the visitor enters this plaza from the southwest, his or her eyes are instantly drawn to the rock and the mountains behind it by the converging framing device of the wayronas. Therefore the Sacred Rock qualifies as the most conspicuous and best articulated image stone at Machu Picchu.

So far, I have introduced the sacred and carved rocks in the core area of Machu Picchu. Although it now appears that access to this core area was guarded and controlled, Machu Picchu was not isolated; it was connected to other sanctuaries and smaller centers by the Inka road system. We will now briefly turn our attention to outliers on Wayna Picchu mountain and on the Inka trail to Cusco.

Carved Boulders in Machu Picchu Outliers

Wayna Picchu is the sacred mountain that overlooks the site in the north, and the Inka built a narrow and steep trail to access it. In the summit area of Wayna Picchu, there are terraces and architectural constructions. At the very top, there is a jumble of huge granite boulders forming crevices. One of these rocks projects to the south and shows a rounded seat-like carving. The back of this seat is triangular, forming another arrow that points south to Mount Salcantay. This constitutes the fourth arrow stone at Machu Picchu and is well noticeable (see Conjunto 5/Intiwatana above and the Wayna Picchu–Intiwatana–Salcantay link).

An alternate return trail passes by the Temple of the Moon on the north side of Wayna Picchu. This complex was never discovered by Bingham but was first reported in 1936. Reinhard (1991:58; 2007:82) mentions that a full moon illuminates the cave interior and that it looks out at the equinox setting point of the sun on the San Miguel ridge, which coincides with the point on the horizon where the sun sets on the June solstice as seen from the Intiwatana (see above).

The site of the Temple of the Moon consists of an overhanging cave with carved rocks and high-quality Cusco-style stonework, an attached structure and a very tall double-jamb doorway, and farther below, more structures, caves, and carved rocks. The large main cave shelters several sculpted rocks cut into seats and horizontal planes. Five tall double-jamb niches framed by smaller niches constructed of precisely fitted Cusco-style masonry are built into the back walls of the overhang. Outside the cave to the west, there is the tall double-jamb doorway leading up to stairs, walls, and eventually to the long narrow structure attached to the west side of the cave (figure 5.6).

There are no direct historical data which would support any kind of interpretation with regard to the use and purpose of this cave except following Reinhard (above). The Temple of the Moon was special in the more general context of caves in Inka thought. Caves were often perceived as origin places where the first ancestors emerged (see the Inka creation stories in the Lake Titicaca and Pumaurqu chapters 3 and 4) and the bodies of dead ancestors were frequently returned to caves. Caves are still seen as entrances into the mountains where the mountain gods or apus reside. Thus the Temple of the Moon constituted a liminal place where the substance of Wayna Picchu Mountain could be entered and his apu approached.

Farther below lie more cut stone blocks, there is an upper building and a lower one constructed around a cave. The interior of this lower building

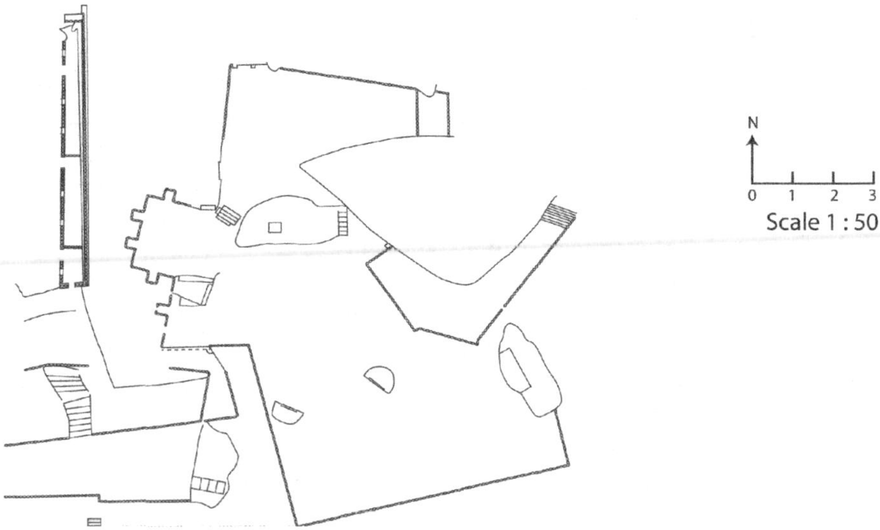

Figure 5.6. Machu Picchu, Temple of the Moon, Plan. Drawing by the author and Brian Garrett.

exhibits multiple niches; in each one I found a stone and sprouting maize kernels in 2003—a sign that people still leave their offerings to the apus. The trail does not end here but descends down to the Urubamba River. In 2003, I explored this trail further together with four workers from the Ministerio de Cultura. We located two more beautifully carved boulders with straight vertical and horizontal cuts in association with caves. In one cave, we noticed holes dug by looters who had possibly removed burials. The other cave displays stonework, yet not of the same quality as that in the Temple of the Moon.

We now move to the southern end of Machu Picchu. The Main Gate is located where the road from Cusco entered the core area. This road slowly rises up along the lower slopes of Machu Picchu mountain to the structures of Intipunku which served as a checkpoint. Along this stretch of the road lie two special rocks of interest in this discussion. The first when walking outbound away from Machu Picchu is a large rock measuring about fifty feet in height and slightly arching toward the trail. Low walls have been constructed in front of it, giving the impression that it sits on its own terrace. Bingham and his team identified this area as a cemetery. Underneath the rock overhang, they encountered five burials numbered 23–27. Burial No. 26 is one of the richest at Machu Picchu. It was the grave of a woman between forty-five and fifty-five years old who had been buried together with her dog, fine pieces of pottery, and bronze objects (see Wright and Valencia Zegarra 2001:118–119; Miller 2003:61–63). Miller suggests that she was the head of a lineage which maintained the camelid herds needed for ritual and the production of textiles. Thus although the huge rock itself remained unmodified, it created a cave-like overhang which was used for funerary purposes. The general symbolic context would have been similar to the one described for the caves and carved rocks on the lower north face of Wayna Picchu.

The second rock is carved into horizontal planes and seats and is associated with architecture (Wright and Valencia Zegarra 2001:120, fig. 149). A row of flat stones between the rock and the trail may have been the foundation of an encircling wall. If such a wall once existed, the sculpted boulder may not have been visible from the trail and might have been treated as a privileged and probably sacred object. The site appears to have been a checkpoint at a crossroad since a broad stairway leads into the forest to what is likely a connecting trail.

The Inka road to Cusco continues to the buildings of Intipunku, translated as Gate of the Sun, on top of the ridge from where travelers see Machu Picchu for the first and last time. Numerous *tampus* (way stations) and smaller sanctuaries—some containing special and carved rocks—are situated along this road. As indicated earlier, these smaller sites were most likely included in Pachakuti's royal estate. Margaret MacLean has analyzed eight sites (1986). Besides Machu Picchu, MacLean investigates Runcu Raccay, Sayaq Marka, Phuyu Pata Marka, Winay Wayna, Inti Pata, Choquesuysuy, and Chachabamba, all located within the area of the present Historic Sanctuary of Machu Picchu.[16] MacLean's argument that her eight sites are functionally related is partly based on the road system. The main transit routes to Antisuyu lead to the lower Urubamba, to Vitcos, and to Vilcabamba, bypassing Machu Picchu. Two main regional roads ran through the area with the outliers, and the final destination of either road was Machu Picchu (MacLean 1986:126–127, fig. 1).[17] Therefore, Machu Picchu and its satellite sites were not stopping places on the way to Antisuyu but existed in their own right.

MacLean calls attention to certain similarities between Machu Picchu and the smaller sites with regard to site selection on high outcrops and site construction which included elaborate water systems with fountains, special and carved rocks, caves, viewing platforms, house groups in approximate kancha arrangement, and terraces. I will briefly single out the sites and features relevant to the present study.

Winay Wayna, the first major tampu after leaving Machu Picchu, consists of an upper and lower house cluster connected by a steep stairway lined by ten beautifully carved fountains. Both house groups, and the stairs and fountains are surrounded by well-made terraces. The fountain system is the most outstanding feature at this site. Water could have been diverted to the residences in simple channels, but instead the Inka chose to sculpt and build ten geometric basins and line them up along the central axis of the settlement. It becomes a visual and auditory pleasure to watch and listen to the cascading water. MacLean (1986:30) has located at least four more fountains in the upper house cluster and two in the lower house cluster. The total number of fountains may be as high as nineteen, which would make Winay Wayna the Inka site with the most extensive fountain system known at this time (MacLean 1986:71). The water source is the Choquesuysuy River, which formed the ravine in which Winay Wayna is situated. Interestingly, there are no carved rocks. Thus the water moves alone in its cycle of bridging the above with the below.

One special boulder sits away from the water channels at the point where the Inka trail from Chachabamba enters the area of the lower house group. A small masonry platform abuts it, but the boulder itself appears unmodified (see chapter 1). It clearly served as a wak'a and may have marked the entry to Winay Wayna because of its location. The boulder faces the terrace system, and offerings for agricultural abundance may have been presented on its platform. It was surely also a *huanca chacrayoq* (Duviols 1979b) or a guardian of the fields and related to local ancestor cults apart from the state system.

Phuyu Pata Marka is located where the road from Machu Picchu bifurcates, and it stands out for its high viewing platform. This platform constitutes the flattened top of a bedrock outcrop and is framed on three sides by a wall constructed of neatly fitted high-quality geometric masonry blocks. The outcrop below appears to have been worked and shaped into a curving and undulating outline sided with walls in a coarser, more irregular masonry style using thick mortar. Some of the rounded projections support platforms and small structures. One might view the entire outcrop on and around which Phuyu Pata Marka was built as one macro-carved rock modified on varying levels.

Significant in the context of this study is the eastern part of the site comprising the fountain system, a plaza, and its central carved rock. The Choquesuysuy River originates in a spring above the first fountain. Five additional fountain basins follow below on the west side of the plaza whose center is foregrounded by a sculpted boulder. The carvings are vertical and horizontal cuts forming seats and platforms. Again, water, open space, and a carved rock unite to form a spatial setting encoded with symbolic layers. On one level, the Choquesuysuy River physically connects the upper altitudes of the mountain range, the realm of the apus, with the bottom of the Urubamba Valley below. Reinhard (1991:64) adds that this particular mountain ridge extends down from the nevado Salkantay, which would have heightened the sacred quality of the river. The six fountain basins slow its course and bring the process to the attention of the observer as the water cascades from basin to basin. On another conceptual level, this downhill movement is echoed in the stepped carvings of the boulder. I concur with MacLean (1986:125) that this eastern area of Phuyu Pata Marka was a ceremonial zone; of course, it also procured the water supply for its residents.

On the trail between Phuyu Pata Marka and Sayaq Marka, I noticed a so far unpublished, overgrown carved rock which clearly displays multiple cut seats. It might have functioned as a trail marker or simply as a minimal tampu which only provided a seat for rest. Yet from the position on the rock carvings, one enjoys a magnificent view of the three nevados—Salkantay, Umantay, and Pumasillo—and herein lies its broader performance context. Visual connectivity established multilevel relations between the sculptures, the traveler, and the apus of the wider geographic range, the material essence of which this rock shares (this carved rock is discussed and illustrated in Christie 2008).

Sayaq Marka exhibits several outcrops and projections of bedrock which were modified to adapt to the architecture but not sculpted. MacLean (1986:125) interprets Sayaq Marka as a multifamily settlement.

In contrast to the three outliers just discussed, Chachabamba is a satellite site built in the valley on a relatively flat area near the Urubamba River. Chachabamba was first investigated and documented by Paul Fejos (1944). The focus of the site is a large carved boulder to which the architecture seems to be oriented. It sits next to the river and measures approximately 12 meters across and 3 meters high. The carvings include steps, two large seats or platforms facing north and south respectively, and traces of possible canals. A structure was built around the carvings and continuing on top of the boulder's lower portion as if to isolate the sculptures from outside viewers. The architecture faces this boulder in a somewhat symmetrical fashion. The boulder constitutes the north side of the first kancha courtyard, followed by a second kancha courtyard to the south which is framed by three structures. Intriguing is the double-sided building which connects both kanchas: its north side opens into courtyard 1 and its southern side into courtyard 2. Another highly important feature are the fourteen fountains arranged in groups of three and four in the southwestern, southeastern, northwestern, and northeastern quadrants of the small site. They were fed by a stone-lined water channel that originates at a cistern near the slope to the south.

The calculated formal layout in accordance with the cardinal directions, the focal points of the sculpted boulder and the fountain groups, and the strategic location literally at the edge of the Urubamba River make Chachabamba a well-planned ceremonial shrine site. To my knowledge, there is no evidence that there was a larger settlement. I think Chachabamba performs the linkages of the carved rocks and water systems in a concentrated and especially beautiful way. The site, and most notably the carved boulder, mark the point where the water that comes down the mountainside joins the Urubamba River and the cycle between the above and below is made complete. This point of connection is highlighted and symbolically controlled by the symmetrical design of Chachabamba.

Choquesuysuy is also situated in the Urubamba Valley about two kilometers downstream from Chachabamba; however, it does not occupy a flat area. Instead, its houses appear to climb up the mountain slope. The name Choquesuysuy was given to this site by Bingham in 1915, and it appears to be connected with Machu Picchu, Phuyu Pata Marka, and Winay Wayna via the road system (see Fejos 1944:41–49). Choquesuysuy is considerably larger than Chachabamba, but it contains no carved rocks. Nevertheless, the settlement exhibits certain features characteristic of Machu Picchu's satellite sites and therefore is pertinent to this discussion: Choquesuysuy is divided into two halves creating a dual division by a stream which comes down from Phuyu Pata Marka in a steep gorge and forms a waterfall above the site; the western half has a fountain system with five basins. Architecture and bedrock are integrated and merge to various degrees.

Although the satellite sites share architectural features with Machu Picchu and the way water, bedrock, carved rocks, and caves are integrated into the site layout, there are also marked differences, most noticeably with regard to scale and selection of location. MacLean (1986:124–126) concludes that each of the small sites had a purpose within the system—which could range from ceremonial to purely agricultural—but that they were all functionally related to Machu Picchu.

INKA LANDSCAPE AND CARVED ROCKS—APPEARANCE, EXPERIENCE, AND PERCEPTION

As outlined in preceding chapters, the first section of the analysis will gather different perspectives of seeing carved rocks at Machu Picchu and register changes as well as continuities in the ways boulder sculptures have been experienced and perceived in relational networks between various agents and the rock art as theorized by T. A. Smith (2003). It will close with one example of altered appearances.

Fields of Vision

The *Groundwork* section was written from the viewpoint of an educated outsider who assumes the vantage point of historical knowledge. Here we will briefly view selected carved rock complexes from the human level of farmers and commoners during Inka times as well as from the perspectives of local traditionalists and New Age spiritualists today.

The architectural group of the plaza with the Ceremonial Rock and the kallanka is located outside the urban sectors in between the agricultural terraces. As Richard Burger (personal communication 2003) has pointed out, the Inka clearly tried to separate public events associated with this kallanka from the more private activities carried out in the urban sectors. Wright and Valencia Zegarra (2001:12) reinforce that this was a place where rituals and festivals for the farmers and workers living in the surrounding areas and not in Machu Picchu proper were held. They note the large number of potsherds from drinking vessels found in this area (Wright and Valencia Zegarra 2001:12). Feasting and drinking chicha provided by the ruler for his subjects to cement loyalties were part of such festivities.

Thus the plaza, the Ceremonial Rock, and the kallanka with the enclosing wall and entrance gate to the estate in the background composed the view of Machu Picchu for most commoners. In this setting, Pachakuti cited his elemental vocabulary in architecture and sculpture. The Ceremonial Rock may have served as a seat and/or space where offerings were presented, but more importantly it communicated the multirelational network between Pachakuti and the human and natural landscapes in which Machu Picchu played a key role. On one level, when sitting on this petrous seat, Pachakuti became a human wank'a claiming ownership of all land in his field of vision. Locals had ancestral field wank'as, un/modified boulders, staking ownership of territories, and Pachakuti appropriated and personified them in the name of the state (Dean 2010:108).

The Temple of the Moon and its nearby structures continue to be viewed as liminal places by spiritualists: each time I have visited the Temple of the Moon, I have found offerings of coca leaves, flowers, and at one time, a crystal stone. Selected stones and sprouting maize kernels had been placed in the niches of a lower building. While some of these gifts were clearly left by New Age tourists, others probably come from local traditionalists. The Museo Manuel Chavez Ballon displays pebbles of quartzite, sandstone, and granite found in and at the Temple of the Moon. The pebbles exhibit differing sizes and shapes, some of which resemble figures and animals. The museum interprets them as *illas* or personal offerings. Although the illas are not explicitly dated to the Late Horizon, the museum apparently does not consider them recent due to the fact that they are on display. When we explored the caves, mentioned above, situated further down the slopes of Wayna Picchu, my four companions offered coca leaves and smoked one cigarette to reconcile the spirits of the cave since we entered their place without having been invited.

Clearly, in the Andes, caves were, have been, and are perceived as potentially dangerous locales where spirits and ancestors reside. Even my Cusco assistant, Henry Conza Leon, is at times scared to enter a cave or at least makes all of us rub a little alcohol from the medical kit in our faces to ward off spirits—"just to be sure." The view and perception of caves as types of liminal rock formations thus constitute one cultural continuity. Of course, the protective measures individuals take today are personal choices and not in any way mandated by political authorities, but they still are the outgrowth of a long cultural heritage.

Cultural continuities can further be observed at Machu Picchu's contemporary rock wak'as. At Aguas Calientes, now called Municipalidad Machu Picchu, the tourist town below Machu Picchu, there sits a prominent boulder crowned by a crucifix next to the main commercial street which leads from the plaza up to the popular hot springs. The boulder can be accessed by concrete steps, and the cross is decorated with thick velvety cloth with gilded embroidery and regularly receives offerings of candles and flowers (figure 5.7). In recent years, the area between the shrine and the Urubamba River has

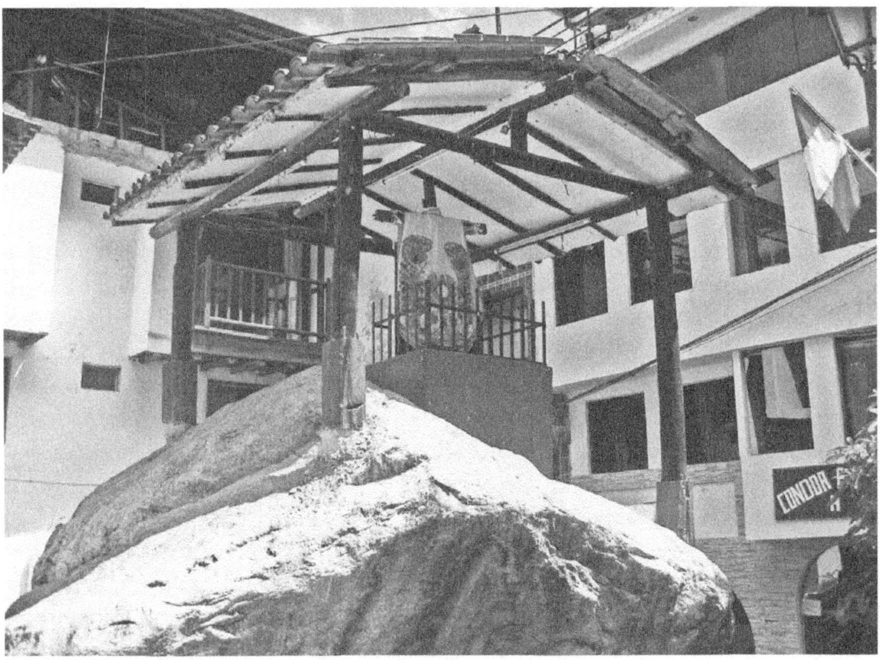

Figure 5.7. Aguas Calientes/Municipalidad Machu Picchu, Cross shrine on boulder. Photograph by Brian Garrett.

been remodeled into a small park which integrates other boulders forming an impressive fountain. In 2001, one storekeeper told me a story in which this cross protected the town from landslides, the major and most feared natural catastrophe due to its location, and that this would constitute the reason for the existence of the shrine. In 2004, there was a new crucifix on top of a rock on the road from Aguas Calientes to Machu Picchu. It is well decorated, accompanied by offerings, and protected by a roof. An emerging pattern seems to be that crucifixes and Christ are the Catholic icon brought into conjunction with rocks (see Christie 2008). Physical appearances may be altered, but boulders and rocks continue to be set in liminal contexts.

Experiencing and Perceiving a Multirelational Landscape

Here we will experience the element of water and certain stones and perceive their relations to the spectacular landscape setting. Although our physical senses are human biological facts and should not have differed from those of the Inka, our perception results from the individual interaction of our own bodies with an object and is culturally biased. Ethnography has to guide us to explore Inka views.

Water had a profound and symbolic ritual function in the Andes. It originates in the snow-covered peaks and comes from high altitudes—here Machu Picchu mountain has been identified as the source of the canal feeding the fountain system around the Temple of the Sun—and flows downhill to join the rivers, establishing connections between the up and down, between the upperworld and the underworld. Salomon (1991:15) has so cogently described as "hydraulic embrace" the circling of water from the ocean to the sky, as rainfall to the earth and into rivers and back to the ocean. Machu Picchu is positioned to experience such connections, and I present it as a case study in which the principles of phenomenology prove useful. This means that our lived bodily movement and perception in this midlevel mountain setting provide meaningful parallels with the way the Inka would have felt their human condition in the Andean cosmos. Machu Picchu is surrounded and enclosed by taller and partly snowcapped mountain ranges, and the narrow Urubamba Valley below can easily be conceptualized as the slithering body of a serpent as Andean people have done.

A similar context applies to the Temple of the Sun: water becomes the physical moving link that fuses the upper level of the temple to the lower level of the sculpted cave. Andean people have always seen and understood oppositions, such as above and below, as dual but integral parts of the whole experience of life, the world, and the cosmos. All elements in the Andean universe possess a life force, a vital essence which may strengthen or weaken due to various causes, resulting in a state of imbalance (Ziolkowski 1966:27–30). Balance has to be restored through ritual intervention, and water is a necessary component of this life force. The complex of the Temple of the Sun was likely perceived as a conceptual Inka cosmogram and balancing act in the ongoing dialogue and circulating exchange of vital energy between social beings which Western thinking divides into man and nature. It is a work of art as defined by the Inka that does not send direct figurative messages but presents its statements in a conceptual, somewhat abstract, and intriguingly aesthetic manner.

Machu Picchu offers an expanded formal vocabulary of carved rocks which can be visually experienced and culturally perceived. The stepped stone in front of the Temple of the Three Windows is one case of stepped motifs, which are extremely common iconographic elements throughout Andean history.[18] Since directly associated archaeological materials are usually lacking, context must be approached through form.

From a practical point of view, the steps are too small to assume a useful function. Therefore we should turn to symbolic connotations. The steps present a condensed profile view of the larger Inka bedrock carvings consisting of vertical and horizontal cuts. Like the latter, they mirror terraced mountain slopes. Van de Guchte (1996) and others (Ziolkowski 1966) have discussed that Inka sculptors created "doubles" of their emperors, which were not portraits in the Western sense in that they would have aimed at realistic likeness, but rather, they were stylized reproductions of the emperor usually executed in precious materials, such as gold, which absorbed his spirit and essence. This concept can be extended to features of the landscape and to the stepped stones. Thus the latter become miniature models of mountains with terraces. In a symbolic context, the stepped stones refer to the mountains and their gods (apus) but also to agricultural terraces; they bring together natural and man-made elements in one stone work forming an organic whole. The unity of the design suggests harmonic coexistence as if the man-made modifications which are necessary for human survival would not weaken the life force of the land. Thus the message of the stepped stones is similar to that of the cave underneath the Temple of the Sun, in which natural bedrock and stones of staged human modifications are placed together to compose the unity and variety of the spatial design. Therefore, I would argue that the stepped stones represent sophisticated and concentrated models of the multiple layers of symbolic mediation Inka rock art reflected.[19] Their role was not functional but purely ritual, which may be indicated by the large number of potsherds which were found outside and beneath the Three Windows. Many sherds came from ceremonial pots and *keros* (ceremonial beakers with flaring rim), which point to feasting and ritual on the Sacred Plaza rather than to domestic activities (Wright and Valencia Zegarra 2001:44).

Another type of carved rocks are the image stones discussed above in Conjunto 5/Intiwatana, Conjunto 12, and as the Sacred Rock. From the perspective of phenomenology, the relationship between human beings and real mountains in the distance versus miniature mountains on site changes in significant ways. The lived bodily experience of the actual peaks remains distant and abstract, while their miniature models were formed by human hands and can be touched as well as viewed at or below eye level. Such interaction would suggest that the ritual context of image stones dealt with controlling space and the vertical divisions of the cosmos. The material of stone is the substance of mountains, and we know from many examples and studies that the Inka as well as Quechua people today have considered replicas or pieces of the original as equivalent to the original (for example, see van de Guchte 1996 and Bolin 2006:116–117). From an expanded art historical point of view, the image stones of Conjunto 5 are not just simple portraits of the two mountain peaks—they also share their substance, essence, and spirit. Sacred liquids, such as chicha and water, may have been poured in the crude canal during ceremonies. If this was so, water would have flowed from the carved mountains just as the snow mountains (nevados) provide water in the form of rivers and streams.

A third formal category that has only been documented in Machu Picchu are the arrow stones. These carvings constitute primary examples of the relational space Pachakuti constructed between his estate and the surrounding mountainous setting as well as between the Inka and the highland landscape in general.

Applying the three dimensions formulated by Smith (2003:69–75), the built environment of Machu Picchu provided the spatial experience. To direct spatial perception, Pachakuti interspersed carved rocks—and here in particular image and arrow stones—at strategic places within his estate to set up evocative spaces in which humans interacted with the mountain landscape. This kind of interaction produced spatial imagination channeled by the placement and form of the—primarily geometric—sculpted rocks, which instilled Pachakuti's programmed stone ideology to which we now turn.

INKA LANDSCAPE AND CARVED ROCKS—ESSENCE, IMAGINATION, AND STONE IDEOLOGY

A primary argument of this study is that Pachakuti initiated the Inka style of rock carving at Machu Picchu and at the wak'as on the Cusco zeq'e lines, which were associated with his panaqa. He devised the formal elements into an ideological vocabulary which made statements about the power of the state and conveyed messages about his own discourse with the landscape. Thus Machu Picchu becomes an origin site of stone ideology and by means of selected examples, I will examine how such a language was authored and exported.

Authoring Stone Ideology

The cave underneath the Temple of the Sun may be understood as a sublime example of harmonizing stone in its varying stages of man-made modifications: there are the untouched bedrock surfaces; the sculpted and stepped rocks have been modified in situ whereas Inka builders removed the stone masonry blocks from the quarry, imposed upon them a geometric form, and reassembled them to form a wall. Stones in these three stages work together to create the space of the cave, which is at once natural as it is based on an opening in the outcrop and man-made as it is enclosed by walls. I argue that Inka artisans as commissioned by Pachakuti present to us a celebration of the reciprocity and dialogue between the Inka and the surrounding land as evidenced by the stony manifestations of the nuances and gradations of their timeless interplay. Rebecca Stone-Miller (2002:205) sees the workmanship of the cave space as a refined materialization of the Andean notion of essence. Reinhard (1991:59) and others show that in the Andes caves were and continue to be understood as entrances into the mountains where the mountain gods reside (see also Temple of the Moon above). Thus mountains and caves constitute different aspects of the same holistic entity. Again, the element that links them together is the material and essence of stone.

As we have heard in the origin narratives, stone is also related to the Sun and to ancestors (see chapter 3). The Sun deity acted as patron god of the Inka dynasty and as father figure of Inka rulers. I reason that Pachakuti called up memories of these relations in the Sun Temple above the cave. The entire settlement of Machu Picchu extends along the eastern slopes of the ridge and is thus oriented toward sunrise in a general sense. This may be partly planned and ideological, but at the same time, it has a sound practical reason: the western slopes are much steeper, resulting in narrow terraces which provide limited space for buildings.

All carvings, constructions, and bedrock at Wayna Picchu in particular emphasize ongoing interactions between the Inka and the land. It seems very impractical to build terraces high up the steep mountain slope; trails lead over bedrock surfaces carved into steps and at one point, such a sculpted stairway passes through a cave; boulders display carved depressions as bedding stones for walls which do not have any functional purpose whatsoever. Thus the man-made modifications of Wayna Picchu are not the result of practical planning but the outcome of other, perhaps ritual considerations.

In a purely speculative scenario, one may visualize that Inka builders and artisans would have climbed this peak to metaphorically converse with the mountains through the medium of stone, which is at once the material essence of the apus and the medium of the artists. Such "conversations" forged calculated ideological relations and memorialize processes of ordering which can still be experienced; they are ongoing and can be repeated in innumerable slight variations visualized in the degrees to which stone was altered and worked. Perhaps it was a dialogue over control of natural forces and resources. One particularly nice example is a finely carved seat which I found on the trail descending on the northern side of Wayna Picchu to the Temple of the Moon. The seat has its own platform which aligns with the trail and it looks out upon the mountain scene.

The Temple of the Moon and the lower two caves and carved rocks on the slope of Wayna Picchu should be understood in a related metaphorical context. In Andean myth, the high mountain peaks are usually personified as male while the earth, water sources, and valleys are female. Water washing down the snowy peaks in the form of glacial runoff, rainstorms, and irrigation canals unites the male and female elements in an act of what Frank Salomon calls "hydraulic sex" (1991:15). Similar ideas apply to Wayna Picchu: the summit with the jagged boulders and the arrow carving, which sets up long-distance relations, was male whereas the interior caves at the lower altitudes and closer to the Urubamba River were female. The main water source is the river, but during the rainy season, water also comes down the mountain slope. The female spaces were caves and womb-like, cut into or formed by the stone substance of the male mountain. This metaphor implies again that the male and female principles were not seen as separate and distinct entities but as part of the same whole, the mountain complex (see also the step-fret design as explained by Grieder 1982:134). In this sense, Wayna Picchu Mountain becomes a cosmogram which includes the Urubamba River that encircles it on three sides.

Further, the Inka often perceived relationships based upon resemblances of essence and spirit more so than visual similarities of form, which is a fundamental point to which we will return several times. Standing at the second Intiwatana, the viewer sees the top of the Intiwatana complex on site above the rim line of a towering mountain, and thus a link of visibility clearly exists. In a conceptual context, the two Intiwatanas may constitute a pair of model and replica, which can be explained through the semiotic model developed by Peirce as legisign and sinsign (see chapter 2 and Christie 2012) and possibly in its Andean setting as Hanan and Hurin. The Intiwatana at Machu Picchu center defines the type category or legisign, and the low-lying second Intiwatana is one of its replicas or sinsigns.

Certain essences of the Hanan–Hurin model may have been materialized in relationships between the Temple of the Sun, the Royal Residence, and Conjunto 17 or Condor Temple. Looking at the spatial layout, the three complexes are positioned along an approximate west-east axis formed by the fountain system and a major stairway. The temples of Hanan and the residential kanchas (courtyards framed by houses creating a four-partite design) of Hurin Machu Picchu lie further north and are clearly set apart by stairs and passageways. The Inka may have conceptualized the Temples of the Sun and of the Condor as two stony religious complexes that stood in a reciprocal relationship related to Hanan and Hurin. Formally, they use similar combinations and infinite subtle gradations of natural cave formations, bedrock, carved and sculpted rocks, and cut building blocks aligned in walls. On a metaphorical level, both groups refer to deified natural forces associated with the sky and upper world: the Sun and the Condor. At the same time, they link these sky deities with the underworld by means of caves. Furthermore, the Condor shrine in Conjunto 17 may allegorically project back to expanded versions of the origin narrative which mention that Ayar Uchu, Ayar Awka, and Ayar Kachi, all turned into birds (Steele and Allen 2004:59; Elorrieta Salazar and Elorrieta Salazar 1996:43). Another bird was the idol or brother statue (*wawqi* in Quechua) of Manqo Qhapaq, which empowered him. Thus the condor figure may imply remembered interactions with the Inka dynastic ancestors and associated territorial claims (chapters 3 and 4).

These two temple complexes constitute the most ingenious material and ideological expressions of Inka relations with the physical setting of Machu Picchu: mountain- and valleyscapes were continuously ordered and constructed through the numerous ways in which the Inka sought to access and control their animated and sentient universe. At its root, this dialogue was individual and may have begun as Pachakuti's personal discourse at Machu Picchu. But Inka stonework was quickly turned into the formal language of the state and in its practice as stone ideology defined and validated a cultivated and ordered Inka landscape in expanding state territories to which we now turn.

Exporting Stone Ideology

Pachakuti's rock art was not simply his localized discourse with the mountain setting but was turned into an ideology and political power tool by exporting it along roads to outlying settlements. Gary Ziegler (personal communication

2005) understands the plaza with the Ceremonial Rock as a staging area for travelers on the Inka road from and to Intipunku (the point—marked by buildings—on the Inka road to Cusco from which the traveler sees Machu Picchu for the first and last times). From the perspective of travelers, the carved boulder would have functioned as a roadside shrine where visitors left offerings. Carolyn Dean (2006b:102) has characterized this rock as an *apacheta*, which in a narrow sense is a heap of stones accumulated on a mountain pass by travelers who leave stone offerings and in a wider sense includes many carved rocks and boulders. Ziegler (personal communication 2005) observed that many of the rocks on the plaza floor were perhaps deliberately arranged and are of varying stone material such as andesite, limestone, and metamorphic rock carried in from other regions.[20] This makes the Ceremonial Rock a very literal apacheta in its narrow sense. Dean (2006b:95) thinks that apachetas including this finely sculpted Machu Picchu rock are representations of mountains as well as offering places. She and Ziegler have noted that the triangular carving of the rock closely mirrors Mount Yanantin visible in the background line of mountains.

Thus style and context bind together important binaries in Inka political and social praxis. The style of carving and most notably the large flat surface suggest that the Ceremonial Rock was used as a special seat and offering table. In both scenarios, the carved rock assumes the role of a mediator. When Pachakuti himself or a high member of his court took a seat upon the rock, they likely presided over ceremonies which in some form established relations between the people, his subjects, and the deities. When people placed offerings on the flat surface or next to the rock, such objects rested on a middle ground below the actual framing mountains, replicated by the triangular back part, and the Urubamba River deep below called into presence by the river rocks on the plaza. Form and context established a political communication between ruler and subjects; an ideologically manipulated spatial discourse between the specific physical place amid the terraces of Machu Picchu and the wide and abstract space materialized by the mountain ranges extending in all directions, as well as economic relationships between the agricultural terraces and the urban sector and between Machu Picchu and Cusco facilitated by the location of the Ceremonial Rock on the Inka road and its role as apacheta. We have seen in other examples, such as the image and arrow stones, that this role of mediation in the sense of setting up ideologically constructed spatial relations in the spirit of Adam Smith's (2003) model is a primary and most eloquently expressed function of the carved rocks at Machu Picchu. The boulder sculptures in the satellite sites carried such notions beyond the site core. It is significant that boulders carved in the geometric style are sparse in the Machu Picchu outliers, with the exception of Chachabamba. During the course of state expansion, stone ideology was exported by strategically placing geometric carved bedrock outcrops along roads and in selective governance centers to mark Inka presence and state concepts of a civilized Inka landscape (see chapters 3 and 7).

CONCLUSIONS

Based on the 1568 document discussed above, Machu Picchu is now understood as a royal estate of Pachakuti Inka Yupanki. The document states that Pachakuti owned a large portion of the bottom lands in the Urubamba Valley and thus valley satellite sites, Chachabamba and Choquesuysuy, clearly belonged to his estate as well.

This context has very important implications for our understanding of the carved rocks. As explained earlier, royal estates were private commissions by the monarch, they belonged to and were maintained by the ruler's panaqa after his death, and they had nothing to do with state government. Therefore, the architecture and carved rocks of Machu Picchu and its surrounding sites are not directly linked to state governance but express, first and foremost, the thought processes of Pachakuti Inka Yupanki. I reason that he formulated ideas and aesthetics at estates, such as Machu Picchu and its satellite sites, and then translated them into a politically expressive stone ideology. If we lend credit to Betanzos's (1996:69–73 [1557:part I, chap. XVI]) description of how Pachakuti rebuilt the city of Cusco, we have to assume that an estate, like Machu Picchu, could not have been constructed without his direct intervention and that what remains today still, by and large, reflects his design. Machu Picchu and its satellite sites are relatively pristine places to investigate carved rocks[21] and therefore have been chosen as case studies.

A second link between Machu Picchu, Cusco, and Pachakuti as the author of both may be found in the urban design of Cusco. The dual, fourpartite, and radial or zeq'e divisions of imperial Cusco are generally credited to Pachakuti and were partially repeated in many settlements throughout the empire. Machu Picchu and its outliers may well constitute another instance of a zeq'e system. The core area of Machu Picchu was virtually surrounded by shrine sites on numerous mountaintops and ridges, and the Ministerio de Cultura continues to find more. Reinhard (1991; 2007:18–53, 130–132) contextualizes Machu Picchu as a sacred center strategically placed within its mountainous setting below Salcantay and above the Urubamba River and aligned with celestial phenomena. Pachakuti may have envisioned sight lines radiating out from the center to these natural and man-made outliers. Of course, the ritual and social contexts would have differed from those of the Cusco zeq'e system because some of the outliers were difficult to access and all satellites would have been maintained by Pachakuti's yanakuna and not by various ayllus.

The most consistent and the most powerful message of Machu Picchu stonework is that of connecting the above with the below, the upper world of *Hanan Pacha* with the underworld of *Uku Pacha*. In a recent article, Catherine Allen (2002) reconstructs a local Andean cosmogram that is based on circulating water with the help of her consultants from Sonqo. I suggest her perceptive observations may also be applied to Machu Picchu and its setting: people in Sonqo (and in other communities) explain that water in lakes builds up and flows out in streams. These streams create local springs and return underground back to the lakes to start the cycle anew. Altitude does not matter in this process and—at least—below ground, streams can flow uphill. The cycling of water is accompanied by a cycling of life: human beings first emerged from subterranean water sources, and after death they travel back through the water to their places of origin. "Carved with rivers and irrigation canals, *Pachamama* (Mother Earth) herself is a kind of *paccha*, a vessel for the passage of water" (Allen 2002:197). As a living body, she "opens" in August to receive the first rains, and her productivity depends on replenishing the supply of internal water and properly circulating it (see Allen 2002:196–198). On a cosmic level in the sky, the Milky Way is likened to a celestial river and called *Mayu*, meaning river in Quechua. Mayu is geographically linked to the Urubamba/Vilcanota River from which it draws water. This water falls as rain and eventually flows back into the Urubamba River to renew the cycle (Urton 1981:60). I argue that this cosmogram of circulating water is visualized in the dramatic landscape setting of Machu Picchu and satellite sites, their fountain systems, and in their carved rocks. Streams, such as the Rio Choquesuysuy, naturally tumble down the steep mountain slopes, but Pachakuti's builders diverted and rerouted some of the water through manmade channels to complex fountain systems which slowed the circulation and forced it to the perception of the eye and ear as well as use of the residents. The predominant vertical and horizontal cuts of the boulders form steps and platforms which mirror terraces and when viewed in profile, resemble mountain slopes (see, for example, the stepped rock in the cave below the Temple of the Sun at Machu Picchu).

I have shown that the primary role of carved rocks was to provide places of *tinku* which, in a general sense, refers to any encounter of opposing yet complementary forces. Allen (2002:192–193, 198–200) discusses a variety of tinkus, and the term may be extended to the functions of carved rocks as well. Boulders were strategically sculpted where elements of the above and below, of the sky and the underworld, touched. The most breathtaking example is the Sun Temple at Machu Picchu in association with the cave below, as was described above. Water as the circulating connecting force flowed from fountain basin to basin immediately next to these structures. Rocks were often sculpted next to water sources: for example, the main boulder at Chachabamba. The cut horizontal planes and platforms served as offering tables where the watery veins of pachamama and the apus could be accessed in ritual. Ritual activity provided the strategies to approach opposing supernatural forces and hold them in a state of temporary equilibrium. Such equilibrium is never permanent but has to be continuously readdressed in relations of reciprocity between human and supernatural actors and forces which are not equal. The flowing water becomes the material image and occasionally the venue of this kind of ritual process.

I further suggest Machu Picchu may have functioned as a training workshop for craft specialists in which Pachakuti and his carvers experimented with innumerable relations between stone, sculpture, integrated outcrops, architecture, and the mountain setting to refine a formal language and sacred geography.[22] Pachakuti then translated this vocabulary of form and space into a state-directed stone ideology intended to mark a specifically Inka political landscape. When he commissioned sculpted rocks and Cusco-style architecture in outlying territories, broader tinkus were performed between the political center and periphery, between Inka and non-Inka.

With regard to form, Pachakuti employed a hierarchy of style in Machu Picchu. The most carefully sculpted rocks displaying vertical and horizontal cuts are all situated in the Hanan or Western Urban Sector. The Sun Temple and the Royal Residence are particularly foregrounded by the use of carved rocks and stone masonry exhibiting the highest quality of craftsmanship. In the Hurin or Eastern Urban Sector, the stonework in general is simpler, more modest, and integrates more natural rocks.

Furthermore, there are two categories of carving unique to Machu Picchu according to my surveys: the four arrow shapes discussed above, which all appear to point south to the dominant nevado Salcantay, and the image stones. Both categories powerfully restate the intimate relations between carved rocks and mountain apus.

Today a new vision of Machu Picchu has been constructed at the core of a contemporary national ideology which again is built on its stones and has been exported globally through innumerable tourism advertisements which visually focus in on Machu Picchu stonework (see also chapter 7, "Discussion and Conclusions"). According to Yazmin Lopez Lenci (2007:83–170, 348–354), it all began with Hiram Bingham's discovery of Machu Picchu, which is recontextualized by Peru in the sense that Bingham was only the first to bring international attention to Machu Picchu while the ruins themselves had been known to local Peruvians all along. According to Lopez Lenci (2007:153), in 1912 U.S. President Taft obtained from President Leguia of Peru an exclusive permit for Yale University to explore and excavate any ancient site in Peru for ten years and to export recovered materials to the United States. This permit has been the battleground of a lawsuit filed by Peru against Yale in 2008, asking for the return of all artifacts taken from Machu Picchu. Many Peruvians argue that the permit issued by Leguia violated two earlier Peruvian *decretos* (officially approved laws) which outlaw the very activities the Yale permit granted and that the latter was never approved by the Peruvian legislative body. U.S.–Peruvian relations escalated in 1914 and 1915 when Cusco newspapers reported that Americans exported or smuggled excavated materials imagined as large treasures from the Urubamba Valley via long mule trains to Bolivia and on to the United States. In Cusco, Bingham was told to cease excavations, which he ignored.

The discourse over Bingham and the meddling of U.S. institutions in Peruvian cultural heritage has kindled passionate feelings of regionalism in Cusco. Traditional Cusquenos perceived that outside entities in the name of universal and ever-advancing science threatened local knowledge and wisdoms, and they resorted to propagate the local cultural heritage and promote a regional economy. At the same time, Cusco regionalism has had to be counterbalanced with the city's interests on the national scene and now with the global world which has brought tourism as the primary economic factor. The extensive celebrations of the one hundredth anniversary of Bingham's rediscovery of Machu Picchu held through July 2011 appear to have reflected a cosmopolitan and conciliatory spirit, including the opening of a new museum in Cusco which now displays artifacts collected and returned by Yale.

Tu eres la realidad terrestre y en el espiritu infinita!
Machu Picchu, salvanos
de nuestra vida
sacanos de la persona humana invital
sacanos de la leccion y la mentira.
Piedra sobre piedra
Vida sobre vida.
[. . .] *Pero estamos a tus pies, Machu Picchu!*
Yo y el Cusco, porque tu eres la eterna vida,
el constante estar en suspenso,
aplastados tu y yo de yanquis y autobuses y valijas!
[. . .] *No temas, Machu Picchu!*
que nada te haran los turistas,
Ningun dano te causara Neruda,
Ningun dano te causara la gringa!
(Martin Adan, 1961–1963, cited in Lopez Lenci 2007:83)

I offer my work with respect to the people of Cusco, their cultural heritage and natural environment unrivaled by any other part of the world.

NOTES

1. Perhaps the fact that many terraces were filled with soil from faraway places facilitated the growth of the exotic crops (Villanueva Urteaga 1971:3, 130). However, Villanueva Urteaga never conducted independent geochemical soil analyses to confirm this assumption (Frank Meddens, personal communication 2013).

2. Wayna Qhapaq practiced this notion of centrality by ordering an equal number of mitimaes from Chinchaysuyu and Kollasuyu to serve him at his estate; see above.

3. Many of these objects were returned to Peru in 2011 during the centenary celebration of the rediscovery of Machu Picchu. At this occasion, the Machu Picchu Museum in Cusco opened, where these materials are now stored and on display.

4. Rowe's archival work and Burger and Salazar's analysis of the Machu Picchu artifacts as well as recent archaeological work at Maukallaqta (Pacariqtambo) by Bauer (1992) and at Vilcabamba the Old by Lee (2000) and the Ministerio de Cultura have clearly outdated Bingham's (1979:233) opinion that Machu Picchu constituted Tampu Toq'o, the Inka origin place, or Vilcabamba.

5. For a thorough and well-researched tour of Machu Picchu illustrated and guided by detailed maps and photographs, the reader is referred to Wright and Valencia Zegarra (2001). My discussion will be in the format of a similar tour but only stopping at major carved rocks.

6. Other known Inka Sun Temples with curving walls are the Qorikancha in Cusco and the Sun Temples in Pisaq and Wina Wayna. One must keep in mind, though, that the curved walls of the Sun Temples at Pisaq and Machu Picchu are clearly the result of the desire to enclose natural rock outcrops and "marry the earth" as Dean (2007) has described them. These curved walls are not an abstract formal feature (see also Protzen 2000:204).

7. Lucy Salazar (2004:36) recognizes the form of a feline in the carved outcrop, possibly a puma, whom Pachakuti claimed as a totemic ancestor. Salazar (2004:37–38) further interprets the form of the upper Temple of the Sun—when seen from above—as a rainbow, a celestial phenomenon that in Inka thought united the three spheres of the cosmos: the sky, the earth, and the underworld.

8. In a general sense, of course, Andean people have associated caves with the spiritual world of the ancestors.

9. Bingham (1997:175) was incorrect when he tried to associate this temple with Manqo Qhapaq by citing the Peruvian chronicler Salcamayhua. Salcamayhua's work is dated to 1620, and he mentions that the first Inka, Manqo Qhapaq, commissioned architectural works at the place of his birth and that these should contain a masonry wall with three windows. Nevertheless, the name "Temple of the Three Windows" has stuck, but it should not be connected with creation mythohistory, Manqo Qhapaq, or the three caves of emergence (see chapter 4).

10. There is one example in Cusco that qualifies as an "image stone": the possible landscape carving in the upper section of Tetecaca documented by van de Guchte (1990:94–98). It depicts terraces traversed by a canal rather than mountain outlines.

11. So far, Dearborn has documented two other cases: the relations between the usnu on the main plaza Awkaypata in Cusco and the sun pillars on the horizon line (Bauer and Dearborn 1995) and between the Sacred Rock Sanctuary on the Island of the Sun in Lake Titicaca and two pillars on the Tikani ridge (in Bauer and Stanish 2001:208–212; Seddon and Bauer 2004; see also chapter 3). In 2005 and 2007, J. McKim Malville (2008), Michael Zawaski, and Steven Gullberg identified a new case scenario: two long-known but uncontextualized pillars on the northeastern horizon of the town of Urubamba appear to be astronomically related to the large white granite boulder in the courtyard of Wayna Qhapaq's palace, Quispiguanca (see Niles 1999). When observed from the vicinity of this conspicuous boulder, the sun rises between these pillars on the June solstice.

The implications are twofold: first, it is significant that the observation points are sculpted or special stones, and the function of some carved rocks may have been to serve as seats from which Inka astronomer priests tracked the cycle of the sun; second, the separation of the pillars as well as the spatial organization around the special rocks suggest that the primary object of solar observations was not to determine specific dates but group participation, with separation by status, in the ritual viewing of sunsets at critical times in the solar year, such as solstices (Dearborn 2000). The Intiwatana may well be another case in this pattern.

12. However, there is the possibility that this perceived arrow was nothing more than a pointed corner of the platform. The reason for this is the observation that one carved side of the arrow continues along the platform as if to demarcate its rim.

13. Today this rock has been nicknamed "The Slide" because of its smooth sloping area. There is a very similar example in the northern part of Saqsaywaman, which is also known as a "Slide." It is unknown if the Inka used them for play, but "Slides" represent a narrow typological category of carved rocks.

14. According to information from the Museo Manuel Chavez Ballon, this compound (Conjunto 9) may have housed a metal workshop.

15. One low rock exhibits a petroglyph incised onto its flat surface. This petroglyph consists of a central point with sixteen radiating lines, a composition which recalls the concept of the zeq'e system. This petroglyph was photographed by Bingham and thus is most likely of Inka origin. The problem is that incising rocks in a low relief is not the typical style of Inka rock art. Therefore it is not further considered here.

16. Her discussion leaves out Patallacta on the Rio Kusichaca, another outlier on the regional road from Cusco to Machu Picchu. This impressive site contains a special rock surrounded by walls as well as other carved rocks and was likely another commission by Pachakuti.

17. More recent investigations by Vincent Lee and Gary Ziegler have identified additional roads.

18. Published examples and references in the literature abound for most periods and cultures: for example, Burger 1992:113, fig. 104; Grieder 1978:pl. 4, 195–263; 1982:fig. 66). Terry Grieder (1978) published numerous ceramic examples from

Pashash, most of which date approximately between A.D. 300 and 700. Many of the stepped motifs are paired with frets which Grieder relates to the mountain and cave symbolizing the cosmos to "Third Wave peoples" in his classification of the developmental stages in pre-Columbian art (1982:134–135). This mountain and cave complex leads back to the Temple of the Sun and its cave with the prominent and finely carved stepped stone. In a very general sense, stepped designs have been referred to as stylized mountains (Wright and Valencia Zegarra 2001:44–45).

The stepped stone in the Temple of the Three Windows is only one of numerous examples at Inka sites. At Pisaq, a small stone with three steps on each side sits on a pedestal carved of the same stone block in front of the Sun Temple. At Saqsaywaman, the stepped stone stands east of the circular reservoir area next to passageways through the rock formations. It displays three steps on each side, and a fourth step forms the top. At Quilla Rumi, a stepped stone was placed prominently on one side of a large platform. At the site of Quespehuara in the Chacan area north of Cusco, a long double-sided stepped motif has been cut out of a rock wall; and the Museo del Inka in Cusco houses a stepped stone assigned to the Colonial Period.

19. The stepped design has received special attention from New Age people and shamans. In tourist stores at Aguas Calientes, I have seen posters which depict the stepped motif doubled to form a diamond shape. The general outline has been associated with the constellation of the Southern Cross, and all the corners and steps have been connected with specialized symbolic meanings. While the Inka may have shared some of those meanings, the New Age presentation is purely speculative and unsubstantiated by any historical sources.

20. On the other hand, Wright and Valencia Zegarra (above) source these rocks as river pebbles from the Urubamba River below.

21. Dr. Barreda Murillo (personal communication 2003) reasoned that there would have been pre-Inka settlements because the bottom lands of the valley constituted a significant coca producing area. Dr. Fernando Astete (personal communication 2003) told me that there has been no evidence of a pre-Inka settlement at Machu Picchu proper.

22. This reasoning does not imply that Machu Picchu would predate the redesign of Cusco. Both building projects were processes continuing for decades and most likely overlapped chronologically.

Chapter Six

Chinchero

Chinchero was a royal estate of Thupa Inka Yupanki, the son and successor of Pachakuti Inka Yupanki. I will begin by laying the *Groundwork* with historical and physical data on site referencing ethnography as well as the recent literature. The analysis sections *Inka Landscape and Carved Rocks* will examine certain urban design elements which were formulated in Cusco and reproduced in several royal estates to project messages about state power and its land claims.

GROUNDWORK

Historical Data from the Literature

Principal sources for the attribution of Chinchero to Thupa Inka are the court case of Dona Beatriz Coya and her husband, Martin Garcia de Loyola, against the Crown in 1588, Sarmiento de Gamboa, and Juan de Betanzos. One of the court documents attests that Thupa Inka Yupanki built leisure palaces at "chinchero guaillabamba and Urcos" (Rowe 1997:277, 281–282), and Sarmiento (1942:152 [1572:chap. 54]) repeats that "he had very rich houses for recreation at Chinchero" (see also D'Altroy 2002:127–129). The first source indicates that Thupa Inka possessed additional estates and properties which are further described in the report of a visit by the Mariscal Alonso de Alvarado at Yucay in 1551 and 1552 (Villanueva Urteaga 1970:31–54). Alonso de Alvarado notes that the Urubamba Valley was divided up between Thupa Inka Yupanki, who owned the upper half, and the Sun to whom belonged the lower half (Villanueva Urteaga 1970:34). Thupa Inka's property at guaillabamba or Huayllabamba is now located near a campground on the tourist Inka trail to Machu Picchu, and little remains to be seen of Inka structures. On the other property at Urco/Urquillos (both spellings are used), there are Inka-style buildings enclosing a large carved boulder with a snake motif. It is my understanding that these structures have not been securely dated to the reign of Thupa Inka or the Inka period in general.[1] This leaves Chinchero as the best preserved and documented of Thupa Inka's estates. Betanzos (1996:159 [1557:chap. XXXVIII]) devotes one entire chapter to the building of Chinchero. He says that Thupa Inka decided "that it would be good to build a town to be populated by the growing number of people in Cuzco, so that he would be better remembered, so he and the lords of Cuzco would have a place to go for recreation, . . ." (1996:159 [1557:chap. XXXVIII]). Thupa Inka himself prepared a plan which was executed by his master builders and artisans within only five years. Each Cuzco lord had to supervise construction of one section of the new town, and Betanzos seems to imply that they also had to fund it. In return, each received a house in Chinchero. This interesting account implies that Chinchero was intended to function as a private estate and town. The Cuzco lords (*orejones*) have to be distinguished from the yanakuna and kamayuq who lived at royal estates to maintain them. Instead, the lords assumed the role of courtiers who moved together with the Inka as part of the court.

Chinchero's Inka settlement includes a number of significant carved outcrops, and therefore it is a focal point in this study. Unfortunately, these are not mentioned in the historical sources except for one vague reference by Cristobal de Albornoz (1967:27 [1580?]). In his list of Inka wak'as, there are two entries that seem to pertain to Chinchero: one speaks of *Guaypon guanacauri*, a rock close to a lake. Here pierced the Cusco Indians their ears; and the other names *Chinchero guanacoari*, which is a rock close by the just mentioned wak'a at Lake Guaypon. It is linked to many other wak'as (1967:27 [1580?]). One of the two lagoons near Chinchero is known by the name *Huaypo*, which must be a spelling related to *Guaypon*. According to Alcina Franch (1976:20), one can see *Huanacaure* mountain from Chinchero, which must refer to the mythologically charged Wanakauri peak just outside of Cusco, featuring in the Pacariqtambo origin account (see chapter 4). Although we cannot identify the specific wak'as Albornoz is talking about, his information should be seen as additional evidence of how the Inka exported features of sacred geography as defined by Pachakuti in Cusco to more distant regions of the empire. Visibility, naming, and social memory were always important factors in claiming multilayered relations between the center and more peripheral areas. It will now be shown that sculpted rocks were a characteristic feature of the Chinchero landscape and that Thupa Inka surely, and intentionally re-created stone ideology at his estate at Chinchero.

Chinchero is located about ten kilometers to the northwest of Cusco on a highland plain with small lakes which is bordered by the rivers Urubamba, Anta, and Urquillos and overlooked by a series of towering mountains. Today it can be reached by the road that connects Cusco with Pisaq, which passes through the populated town adjacent to which the Inka ruins are located. From 1968 to 1971, Chinchero was the object of an extensive project by the Misión Científica Española en Hispanoamérica under the direction of Jose Alcina Franch. Their work included survey, excavation, analysis of artifacts, as well as ethnohistorical studies, and resulted in the publication of four volumes. In 2000, Carmela Caceres Silva and Carmen Gladys Jurado Carrasco completed their dissertation about the iconography of Chinchero at the Universidad Nacional de San Antonio Abad del Cusco, which is particularly insightful to my study of the carved rocks. Stella Nair's fieldwork, dissertation (2003), and publications have since expanded their data.

Alcina Franch's team as well as Nair found Killke material at Chinchero and its environs which indicate that the region was inhabited before Thupa Inka founded his estate. Killke people, now more narrowly identified as the Ayarmaca (Bauer 2004:79–80), probably lived throughout the area of Chinchero in hamlets of up to six houses before and during the rise of the Inka (Alcina Franch 1976:15; see also introduction, Of the Inka: A Brief Cultural History). Traces of one such Killke settlement were documented underneath the Great Plaza of Inka Chinchero (Alcina Franch 1976:11). How the presence of Killke people impacted the establishment of Thupa Inka at Chinchero is not entirely clear and remains a subject of ongoing investigations. This raises the question that Killke people could have carved rocks at Chinchero as well as in the immediate surroundings of Cusco. This remains an open possibility. However, I will present sound evidence that the sculpted rock outcrops constituted one defining element of Thupa Inka Yupanki's royal estate.

The Carved Outcrops of Chinchero

Alcina Franch (1976:23, fig. 6) and his teams divided the settlement of Chinchero into twelve sectors. Nair drew up four sections for her investigation. Alcina Franch's Sector XII covers the streets of the inhabited town, and Sectors IV and V include the church and the Plaza de Armas since a large portion of the town was built over the Inka site. The present town and these archaeological sectors occupy a hilltop from which the other sectors gradually descend north toward the Inka road from Cusco to Yucay and to the Urquillos River. Sector VIII contains the Pumaccacca rock, and the sculpted outcrop Titicaca falls into Sector X. Sector XI defines the large area of agricultural terraces which describe three Us or *anfiteatros* as they outline and adapt to the natural curvature of the hillside. The second carved outcrop to the northeast of the terrace system was excluded from Alcina Franch's work area. This outcrop was mapped and studied by Caceres Silva and Curado Carrasco. Here I will focus on the sectors and areas with the carved rocks.

Sector CH-VIII encompasses Structure 11 and several carved rocks which form the southeastern corner of the Great Plaza (for a detailed analysis of its architecture, see Alcina Franch 1976:100–114). Structure 11 was built over a natural outcrop, resulting in a pyramidal volume which consists of four platform levels. Platform 1 forms the entrance area in which Alcina Franch (1976:105) documented wall remains from the Colonial Period. He interprets these walls as possible leftovers of a colonial chapel, indicating that perhaps the Spanish were aware that the Inka ascribed a sacred character to this building rooted in its carved rocks. It was a general Spanish custom to build religious structures over precontact sanctuaries. The second platform is accessed by a stair framed by a double-jamb doorway. The third platform has a long passageway which runs along the backside of the carved rock Pumaccacca. A drainage canal lines its eastern wall and descends to the lower platforms. There was no identifiable access from the third platform passageway to the upper part of Pumaccacca (Alcina Franch 1976:103–105).

Pumaccacca constitutes the focal point of Structure 11 and sits approximately at its center. It towers about six meters above the level of the Great Plaza, and Alcina Franch (1976:106) likens its form "to an eruption of the mountain coming out of its interior, which turns over on the outside and overflows." This verbal image may well describe part of the sacred character the Inka most likely saw in this rock. The carvings are vertical and horizontal cuts forming planes as well as two sculpted pumas with crossed legs (figure 6.1). Nair (2003:72) cautions that traces of a foundation indicate that Pumaccacca was surrounded by walls and embedded into the terrace and as a result, would barely have been visible from the plaza. Her reconstruction differs from Alcina Franch's (1976:fig. 32), which depicts the carved rock as well visible and overlooking the plaza. Since we cannot know how tall the surrounding walls were, the evidence that this sculpted outcrop might have been concealed from public view is inconclusive.

Pumaccacca has to be understood in relation to four other carved rocks situated nearby: first is a smaller rock with a seat-like carving which is located at the foot of Structure 1 and in the southeastern corner of the Great Plaza; the second and third sculpted rocks sit on the third platform; the fourth carved boulder is on top of the outcrop on and around which Structure 11 was built.

Most walls pertaining to Structure 11 were made of well-fitted and high quality, imperial Cusco-style masonry. Alcina Franch (1976:106–112) distinguishes two construction phases: the structural retaining walls of the three

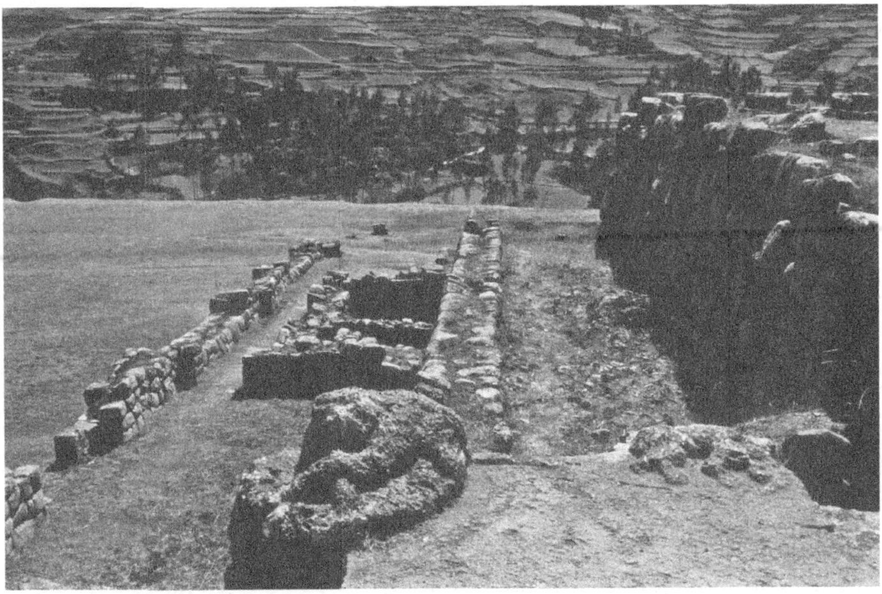

Figure 6.1. Chinchero, Pumaccacca in Structure 11. Photograph by Brian Garrett.

platforms were built of large and carefully fitted blocks, dating to the founding of Chinchero by Thupa Inka Yupanki around 1480; the walls which define rooms winding around the outcrop of Structure 11 were made of smaller stones held together with mortar and probably constitute a remodeling effort which occurred later during Inka occupation but still before the Conquest.

The south side of the Plaza is lined by Structures 1, 2, and 3. Structures 1 and 3 form long rectangles which open to the Plaza with six (Structure 1) and seven (Structure 3) double-jamb windows or entryways. Structure 1 measures 48 meters in length, Structure 2 17.80 meters, and Structure 3 is 42 meters long (Gasparini and Margolies 1980:214). They are built on a terrace and their floor level is higher than that of the plaza; and therefore access is accomplished from the short lateral sides via two passageways between the buildings. Elongated rectangular buildings with many door openings facing the main plaza are the general diagnostics of kallankas, and I suggest Structures 1 and 3 fall into this category. Chroniclers have described kallankas as public shelters for audiences of ceremonies held in rainy weather (see more in the Introduction; and for a detailed discussion of the building type kallanka, see Gasparini and Margolies 1980:196–219). The elevated level and the lateral access to the Chinchero kallankas might have made it difficult for large numbers of people to enter at one time, but their elevation is low enough that the openings to the plaza could have been used as additional doorways.

The combination of a foregrounded rock—often carved but sometimes uncarved—kallanka/s, and a large plaza form an architectural pattern which has been documented in many Inka settlements. The analysis section Inka Landscape and Carved Rocks—Essence, Imagination, and Stone Ideology will explore it further (to review the full discussion, see Christie 2007a).

Northeast of Structure 11 and several terraces below sits a medium-sized carved rock formation known by the name Condorqaqa on an extended platform surface. The name is derived from the weathered relief carvings of three animals found at ground level on the northwest side of the rock. The upper ones appear to be four-legged animals, and the lower one may be a bird with outstretched wings, hence the name Condorqaqa or Condor Rock. These reliefs form a row in descending order as if marching or flying into the ground. Caceres Silva and Jurado Carrasco (2000:131) conclude that the animal carvings might refer to the ancestors of certain ayllus as suggested by Cristobal de Molina El Cusqueno.

The rock formation itself is not a solid mass but a clefted limestone. Visually it looks like a bubbling liquid thrown up by the earth and frozen in its act of undulation. Of course, weathering has been increasing this visual effect; and though it may not have been as poignant in the fifteenth century, it was surely unusual enough for the Inka to notice. Caceres Silva and Jurado Carrasco (2000:125) liken the whole rock of Condorqaqa to a bird in flight with wide wings. Such associations are likely a primary reason why Thupa Inka's artisans carved this limestone formation into innumerable seats, platforms, and niches. Most of the cuttings form right angles; others are rounded. The largest, most finely sculpted, and least weathered seats and platforms are found on the north and northeast sides.

I suggest Condorqaqa or Condorqaqa a, as I name it, has to be understood in conjunction with a smaller carved rock, Condorqaqa b, situated a short distance higher up against a retaining wall. Condorqaqa b is formed of a similar undulating and clefted limestone but much smaller in extent. It displays two carefully crafted seats facing in a general northern direction. Seated upon these carvings, one enjoys a comprehensive view of the U-shaped terrace complex below.

Condorqaqa a and b are located on peripheral terraces to the southeast of the third U—or the third anfiteatro (amphitheater) as Alcina Franch (1976:131–140) calls it—of the terrace system with open space between them. One pronounced seat on Condorqaqa a also looks toward the U-shaped terraces. I reason that Condorqaqa a and b should be understood in relation to the terrace complex and agricultural productivity. Their visual qualities would make them special places charged with power to contact the earth. Van de Guchte (1999:165–166) has commented on the principle of alterity that drew the attention of the Inka to rocks with unusual visual characteristics and motivated them to sculpt those over plain and regular-looking stone formations. A particular connection with the earth is reinforced by the placement of the relief carvings, which seem to move into the ground, and the bird may be flying out of it. They are difficult for human beings to see and appreciate. Thus they may have been intended as a type of offering to the earth. In the worldview of the Inka, the earth was alive with formidable powers to bring up plants and crops to sustain people. Condorqaqa a and b may have served as wak'as on whose carved seats and platforms offerings to the earth were placed and perhaps sacrifices conducted. Accompanying rituals to increase the earth's fertility may have been acted out in the open space between them. The pronounced seats facing the U-shaped terraces could also have functioned as thrones for high-status religious specialists overseeing such ceremonies. Alcina Franch (1976:137) notices a usual and mysterious construction at the eastern end of the upper terrace, which may also have stood in a relation with the Condorqaqa carved rocks. It consists of two rooms composed of retaining walls exhibiting well-cut and precisely fitted polygonal masonry but no doorways.

Further northeast, a group of carved rocks spread around the eastern periphery of the site. One outcrop, in particular, stands out for its exact vertical and horizontal cuts creating seats, platforms, planes, and steps. Some of the platforms and planes are arranged in approximate rows of three grids. I suggest that this boulder might possibly have been used as another counting device as explained in chapter 2. The grids are reminiscent of the boxes of a yupana which, in turn, can be translated into a khipu. So far, I have emphasized that royal estates were private properties and thus more or less detached from the political, administrative, and economic spheres which depend on accounting and record keeping for their proper functioning. In support of this reasoning, I have found very few grid-like carvings on royal estates. This Chinchero outcrop is one notable exception, and I would explain it by its proximity to the U-shaped terraces. The U-shaped terraces constitute a central feature at Thupa Inka's estate and must have produced large quantities and possibly varieties of crops and agricultural products which may have been sorted and counted on this sculpted rock. It is large enough for people to stand on, sit on, and heap objects on. After their quantities had been established, they were officially recorded on khipus.

The northeastern boundary of Chinchero is formed by a very large and fancifully sculpted boulder that sits below the terraces and at the Inka road which leads to Yucay in the Urubamba Valley (figure 6.2). Some local people

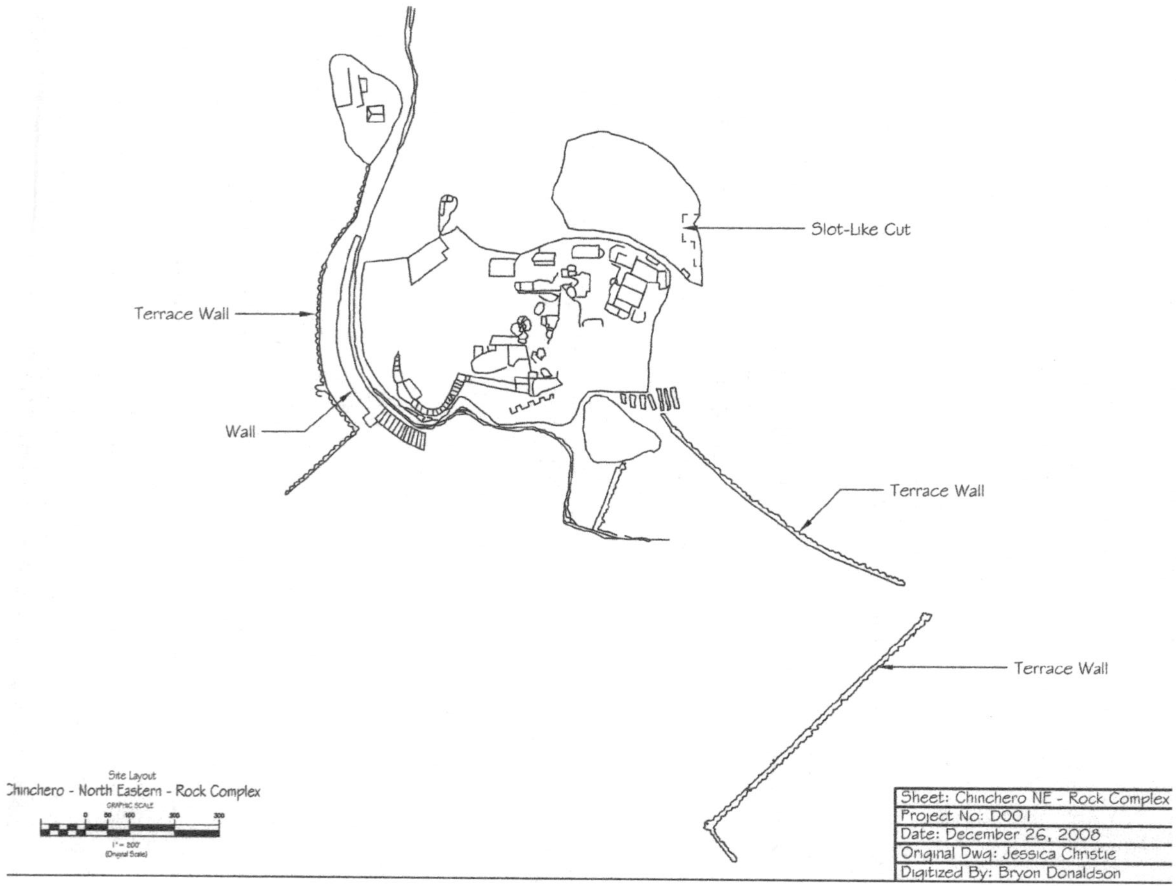

Figure 6.2. Chinchero, Northeastern carved rock complex. Drawing by the author.

have nicknamed it *Chinkana* due to an interior room. This room can be entered through a thin opening followed by a long passage from the top northern face of the outcrop. The walls of this room have been smoothed, the floor plastered, and the south wall exhibits two niches (Nair 2003:73–74). I prefer to designate this boulder as northeastern carved rock complex as opposed to Titicaca, which forms the northwestern carved rock complex. Caceres Silva and Jurado Carrasco (2000:124) call it Q'ellomachayniyoq.

This northeastern carved rock complex is one of the most magnificent examples of Inka rock art because it constitutes a true summary and cosmogram of all its forms, associations, and symbolism. First of all, it stuns by its sheer size and height and by its rounded outline. It looks like a gigantic ball, which the Inka would surely have noticed. When approached from the Inka trail, the visitor has to climb around it to the back side to gain access to the top. The upper section is lavishly sculpted into seats, platforms, and planes, some of which appear to be rounded. To some extent, the rounded forms were heightened by erosion since the rock is limestone as that of Condorqaqa. Most intriguing are the carvings on the west side. A narrow and at times deep stair winds precariously along the western edge of the outcrop. It provides access to a cut seat from which a person could overlook the Inka road. On the other hand, it is conceivable that the steps functioned as foundations for a wall which would have protected the sculpted top from the eyes of anyone walking on the road.

In the lower section, a stream named Parq'o which originates nearby (Caceres Silva and Jurado Carrasco 2000:127) and thus has no connection with the anfiteatros of the terrace system accompanies and follows each contour of the northeastern carved rock complex. In the southwest, the water flows by and under a rock overhang the ceiling of which displays a finely carved crenellated motif. The stream continues repeating the curvature of the outcrop. Toward the outside, it is framed by steps and what is probably a recent low wall. Finally the Parq'o stream departs from the rock complex in a northerly direction crossing the Inka road. On the north side, the bottom part of the outcrop exhibits a shallow cut cave.

Several smaller carved rocks seem to be related to the northeastern carved rock complex even though they are physically detached from it. To the north, three carved rocks stand in very close proximity: one large rock is only separated by a very narrow passageway and displays a slot-like cut; a small rock with a carved seat sits in front of the cave; and a medium-sized rock with a highly unusual, finely cut triangular sculpture which resembles an open book is located a short distance down the stream. Another modified boulder can be found roughly 30 meters to the northeast, and one carved rock stands approximately the same distance southwest next to the Inka road. This rock, northeastern carved rock b, abuts a terrace wall and is oriented north.

It contains a number of vertical and horizontal cuts forming platforms and planes. The sculptures demonstrating the highest quality of craftsmanship are a double seat facing west onto the Inka road. A person in this position could easily have observed and perhaps controlled traffic on the road, and this rock might have functioned as a type of guard station.

Today the northeastern carved rock complex is surrounded by terrace walls which branch off in various juxtaposing directions. Crops are grown and thrive immediately below it, benefiting from the water channeled off the Parq'o stream. Although these irregular terraces are recent and do not form part of the U-shaped system laid out in three anfiteatros, one would expect that during Inka times as well agricultural production extended to the northeastern carved rock complex, taking advantage of the stream. It is most likely that one symbolic aspect of this sculpted outcrop focused on rituals dedicated to fertility and agricultural growth. Originally it was probably treated as a *huanca chacrayoc* (see below). People may have laid offerings to pachamama on the platforms carved into the top; the whole outcrop may have been a double or stone image of her body which could be accessed by means of the interior room and the cave below. The vicinity of the Parq'o stream and its spring proved that hers was a fertile body with the innate power to bring forth water and nurture crops.

Yet there was a second function which was more authoritative in the context of Thupa Inka's royal estate: the northeastern carved rock complex marked a strategic entrance to and exit from Chinchero on the Inka road. It was truly the first or last monument any visitor traveling on this Inka trail would have seen. In this sense, the outcrop sculpted in Inka style comes to embody Inka-ness and an Inka identity which, I postulate, Thupa Inka defined at least partly through the rock art.[2] Having this rock complex carved in the style promoted by his father, Pachakuti, proclaimed his descent line and his affiliation with his father's heritage, which the chroniclers in general confirm. One important difference, though, in the way Pachakuti and Thupa Inka employed carved rocks lies in the public settings emphasized by Thupa Inka, while his father seems to have preferred more private contexts (the majority of the examples at Machu Picchu, Ollantaytambo, Pisaq, and Vitcos). A third but minor function of the northeastern carved rock complex and northeastern carved rock b could have been as guard stations where watchmen overlooked the trail and reported visitors to the administration of Chinchero.

The northwestern carved rock complex is known as Titiccacca. It is situated below the Great Plaza, west of the U-shaped terrace system, and near the main entrance stairway from the Inka trail to Yucay. Titiccacca is more massive in size than the northeastern carved rock complex and divided into two halves by a narrow cave-like passageway. Only the slightly larger half is sculpted (figure 6.3). This part of Titiccacca is entered by means of a finely carved stairway

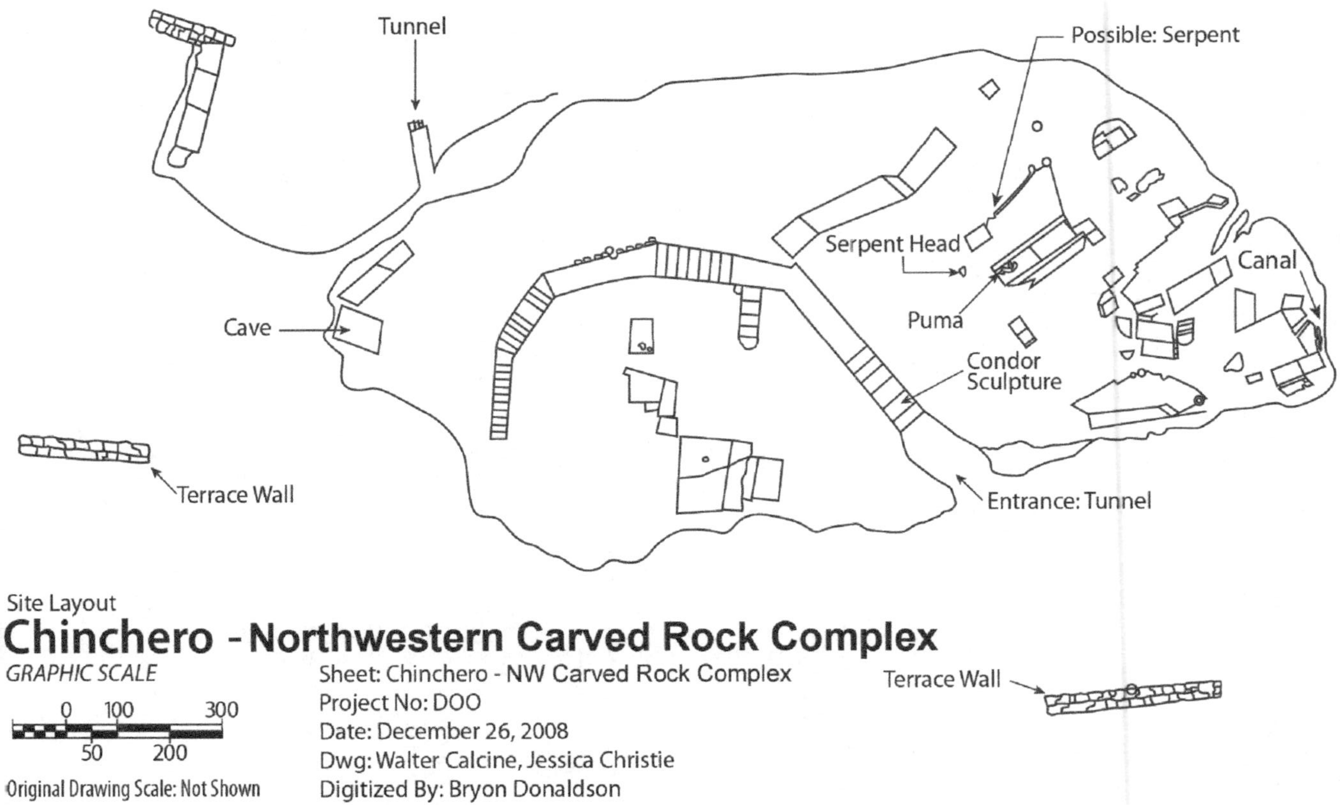

Figure 6.3. Chinchero, Northwestern carved rock complex. Drawing by the author.

which leads in and through the rock. On one wall of the passage formed by the stair, there is a form which some people identify as the relief carving of a condor with its head turned and its beak tucked under one wing. Leaving the passage, the visitor climbs a few more steps which lead him or her to the top, which is richly carved with many vertical and horizontal cuts creating more steps, platforms, and planes. Focusing the eye, one can recognize two damaged high-relief carvings of pumas and one of a snake. Looking east, one sees the ball-like mass of the northeastern carved rock complex. A continuation of the entrance stairs curves west and leads to the west side of Titiccacca where more well-crafted seats and platforms appear. At least two of them face west, and from this position a guard could have watched the Inka road and reported all traffic to the administration of Chinchero. On ground level in the passageway between the two halves of Titiccacca, there are a few minor carvings consisting of some platforms, a shallow cave, and a short tunnel in both halves of the rock complex.

The functions of the northwestern carved rock complex were probably similar to those of the northeastern one. On one hand, it served as a location for rituals intended to increase agricultural abundance. Titiccacca overlooks the terrace system and is physically separated from it by the main stairway. While it lacks a direct water source, such as the stream at the northeastern carved rock complex, there are isolated carved canal segments which may have been used in libation rites. In general, an underworld connection would have been evoked by the cave-like passageways of the entrance stairs and between the two halves of the rock complex. The human world existed outside the rock and was transported to an upperworldly realm as a person emerged from the dark entrance tunnel. The sculpted top section is bathed in sunlight and offers multiple shelves and platforms to place offerings and perform ceremonies. The animal relief carvings may have played a role in such performances since the snake is often associated with the underworld and the puma with the terrestrial realm in Andean thought. The top further provides strategic views of the terraces and the Inka trail, imparting upon the viewer an experience of visual control. At least the western side of Titiccacca could have been a type of guard station similar to those on the northeastern carved rocks to control the western entrance and exit of the settlement via the main Inka road.

INKA LANDSCAPE AND CARVED ROCKS—APPEARANCE, EXPERIENCE, AND PERCEPTION

Fields of Vision

Our experiences and perceptions of appearances are culturally influenced by language: the words we use to describe and name things direct our perceptions of them. As explained above, this lay at the root of much confusion in the Andes: since Quechua was only a spoken language, the Spanish misunderstood many issues and misspelled even more terms, and their versions of Andean history and culture have been transcribed into contemporary Spanish and English. Here I will explore two cases in which terminology has conditioned multiple fields of vision.

Nair (2003:127; 2010) has brought to attention that in colonial Quechua the term *kallanka* did not signify the building type of a great hall but rather the construction technique of using large worked stones for foundations and lintels. Thus hearing the word kallanka, the Inka would have perceived any structure built of large ashlars. The Qorikancha would have qualified as a kallanka-*wasi* (house), but the Chinchero large halls would not (Nair 2010). In the context of this study, the Quechua term kallanka would have been an addition to the vocabulary of stone ideology as it described architecture. Although this clarification is important, I will continue to use the word *kallanka* meaning "great hall" the way it was first used by Morris and Thompson (1985) since it is firmly established in the literature.

The northeastern carved rock complex as well as Condorqaqa may originally have been viewed as *huanca chacrayoc*. As Duviols (1979b) amply demonstrates, huanca chacrayoc were large monoliths placed in fields to protect the crops. At Chinchero, the Killke people/Ayarmaca had likely noticed the unusual forms of these outcrops and venerated them as agricultural *huancas* (field guardians). Quechua dictionaries remain vague on the point of whether huancas were natural stones or modified by man (Duviols 1979b:26). Duviols (1979b) weighs the sociopolitical contexts of agricultural huancas consulted for fertility of the crops versus the huancas used by herders and nomadic people in the Andes, which marked the territories they occupied and referenced their culture heroes and ancestors. This is precisely a transition stone ideology elucidates: the Inka appropriated many earlier huancas which acted in agricultural as well as social contexts since they belonged to specific ayllus but otherwise were apolitical. They turned them into political wak'as through the geometric vocabulary of Inka rock art, which became the Inka way of marking sacred geography and claiming land.

INKA LANDSCAPE AND CARVED ROCKS—ESSENCE, IMAGINATION, AND STONE IDEOLOGY

The Plaza—Kallanka—Foregrounded Rock Paradigm

In the second part of the analysis, I will summarize the architectural-sculptural pattern defined by a plaza, kallanka, and foregrounded rock (full argument in

Christie 2007a). The Chinchero candidate is Pumaccacca with Structure 11, Structures 1 and 3, and the Great Plaza.

The archetypal example of the pattern plaza—kallanka—foregrounded rock existed in the center of Cusco itself. The urban design of Inka Cusco as authored by Pachakuti was described in the Introduction (Of the Inka: A Brief Cultural History) based upon archaeological evidence and primary ethnohistoric sources (Rowe 1967).[3] The central plaza had two sectors separated by the Huatanay or Saphy River: Awkaypata and Kusipata. Several kallankas lined the main plaza Awkaypata (see D'Altroy 2002:117), which were found in the Qasana compound, on the east side where today the cathedral stands, as well as within the Amarukancha and Hatunkancha compounds (for a detailed description and reconstruction of the Cusco main plaza, see Hyslop 1990:34–44; Bauer 2004:111–135; Rowe 1967).

The Cusco version of the foregrounded rock is referred to as *usnu*. Angles Vargas (1988:76–77) and others place the usnu exactly in between the Awkaypata and Kusipata sectors on top of the canalized Saphy River. In this interpretation, it had direct access to underground waterways (see Zuidema 1980:357), and its mediating qualities were physically brought out by its very position between Awkaypata and Kusipata. Nobody knows exactly what the usnu looked like because descriptions by the chroniclers are not consistent (see, for instance, D'Altroy 2002:115, 329; Bauer and Dearborn 1995:36). John Hyslop (1990:69–101), Tom Zuidema (1980), and Frank M. Meddens (1997) have investigated the Cusco usnu, other usnus, and the complex symbolic connotations associated with them. They conclude that usnus could assume a number of material forms: they could be stone pillars, stone seats, stone basins or fonts linked to underground channels, platforms, as well as truncated pyramids. Depending on their forms, usnus could have a variety of functions (Christie 2007a).

One important issue in the usnu discussion which bears upon the pattern under study is the question of its origin. Zuidema (1978 cited by Hyslop 1990:72) thinks that the usnu as a platform developed from the seat of the Sun or the stone set in the Awkaypata plaza of Cusco. In this scenario, the Cusco usnu would represent the earliest prototype for all later usnus. I argue that the usnu concept evolved from the shared pan-Andean symbolic associations surrounding stone. Most likely, there had been some kind of rock wak'a in the Cusco plaza before the rebuilding of the city under Pachakuti took place. Pachakuti and his successors devised the usnu as a special category of foregrounded rocks and as a state symbol and manifestation of Inka stone ideology. The formal vocabulary of the usnu was expanded from a single stone to include the basin and drain complex, seats, and the solid platform.

The pattern of a special rock, kallanka/s, and a large plaza was repeated throughout the empire at numerous Inka settlements with varying functions, such as royal estates, administrative centers, border towns, and the new capital in exile Vilcabamba the Old (known as Espiritu Pampa). It is absent from other settlements with the same functions. In general, the pattern may show up at any settlement where the Inka state had a presence and Inka identity was publicly performed. Its absence may be explained simply by the lack of archaeological excavation and by the geography of some sites. In the following, we will briefly discuss some of the sites exhibiting the pattern in relation to Chinchero.

Closely linked to Chinchero are other royal estates adopting the pattern. The earliest case is Juchuy Qosqo, Wiraqocha Inka's estate, which is situated high above Calca overlooking the Urubamba Valley. The center of Juchuy Qosqo is laid out around the Main Plaza. A kallanka building placed on a terrace overlooks the plaza from the south. In the southwestern corner of the plaza, below the kallanka, and next to the access steps to the upper terraces, sits a prominent and very intriguing sandstone boulder investigated by Kendall, Early, and Sillar (1992:199).

Juchuy Qosqo brings up the crucial issue with regard to chronology. The chronicles make it clear that Wiraqocha Inka built Juchuy Qosqo and that his son Pachakuti gave central Cusco its present form. Would that mean that Juchuy Qosqo predates the archetypal model of the pattern under discussion?[4] Can we ascertain under which ruler the Cusco usnu was erected? As observed in the introduction (Of the Inka: A Brief Cultural History) the pre-Inka settlement on the future site of the capital was large and already had certain spatial divisions which Pachakuti adapted, but there are no data to tell us whether it displayed an usnu-like stone.

Some of Pachakuti's private properties take up the plaza-kallanka-foregrounded rock paradigm. At Machu Picchu, it is articulated by the kallanka overlooking the plaza with the Ceremonial Rock situated outside the entrance gate discussed in chapter 5. Vitcos and Ollantaytambo once exhibited elements of our pattern which were spatially disjointed, and at Pisaq it is absent.[5]

Amaru Thupa Inka, Pachakuti's oldest son who would legally have succeeded as emperor instead of Thupa Inka, used the same architectural pattern in his estate at Callachaca. Callachaca spreads in several sectors over the hillsides east of Cusco. The one of interest is Susan Niles's Sector F (Niles 1987:fig. 1.3, 106–114). What she calls the T-shaped Plaza Group is a plaza in the form of a thick-stemmed T which is fronted by two partly preserved kallanka-like halls in the north and by a rock outcrop in the south. The rear walls of the kallankas were built into a hill, and their doorways open onto the plaza. The outcrop exhibits several man-made modifications which Niles describes as nooks and elaborations of natural crevices, some of which evidence high skills of craftsmanship. On its south side, the outcrop has at least one seat-like sculpture facing south across the Huatanay Valley, which means away from the plaza.

There is one other special rock sitting in the northeast corner and slightly outside the plaza arm. Niles (1987:108) calls it "an unusual item of Inca handiwork." It is a natural stone projection crowned with a cap of Inka masonry (see Niles 1987:figs 3.49 and 3.50). Niles could not find any evidence that this rock was originally part of a larger structure, and she concludes that the masonry topping was an Inka improvement of the form of the rock. In its "improved" form, this rock may have replicated one of the mountains across the Huatanay Valley, recalling the image stones from Machu Picchu. It follows that either the sculpted outcrop or the stone projection with the masonry cap or perhaps both could have constituted the special foregrounded rock/s linked with the plaza and kallankas.

Chinchero follows next in line when proceeding in a chronological order, and its example is formed by Pumaccacca with Structure 11, Structures 1 and 3, and the Great Plaza (see above). Another commission by Thupa Inka exhibiting the plaza-kallanka-foregrounded rock paradigm was likely Kasapata on the Island of the Sun in Lake Titicaca discussed in chapter 3. Wayna Qhapaq's well-documented estate at Urubamba and Yucay also used the pattern, though in a diminished form, in the palace complex of Quispiguanca. Niles and Robert Batson (Niles 1999) investigated and documented the standing architectural remains. Their reconstructions illustrate that the monumental entrance to Quispiguanca lay on the east side of the complex between two gatehouses. Once the visitor had passed through the portal, he or she stood in the main plaza in the middle of which sat a pronounced white boulder with a possible platform construction. North of this plaza, there were two symmetrical kancha compounds fronted by two kallankas. The white boulder probably functioned as a shrine. It is now topped by a Catholic chapel located at the east end of the present cemetery and therefore cannot be reconstructed. Niles also thinks that water may have been channeled through the midline of the plaza, over or around the rock and its shrine, and on to the south terrace wall. Today water drops down the south terrace façade of the palace; during Wayna Qhapaq's time, it would have been collected in an artificial lake described in historical documents (Villanueva Urteaga 1971:38).

The Quispiguanca example reinforces our pattern with a foregrounded but uncarved rock. It seems reasonable to suggest that given the central position of the white boulder and its possible companion platform, the two served as a focal point for ritual activity. This ritual activity could have been primarily political and related to the usnu concept, and it is conceivable that Wayna Qhapaq sat on the speculative platform overseeing events in the plaza. Or the rituals might have been religious in nature, during which offerings were brought to the boulder.[6]

The cases discussed above represent royal estates of Inka rulers which use the same general pattern as Chinchero, but their specifics are quite distinct. For example, in Cusco, Machu Picchu, and Quispiguanca, the usnu and usnu-like stones are centrally positioned in the plaza whereas Pumaccacca crowns Structure 11 in the southeast corner of the Chinchero plaza. These differing positions significantly alter the viewing directions and the resulting relational networks between the ruler/centralized activity and the audience, center and periphery, place and space. They indicate that in Cusco, Machu Picchu, and in Quispiguanca, the usnu and similar central stones were essentially oriented outward: the ruler and/or official on the usnu communicated with the immediate audience in the plaza and metaphorically interacted with the surrounding land and mountains, forging imagined strategic relations. At Chinchero, on the other hand, communication between Thupa Inka on Pumaccacca and his audience in the plaza would have been more inward oriented. The pattern using a plaza-kallanka-and foregrounded sculpted rock is most strongly developed on the properties of Pachakuti and his sons, Thupa Inka and Amaru Thupa Inka; while his father, Wiraqocha Inka, and Wayna Qhapaq used uncarved boulders.

The paradigm of a plaza-kallanka-foregrounded rock was exported to governance sites and administrative centers further away from Cusco. Selected examples are Samaipata in Bolivia (Meyers 1998; 2007; Meyers and Cornelius Ulbert 1997 on a similar case at the site of La Fortaleza located some 50 km air-line distance east of Samaipata); Inkallaqta located east of Cochabamba in Bolivia (Gasparini and Margolies 1980:207–212); Huanuco Pampa in the northern highlands, which was investigated by Craig Morris (Morris and Thompson 1985; Hyslop 1990:27, 203–206, 215–218; Gasparini and Margolies 1980:201–206); Vilcabamba the Old (Espiritu Pampa) documented by Vincent Lee (2000) and more recently by the Ministerio de Cultura (Fonseca Santa Cruz 2009; see further Legnani 2005:36).

These examples fully discussed by Christie (2007a) demonstrate that the combination of architectural and sculptural/natural features plaza-kallanka-foregrounded rock forms a recurring and meaningful pattern. In some cases, the special rock had the form of an usnu with a seating area and thus functioned as a place where the ruler appeared in a prominent and elevated position. Guaman Poma de Ayala (1987:391, 407, 377 [1615:374/384, 398, 369]) represented Atawallpa and Manqo Inka seated on usnus in Cajamarca and Cusco and surrounded by their soldiers as well as a stepped platform as part of an architectural setting in a scene during the reign of Wayna Qhapaq. In these three drawings, the usnu is a stepped pyramidal stone with four platforms, but it is by no means certain that this was the historically correct form of the Cusco usnu. The carved rocks at Chinchero (Pumaccacca) and Machu Picchu (Ceremonial Rock) may have been used as similar royal thrones.

Yet, such a context does not explain the entire pattern since other cases, for example, Quispiguanca, Vitcos, and Vilcabamba, were not carved as seats. I argue that first and foremost, all the foregrounded rocks which fall under the pattern were wak'as with all their symbolic connotations and mediating qualities addressed by many writers (Bray 2014). Frank Salomon (1998:7–17; Salomon and Urioste 1991:16–19) examines wak'as from a linguistic perspective through the language in the Huarochiri manuscript written in Quechua. He shows that wak'as, like people, plants, and animals, pass through several states of being: "from kinetic, fleshy, fast-changing . . . toward static, hard, slow-changing . . ." (1998:9). Since this process appears to be the same for wak'as, people, plants, and animals, it follows that they all would be subject to similar life forces and animating essences, thus establishing a clear parallel between wak'as and humans. The ultimate states of permanence are visualized in deified mountains, other land features, and rock wak'as. In Andean thought, there is a continuum from transitory to durable modes of being, the latter being materialized by bones, seeds, stone, mountains, and sculpted rock wak'as. It implies that the rock wak'as had not always been hard and timeless but had passed through softer, more pliable material stages. Within this concept, the process of sculpting may have reflected the shaping of a softer material stage into its permanent state of existence. At the same time, though, not all natural stones were wak'as. Those foregrounded in the pattern under discussion were thoughtfully selected; some were carve and deliberately placed into a specific architectural and spatial context, meaning that in many cases the rocks stood there first, and architectural space (plaza, kallanka) was designed around them. The differing states of being for humans and the natural world were brought into relationship through ritual at the rock wak'as and in the adjoining plazas and kallankas. While the ritual participants—people and wak'as—interacted asymmetrically, with the wak'as being the more durable and hence more powerful, ritual activities mediated between the multiple and complex inhabitants of the cosmos and states of existence, made them understandable, and reinforced coherence (see also Allen 1998:25).

Van de Guchte (1990:237–271) superbly explores the complex layers of the concept wak'a based upon ethnographic and linguistic sources. He notices that while Spanish writers in the sixteenth century tended to focus on wak'as as objects, in the seventeenth century, there was a greater interest in the processual and ritual characteristics of wak'as, and they are treated in the literature as living, dynamic forces charged with spiritual powers. Wak'as can make sounds and speak in human languages and have relations with weather phenomena and the celestial bodies, and wak'as may have unusual physical features which separate them from normal people and natural objects. Wak'as in all these categories lively interact with humans;

and we recall Guaman Poma's (1987:253 [1615:261]) well-known drawing of Thupa Inka addressing a group of wak'as, many of which are carved stones. Pumaccacca at Chinchero acted as one such wak'a and related usnu. All discourse between wak'as and humans, between the landscape and humans, is based on the principle of reciprocity central to Andean thinking. Reciprocity (*ayni*) cements Andean social and economic lives in the form of a vertically organized trade system between altitude zones; and reciprocity was fundamental in the political obligations between the Inka state and its subjects (Stone-Miller 2002:15–16). Reciprocity also characterizes the relations between man/natural world and man/supernatural beings and always invokes mediation and dialogue rather than dominance, performed in ritual in the plazas of the model under discussion.

Relationships of the highest order may have been called up by the rocks and boulders in this pattern. One fundamental concept that qualified Cusco as the capital and elevated it to the center of the empire was that the division of the four quarters of Tawantinsuyu originated from its main plaza Awkaypata. More specifically, Zuidema (1980:326) interprets Molina's description of the citua festival in the way that the four-partite division indeed began at the usnu. The document of the visit by Damian de la Bandera in the valley of Yucay in 1558 states that this valley was the personal property of Wayna Qhapaq ("como recamara suya"—like his private room) and that therefore the Indians who lived there did not belong to any suyu (Villanueva Urteaga 1971:94; see chapter 5). Witnesses Lucas Chico and Martin Cutipa, mitimaes from the Yucay Valley, confirmed that neither this valley nor Wayna Qhapaq's properties were ever part of any of the four provinces into which the empire was divided (Villanueva Urteaga 1971:129–131). In the context of this study, this establishes an intriguing political parallel between Cusco and the royal estates: they all may have been considered neutral spaces, types of centers, and certain origin places. The division of the four quarters started in Awkaypata and possibly at the Cusco usnu itself. I suggest that the pattern of plaza-kallanka-foregrounded rock at the royal estates evoked similar ideas.

While plazas and kallankas generally have practical functions, the special rocks are clearly the most powerfully charged feature in the pattern (see Christie 2003; 2005 for the mediating powers of rock wak'as). It should be recalled that Salomon (Salomon and Urioste 1991:19) recognizes clear correspondences between the organization of wak'as and human social structure. I speculate to summarize such interrelations as follows: Cusco—ruler—center—plaza and kallankas for people—stone usnu—libations for *ukhu pacha*, offerings to *hanan pacha*; royal estates—ruler—plaza and kallankas for people—stone wak'as. Ritual in the plazas, kallankas, and at the stones brought these interrelations alive through performance and acted them out. I argue that the pattern metaphorically centered in Cusco royal estates and any other settlement using it.

There may have been interesting links between the foregrounded rocks in our pattern, usnus, and Inka origin places (see Christie 2007a). A connection with the origin place Pumaurqu may have been established by Thupa Inka through his carved rock Pumaccacca in Chinchero Structure 11, which exhibits two puma figures among seats and platforms. Thupa Inka may have envisioned to symbolically transfer the memory landscape of the origin site of the Inka dynasty to his personal estate.

This analysis of the plaza-kallanka-foregrounded rock paradigm includes numerous case studies which have brought up multiple thought structures about stone ideology which were acted out and explained publicly on the human stage constructed by the features of the pattern. While there is the general overarching pattern, each occurrence is unique and perhaps personal in the way it was interpreted and manipulated by specific rulers. This is true in particular of the royal estates, which reveal that Pachakuti and his two sons, Amaru Thupa Inka and Thupa Inka, used the pattern in a strict sense, meaning that their special rocks were sculpted. I believe our pattern speaks to the individuality and creativity of Inka rulers and artisans in the way they adapted a specific design to local land formations and changing political contexts. This individuality has been a key challenge in our efforts to define clear models and patterns in Inka culture in general (see Christie 2006a; Hyslop 1990; Kendall 1985; Morris 2004; Protzen 2000, 2010 on Inka palace and royal estate architecture as well as urban design). Perhaps we should concentrate more on the individuality of Inka design, approach case studies from a local context, and by reconstructing from the bottom up find the point at which individuality and local interpretation were balanced and counteracted by state control, which is the very principle of Andean reciprocity or *ayni*. Chapters 5 and 6 are contributions in this direction.

CONCLUSIONS

The analysis of the carved rocks at Chinchero shows that Thupa Inka continued the tradition of rock art initiated by his father, Pachakuti Inka Yupanki. The style of the Chinchero sculpted outcrops is consistent with that at Pachakuti's royal estates and at most of the examples in the immediate vicinity of Cusco. It is intriguing that even in figure sculpture, representations of the same animals keep being repeated: felines, so-called "condors" or more general birds with outstretched wings, and snakes (compare with Laqo, northeast of Cusco, chapter 2).

However, one noted difference seems to be that Thupa Inka commissioned and/or remodeled[7] carved rocks in more public spaces. Pumaccacca seemingly erupts over the southeast corner of the Great Plaza. The rocks of Condorqaqa enjoy their own plaza space. The northeastern and northwestern carved rock complexes impress through their monumentality and their location on the main Inka road. They would forcefully have announced an Inka presence to anyone entering or exiting Chinchero. At the same time, though, from the trail, visitors would have recognized very few carvings since they grace the tops. I argue that the strategic position, size, and alterity of the rock complexes suggested the existence of carvings, that both the material essence of the stone and the carvings expressed concepts of mediation with the cosmos, and that therefore the visitor would have understood without necessarily viewing all the individual sculptures. Such reasoning further entails that Thupa Inka also subscribed to Pachakuti's ideology of stone. Indeed, he interpreted and applied it in more public and political settings whereas Pachakuti had still experimented with the formal language in some more private sectors at Machu Picchu. While the Chinchero carved rocks occupy rather open and public spaces, the ritual-political discourse enacted at them appears to have been internally oriented toward local audiences whereas Pachakuti clearly invoked the surrounding mountains at Machu Picchu.

NOTES

1. The boulder displays a long and meandering channel-like snake body terminating in a sculpted head. In the summer of 2001, I found offerings of flowers and coca leaves on top of this sculpted rock.

There is a circular structure on top of the hill directly overlooking the building group with the sculpted boulder. This tower-like structure looks very un-Inka and raises many questions with regard to dating and chronology.

2. I have discussed the northeastern and northwestern carved rock complexes as trail markers elsewhere (Christie 2008).

3. John Rowe (1967) uses five eyewitness accounts, records of the distribution of house lots to Spanish settlers in 1534, as well as some archaeological data to reconstruct Inka Cusco. These eyewitnesses were Francisco Pizarro and his secretary, Pedro Sancho de la Hoz, who wrote their accounts in Cusco in 1534; an anonymous source known as the *Noticia del Peru* written about 1535; Juan Ruiz de Arce, who produced his report in 1543; and finally Pedro Pizarro, who did not write until 1571 (Rowe 1967:59).

4. The ethnographic evidence is inconclusive on this issue. Betanzos (1996:21, 32–38, 42–43 [1557:part I, chaps. VI, IX, X]) refers to Caquea Xaquixahuana/Juchuy Qosqo as a "hilltop stronghold" and "fortification" before the Chanka wars. After Inka Yupanki's victory over the Chanka, he sent his lords to serve Wiraqocha Inka,

his father. The latter told them he needed help with a construction project: "he wanted to build a town there in that stronghold." The lords asked for a plan of that town using clay models to indicate the buildings. Cieza de Leon (1959:209–210 [1553:chap. XXXVIII]) speaks of a battle between Wiraqocha Inka and the people of Calca. After Calca had been subjected, Wiraqocha Inka ordered tribute payments and that the men from Calca should help him build two palaces—one in Cusco and one in Caqui [Caquia-Xaquixahuana] or Juchuy Qosqo. The archaeological data obtained by Ann Kendall (Kendall, Early, and Sillar 1992:208–210) place most of the buildings into the early phase of Inka reign, which would include Wiraqocha. It thus appears that the site of Juchuy Qosqo started out as a fortified stronghold and was enlarged to a town by Wiraqocha Inka before or in the early stages of Pachakuti's reign. Chronologically these activities cannot clearly be dated before or after the rebuilding of Cusco.

Following another line of reasoning, one might suggest that the eighth Inka ruler made specific political claims by adopting the name of the Andean Creator God Wiraqocha. Sarmiento (1942:91–92 [1572:chap. 24]) tells us that his name was Atun Topa Inga before he succeeded his father in office. Such political claims of creating a new era and order may well have translated into the architectural designs of Juchuy Qosqo and perhaps Cusco. On the other hand, the name *Pachakuti* taken on by Wiraqocha's son upon his accession means "mover of the earth/world" and seems no less ambitious and makes similar claims. Thus the ethnographic and archaeological evidence remains inconclusive as to the question where and by whom the pattern was first employed.

5. Vitcos, which is situated overlooking the upper Rio Vilcabamba near the present town of Huancacalle, has a small plaza lined by kallanka-like buildings and several nearby and closely spaced kancha groups. From this palace complex, an Inka road led south and abutting this road, a system of broad and well-fitted terraces was constructed upon which sit several carved rocks. The final destination of the road and the highest point of the terraces is the building group which encloses the sculpted Yurak Rumi Rock (see Lee 2000:figs. 17–26). This large boulder displays a few carved planes but its main function was clearly not that of an usnu-related seat. Its south side projects over a wetland and a wall constructed in precise Cusco-style masonry which exhibits four seats. It stands in close proximity to smaller carved rocks and is virtually surrounded by walls and buildings. A plaza is missing in the Yurak Rumi Group and has to be found in the palace complex of Vitcos. This makes Vitcos a weak link in support of the pattern since it was devised to address large audiences. If the pattern was at work at all in Vitcos, it targeted small audiences of elites and members of the imperial court. On the one hand, one could argue that the natural topography called for a separation of the elements in the pattern. The palace including the small plaza sits on a ridge, and the Yurak Rumi Group occupies the top of another hill, the side of which has been transformed into man-made terraces. In between, in a valley, extends the Central or Village Sector as Lee calls it (see Lee 2000:figs. 17, 20). During my visits to Vitcos in 1999 and 2004, I never noticed carved rocks on the ridge of the palace group. Perhaps the Inka did not find any suitable rocks on the palace ridge either but located an abundance of them on the hill to the south, which is now graced by the Yurak Rumi Rock, and these natural conditions compelled them

to separate the pattern. Finally, during my investigation in 2004, I identified several large worked building blocks high up on the hill with the Yurak Rumi Group which had seemingly been abandoned. This could indicate that the Inka intended to build more on this hill but were unable to complete additional structures, or that there were more Inka buildings which the Spanish destroyed.

On the other hand, of course, Vitcos may simply not share the pattern which would align it with Pisaq and Ollantaytambo, two additional and very important estates of Pachakuti located in the Urubamba Valley. Pisaq is divided into numerous sectors which crown the mountain ridge line overlooking the valley and the contemporary town of Pisaq. In this case, the topography was not conducive for elaborating the pattern since the ridge locations do not provide sufficient space to construct large plazas. Ollantaytambo as well had different sectors, one in the flat valley bottom of the Patakancha River and the other on the hillside which contains the famous so-called Sun Temple. There would certainly have been space for the pattern, but the Inka apparently did not use it here. This impression is based upon the structures and spaces visible today. But Ollantaytambo has been remodeled more often than Pachakuti's other estates: Manqo Inka used it as a military fortress; the late Inka architectural style of the Araqhama Sector; and the contemporary town which sits on top of part of the Inka site. Jean-Pierre Protzen (1993:50–53) has shown that there was indeed a plaza opening within the kancha zone, and he thinks this plaza was bounded by kallankas. This area has been built over with houses; and if there ever was a sacred rock, it has disappeared. Therefore the pattern cannot be documented for Ollantaytambo.

6. J. McKim Malville, Michael Zawaski, and Steven Gullberg (in press) have shown that the white boulder and its close vicinity functioned as an observation point of June solstice sunrise, which could be seen between the two Inka pillars on Cerro Saywah overlooking the Urubamba River (see also chapters 3 and 5).

7. Again, I am not proposing that Thupa Inka necessarily commissioned all the carved rocks at Chinchero. There was an earlier Ayarmaca/Killke occupation, and some of the sculpted boulders could already have been Ayarmaca/Killke shrines and sacred places. I do, however, think that Thupa Inka gave them the form we see today and established the spatial relations with other features of his estate as discussed above.

Chapter Seven

Discussion and Conclusions

In the preceding chapters, I have examined the formal elements, style, and history of carved rocks as well as the cultural context at a number of precontact and contemporary sites. The discussion has focused on Cusco, the capital where the style of Inka rock sculpting developed on wak'as of the zeq'e lines (chapter 2), on the creation sites on Lake Titicaca and at Pacariqtambo where Inka ruling dynasties legitimized their authority through their stony paqarinas (chapters 3 and 4), and on the royal estates of Machu Picchu and Chinchero (chapters 5 and 6). At Machu Picchu, Pachakuti placed carved rocks into calculated relations with surrounding mountains; and at both estates, Pachakuti and Thupa Inka initiated tactics to employ rock art in peripheral regions as markers of state identity. Archaeological reports, archaeological and anthropological fieldwork, and the Spanish chronicles were used as building blocks to reconstruct some of the functions of Inka sculpted outcrops. Many have been illustrated by drawings from my mapping project of approximately thirty carved rock sites, five of which had not been documented before.

In this sense, the methodology has been a mosaic work of adding data and pieces of information from multiple disciplines guided by theories that the relations between subject–object, the researcher, and the objects of study are never static but always dynamic, changing, and layered with power. My new evidence and findings are condensed in the term *stone ideology* used to characterize the political landscape the Inka formulated, following the model of T. Adam Smith (2003). My approach has been agent-based, attempting to write an "ethnography of the particular" in Abu-Lughod's words (Abu-Lughod 1991:158–159). Ethnographies of the particular focus in on local action by individuals who perform their lives through their bodies, minds, and words instead of reconstructing cultural patterns. This theoretical framework is mirrored in the structure of the preceding chapters in which the traditional narrative of the *Groundwork* sections presenting empirical data is juxtaposed by the segmentation into themes of the Inka Landscape and Carved Rocks analysis sections.

Pachakuti clearly emerges as the individual Inka protagonist. Yet one has to caution that the image of Pachakuti was largely constructed through the lens of Spanish writers and not from archaeological data (see Nair 2003:28). This means I do not know or claim that all personal details truly happened the way they have been described in the texts (see, for example, Betanzos 1996:71 [1557:chap. XVI]) or when precisely he acceded to the highest office. I do concur that the Inka had powerful and influential leader figures in the mid-fifteenth century that were instrumental in building out Cusco as the imperial capital, authoring an Inka style of rock art, reorganizing religious worship, as well as in economic intensification and imperial expansion. For the sake of clarity and by convention, I refer to them as Pachakuti Inka Yupanki and Thupa Inka Yupanki.

Beyond the theoretical emphasis on the particular, there is the pervasive and always recurring trope in Andean history on the timeless and layered qualities of the substance of stone and its superhuman manifestation in the mountains. In this sense, I present three main findings.

MAJOR RESULTS OF THIS INVESTIGATION

1. Pachakuti's Stone Ideology

I argue that Pachakuti institutionalized the Inka style of rock sculpting.

Evidence for Pachakuti's Authorship

Such reasoning is largely based upon documentation and dates for Machu Picchu which tie this royal estate to the reign of Pachakuti and to his panaqa (chapter 5). Machu Picchu and its outliers represent a grammar of Inka stonework and its constructed and shifting relations to the surrounding mountain landscape. The Spanish chronicles and archaeological survey work (Bauer 2004; Bauer and Covey 2002) also tell us that Pachakuti redesigned Cusco into an imperial capital, that he built up the economic infrastructure of terrace systems in the Cusco Valley, that he reorganized wak'a worship throughout the growing empire and instituted or revised the Cusco zeq'e system and others in different towns (Coben 2006). As discussed in chapter 2, four or possibly five of the nineteen carved rock wak'as on the Cusco zeq'e lines were connected with Pachakuti Inka Yupanki, and the fifth zeq'e of Chinchaysuyu was maintained by Pachakuti's panaqa (see also Niles 1999);

another six were associated with his son Amaru Thupa Inka, who was one of two supervisors of the Cusco wak'as and their required worship. Such findings are significant because they link the majority of the carved rock shrines to two individuals, the ninth ruler and his eldest son. Based upon the tight links between Pachakuti and the carved stone wak'as at Machu Picchu and on the Cusco zeq'e lines, it appears reasonable to suggest that he commissioned the Inka style of rock art analyzed in this study.

Yet such commissions were never shallow orders to his artists/artisans to cut certain outcrops into geometric volumes. Pachakuti turned formal elements and associated features into a sophisticated aesthetic language which may have co-opted earlier shrines but now was used to advertise Inkaness and state identity in a growing empire. I argue that Inka rock art became the core of a state-sponsored *stone ideology* spread and performed by Pachakuti but derived from many Andean practices surrounding stone wak'as. The carved rocks came to stand out as the materializations of certain concepts about the supremacy of the Inka state. This reasoning is grounded in Andean origin narratives (chapters 3 and 4), and Inka rulers chose to construct and publicly proclaim their history and origin as closely tied to sacred rock formations.

The Metaphorical Language of Style: Formal Discourse with the Natural and Social Worlds

The specifically Inka style of rock carving using vertical and horizontal cuts and geometric forms was intended to visualize the reciprocity that existed between the Inka and the natural environment. The range of modification reached from untouched natural rocks left in place and worshipped as shrines, to small rocks with a few simple carvings, tall and elaborately sculpted outcrops containing caves, carved boulders associated with walls, and finally to the precisely cut and portable Inka building stones which were put together to create new man-made forms of rock art: that is, walls. I suggest that the degrees to which natural materials were altered allow insights into the discourse the Inka carried on with their natural surroundings. The nature–culture binary opposition coined in Western cultures never applied to Inka ways of thinking. Rather, the land, man, flora and fauna all inhabited and shared the cosmos and, as animated and active agents, interacted in a give-and-take relationship which had to be kept in balance. Rituals did not mediate between dialectic poles but maintained harmony and order within a whole defined by everyday experience. This Inka view is very much consistent with the theory of phenomenology applied to this study, which advocates that the world and our own existence are materially and conceptually intertwined through lived experience.

Carving stony outcrops and thus altering the natural setting is clearly an act of exerting control. Yet the Inka would not have understood it as an artistic "conquest." In Quechua, one verb which describes the process of making something is *camay* (Salomon 1991:16). Camay connotes the act of endowing an object with life force and maintaining the flow of this vitalizing force from a prototype provider (*camac*) to the tangible object (*camasca*). The Inka state chose stone and its encapsulating mountains as their camac which was publicized in their oral narratives of origin and emergence from stony outcrops. State ceremonies and religious ritual invoked the stony camac of memory landscapes to perpetually energize its camasca, which were the boulder sculptures and ultimately the ruler himself who had taken up the responsibility of directing this flow of energy.

In terms of landscape iconography, sculpted rocks mediate between object and image (see van de Guchte 1996). Their place is always somewhere in between these dialectic poles. Some stones and carved rocks become vital nodes on trails/roads and conceptual lines/zeq'es since the latter outline a horizontal direction and stones/rocks access the vertical cosmic divisions. They mark a center and become a third entity in an encounter or *tinkuy* of opposing axes and opposite forces, and offerings have to be placed and rituals conducted to maintain equilibrium in the animated world of the cosmos.

Likewise, Andean people live in opposing ecological zones, and trade and barter their products grown in differing altitudes. This system of symbolic and economic reciprocity/ayni has been fundamental to Andean culture in pre- and postcontact times. It implies that opposing elements have to come together and complement one another in order to form a functioning holistic entity. From a similar perspective, ayni guided Inka interactions with conquered tribes and subdued subjects,[1] and I see a conceptual parallel in the way carved rocks and administrative strategies reflect reciprocity. Varying landscape settings called for particular modifications of stone, and different sites and situations demanded differing degrees of give-and-take relations with local populations (D'Altroy 2002:232–234, 327). This may explain why Pachakuti commissioned sculpted rocks in strategic outlying areas to claim Inka presence and to export Inka mythology to Inka-conquered tribes (Christie 2013b, 2014). In other words, he co-opted the ancient Andean custom of venerating rocks and mountains and turned stone wak'as vitalized by Inka camac into new geometric forms, which he advertised as an Inka visual language to mark Inkaness and the presence of the state. Simply speaking, Pachakuti transformed traditional Andean stone worship into a political power tool of his empire. A parallel is found in architecture. Guaman Poma de Ayala (1987:356–357 [1615:352]) describes and illustrates how Inka noblemen built boundary markers (*mojones*) in the geometric and precisely fitted Cusco masonry style throughout the empire (see also Nair 2003:51–52). I refer to all these practices as *stone ideology*.

In a sense, all rocks share this inherent animistic potential, but not all rocks were or are wak'as. The Inka preferred in particular boulders and outcrops which exhibited cave-like overhangs, were associated with water sources, and/or displayed some unusual visual characteristic. Van de Guchte (1999:151) argues for an Inka aesthetic of alterity or difference in the way they selected wak'as, which was an integral aspect in the Inka cognition of landscape embedded in a state-controlled practice blending mythology, geophysical reality, and political ambitions. Anthropologists have frequently talked about culture as a system of shared meaning. Recent discussions, however, have emphasized that the production of meaning is always the result of power relations (see Sharman 2006:843–844). In our context, I argue that Andean people were used to experiencing stone within their ayllu organization (for example, huanca/field guardian and mojones/boundary marker stones; see Duviols 1979b); the Inka gave the rock wak'as at their state installations a new form of sculpted geometry which mirrored the order and control of the state as well as the energy of Inka camac. As van de Guchte (1999:155) emphasizes, the Inka selected stones, rocks, and boulders to become wak'as but could also change them back into non-wak'as (*wak'a atisqa*). These selections clearly reflect the processes of power involved in bestowing meaning or non-meaning upon rocks, which I will contextualize further.

Lived and Performed Discourse with the Natural World

Exactly how the Inka forged and maintained reciprocal relations with the natural setting through carved rock wak'as is a fascinating but still elusive inquiry. We have gained certain insights from contemporary Quechua and Aymara practices (Allen 1998, 2002; Bolin 1998, 2006). From a linguistic perspective, stone was seen in the context of durable and permanent things; it was linked with mountains which had *camac* or a vitalizing force which enlivened carved rocks, their *camasca* (see above; Salomon 1998; Salomon and Urioste 1991). Early colonial Quechua words for stone, for example, *rumi* and its derivatives (Gonzalez Holguin 1989:319–320 [1608]) carry connotations of hard and precious substances. The word *uma* and its derivatives link the concepts of head and mountaintop with precious stones (Gonzalez Holguin 1989:354 [1608]; see the enqaychu, Bolin 2006). Fundamental are the multiple and complex layers of mediating qualities—between high and low, sky and river valleys, hard and soft—stones/rocks seem to be imbued with a mountain camasca.

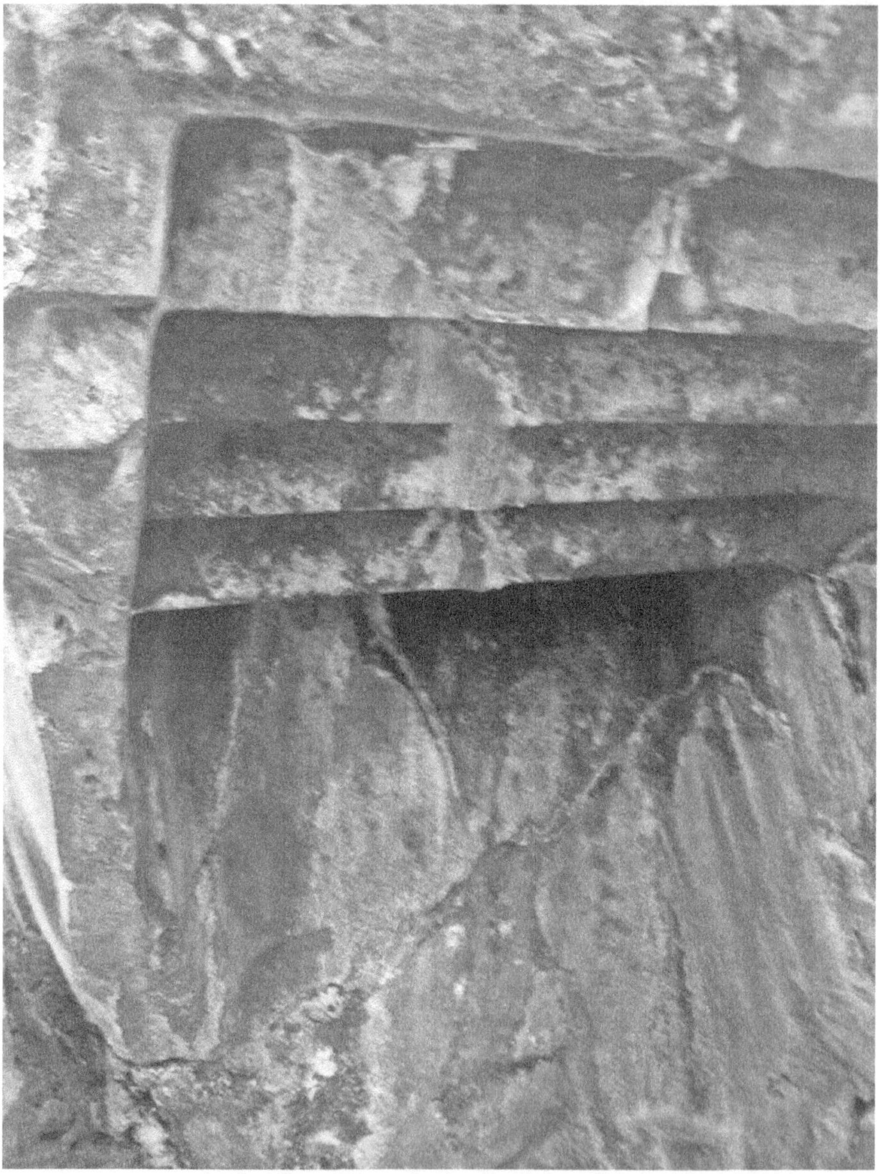

Figure 7.1. Cusco, Saqsaywaman, overhang with carving stepping in and down (as opposed to up and out). Photograph by Brian Garrett.

Nevertheless, such a permanent state of an ordered cosmos was always in danger of being undermined. A scenario of this kind may be alluded to in certain carvings which step in and down as opposed to the common stepping up and out forms which create the steps and platforms. Several such stepping in sculptures can be found at Saqsaywaman near the reservoir area (figure 7.1). Allen (1998:20–22; unpubl. manuscript) has carefully documented relations between people in Sonqo and domesticated animals and utensils, both of which are treated as living creatures. These interactions continue beyond death with the important change that positions of dominance and dependency are reversed. In this liminal world of inversion, it is now humans who suffer from and are at times tortured by the animals and objects which served them for as long as they resided in the world of the living. Steve Bourget (2006:179–181, 233–235) has reconstructed a similar world of ritual inversion consisting of three layers for the Moche.[2] Salomon (1998:9–11) likewise speaks of a third liminal world called *Uma Pacha* where the life forces of the deceased have been stored and continue to interact with the living. Bourget (2006:210–213) argues that the theme of inversion was deliberately expressed in material terms by a stirrup spout bottle which had been carefully placed in an upside down position into a niche in a tomb in San Jose de Moro. I suggest that the step-in carvings constitute the Inka vocabulary for this concept of inversion. Thus the stepped-in overhang at Saqsaywaman marks a liminal place at which the vital energies of natural elements, such as stone, could be approached; it is also a potentially dangerous place where power configurations can be inverted (see below). Archaeologist Sabino Quispe Serrano shared with me in 2004 that the Ministerio de Cultura discovered burials in natural vaults at Saqsaywaman with carvings stepping into the earth, so to speak.

More generally, Spanish writers report that stones (*pururaucas*) turned into warriors to assist Pachakuti during his battle against the Chanka; Thupa Inka spoke with carved stone idols (Guaman Poma 1993:193 [1615:261]); a large stone block—the Piedra Cansada or Tired Stone in chapter 2—cried and resisted to be moved any further (Guaman Poma 1993:123 [1615:159]); and mountains were personified. These selected examples illustrate that the world of stone was alive, endowed with a timeless life force which defied death, and the Inka actively engaged with this world. Nature in all forms was composed of numerous agents with whom the Inka sought to negotiate dynamic reciprocal relations.

Stone Ideology Inscribes the Inka Imperial Landscape

What made Inka engagement different from pan-Andean engagement with stone is that it was channeled by the state in the imperial vision of landscape.

While local wak'a worship continued on one level, the Inka ruler and his officials activated through ritual rocks as strategic wak'as in state-sponsored settlements, on zeq'e lines, and on roads. I have argued elsewhere (Christie 2006b; chapter 3) that the Inka may have been implementing an extended zeq'e system that would eventually have encompassed and integrated the entire empire and reconstructed it as an imperial landscape. This line of reasoning explains the existence of isolated and far-outlying carved rock complexes, such as Sayhuite, Ingapirca, Samaipata, as stations or end points of extended zeq'e lines (for a map of Inka carved rocks, see van de Guchte [1990]). This extended zeq'e system has remained incomplete, and only a few lines can speculatively be postulated based upon outlying carved rock complexes. However, its existence is partly supported by the consensus of numerous colonial and contemporary writers that other Inka zeq'e systems existed outside of the area of Cusco (see Bauer 1998:143–154; Polo de Ondegardo 1916 [1571]) and by the documented facts that the conceptualizing of ray centers and straight ritual lines is profoundly Andean (Christie 2008).

There was an evident political component in the Inka strategy of forming an *inherent* (see van Dommelen 1999:277–278) Inka landscape by dotting roads and conceptual lines with carved and/or unmodified but foregrounded (emphasized) rock wak'as, which was part of the strategy of stone ideology. The recognition that stone has powerful animating essences is essential pan-Andean memory (for Chavin de Huantar, see Burger 1992; for Tiwanaku, see Kolata and Ponce Sangines 1992; and for postcontact stone idols, see Pablo Jose de Arriaga 1968:201 [1621:chap. II] and Cristobal de Albornoz 1967 [c. 1582]). Clearly, stony places, such as special rocks, caves, and mountains, as well as sculptures, had always been Andean locations where religious and political powers were negotiated in some form (see Chavin and Tiwanaku). I argue that the Inka co-opted this general form of Andean stone worship and manipulated it for their own political strategies. What is specifically Inka is the style of carving and the calculated placement of the rock wak'as. These state wak'as replicated aspects of the Inka stone paqarinas (rock outcrop on the Island of the Sun, Pumaurqu) and constructed Inka territory as an Inka version of the Aymara space–time concept of *taypi*, which implies an original concentration on a central place of origin (rock sanctuary on the Island of the Sun/Cusco, the capital) followed by a phase of spatial diffusion and extension (paqarinas of individual ayllus or lineages/wak'as on zeq'e lines and roads) (Bouysee-Cassagne and Harris 1987:19–21; chapter 3). This spatial extension was likely ordered by Pachakuti into macro- and micro-suyu and zeq'e systems in order to reciprocate administrative, economic, and ideological needs (Duviols 1976). Roads and stone wak'as defined many of these divisions. Rostworowski (1999:138–139) reconstructs similar cross-links of physical space and political power in the settlement pattern in and around Cusco: the royal panaqas and ayllus of the nobility resided in the center of Cusco; each senorio (Andean chiefdom in the Late Intermediate Period) annexed by the Inka was required to send one of its dual chiefs to live in the capital to ensure its loyalty; these chiefs occupied the geographical suyu corresponding to their home within the city, and those who had been integrated earlier resided and owned land closer to the center (see also Nair 2003:112).

Sherbondy (1992:61–62) argues that the Inka used the Sea as a symbol that would help unite all the diverse people they had conquered because all Andean groups respected the Sea or the Pacific Ocean as the source and end of all water. To apply this symbol to their political advantage, they developed what Sherbondy (1992:62) calls a water ideology, which included placing the Sea at the religious and political center of the empire by filling the central plazas of Cusco, Awkaypata, and Kusipata with sand two-and-a-half palms thick brought from various Pacific beaches. This sand stood for the Sea, which highland and coastal peoples revered as an origin place. I would add to this the symbol of stone and the concept of stone ideology. Stone in its camac form as mountains was visible to all highland and most coastal peoples; and for all Andean inhabitants, mountains constitute the pole in the great water cycle opposite to the Sea and are paramount sources of water because rivers originate in their high altitudes. Thus water, stone, and mountains are intimately linked in Salomon's (1991:15) image of the world mountain around which water circulates. The Inka thus manipulated the elements of water/sea and stone/mountain to build a political landscape in the sense proposed by Smith (2003), which interwove the natural and man-made landscapes in a network of geopolitical and perpetual memory relations.

2. Grid-like Carvings Functioned as Counting Devices

A second important argument is that some grid-like carvings were possibly used as counting devices or yupanas in an economic context. Primary examples are Sabacurinca or the Throne of the Inka at Saqsaywaman (chapter 2), the boulder carved into grids in the northeastern sector of Chinchero (chapter 6), and Intinkala as well as possibly two rows of grid-like carvings on the summit of Copacati in Copacabana (chapter 3). This interpretation is reinforced by Cummins's (2007:277–278) cogent analysis that "Andean objects cross over or through genres, media, and technologies to perform various meanings in relation to other objects so as to form modalities of expression that become specific by place and arrangement." He attributes this to camay, the vitalizing force in every object (see above). Although Cummins focuses on drinking vessels and textiles, a similar potential is clearly implied for other objects and differing media. Thus a yupana could have been created in different settings and by various techniques, which include sculpted outcrops.

3. Stone Ideology Turned Contemporary

The third major point has been that the veneration of stone has been an Andean tradition with great time-depth documented from the Pre-Ceramic Period to the present. What set Inka cults of stone apart was that the selection, shaping, and performances at stone wak'as were aspects of the political strategy of stone ideology and the resulting political landscape masterminded by Pachakuti.[3] Here I will present two additional cases of contemporary ritual practices at carved rocks and then explore some of the power relations they construct today which appear to have shifted from a state-sanctioned tool of political authority to social processes shaping a national identity and perhaps new political constellations.

Contemporary Rock Wak'as: Tetecaca in Cusco and Mesarumis in Chillihuani

The first example represents an in-town setting known as Tetecaca. This rock complex is located on a hillside within the eastern suburbs of Cusco overlooking the city. The Ministerio de Cultura has designated Tetecaca an archaeological site. It is a large carved outcrop draping over part of the hill. The most spectacular carvings can be found in the upper section: the top and approximate central axis of Tetecaca is marked by the small panel which van de Guchte (1990:96–97) first identified as the depiction of a terraced landscape. Although the conceptual similarities between stepped carvings and agricultural terraces (andenes) have been noted by van de Guchte (1990:95–98, 234–236) and me, this is one of the very few examples in which Inka artists attempted to copy a hillside with straight running and curving andenes as a small-scale landscape. Over this landscape model and continuing below ran a carved canal. It appears to have emptied upon the carvings on the level below.

Other intriguing sculptures in the upper section are a large rounded knob with a small depression on top, a basin with a canal, precisely cut seats and steps. The northeastern side of the rock formation displays carved depressions which likely footed a wall that shielded the sculptures from public view. The lower section of the rock complex mostly displays large niches and large-scale seat-like platforms.

It is most interesting what local people have made out of this Inka site probably only during the last century. The lower section has been turned into a sumptuous Catholic shrine dedicated to Christ on the cross (figure 7.2). Behind a locked glass case rests a large painted cross draped with thick velvety

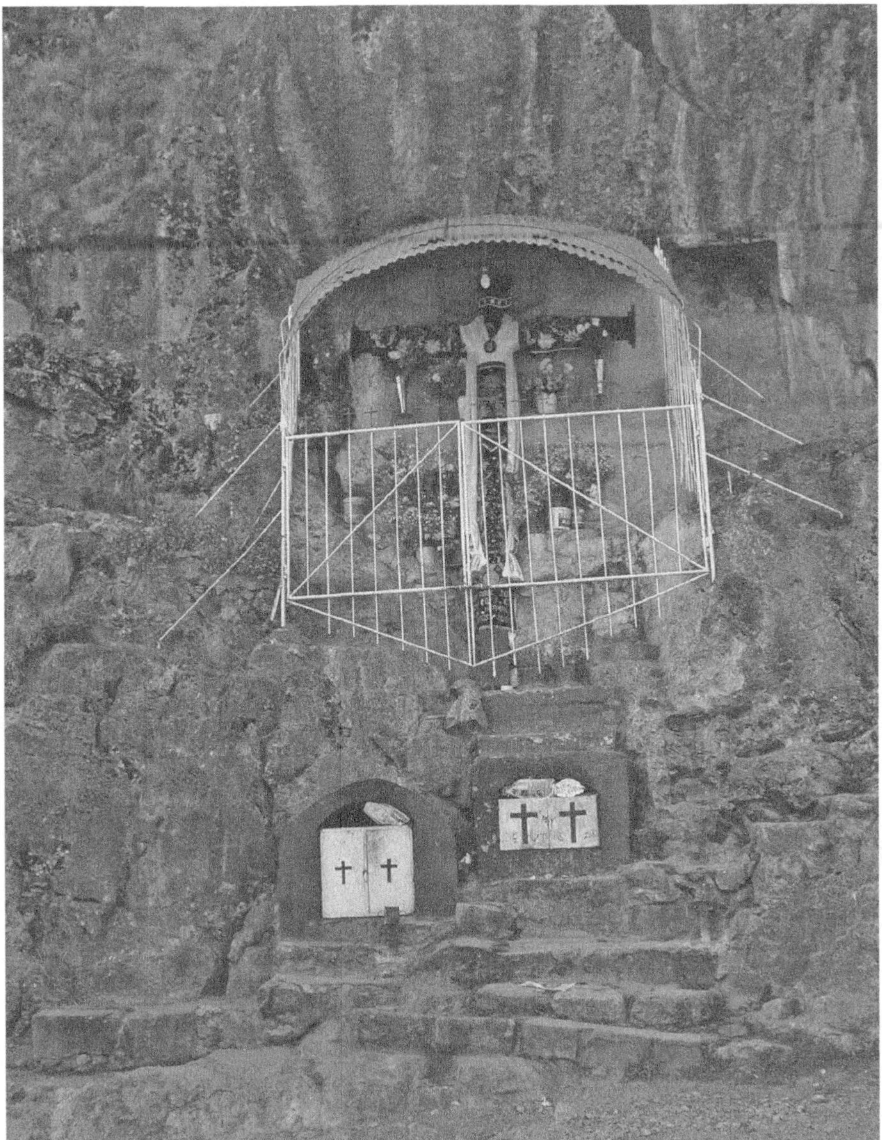

Figure 7.2. Cusco, Tetecaca rock outcrop, Catholic Cross shrine in the lower sector. Photograph by Brian Garrett.

cloth adorned with gilded embroidery. The shrine is always decorated with candles and flowers placed in containers upon the carved shelves. There is also a carved depression to hold holy water; yet it was dry in June 2005. Most Cusquenos who walk by the shrine stop to cross themselves. Every year on May 3, the Day of the Cross, the cross is taken out and paraded in procession (for a description of the annual celebration, see Paz Flores 1992). Further, in the summer of 2001, I observed a neighborhood meeting taking place in front of Tetecaca. This meeting included a certain amount of social drinking in addition to the business part. Paz Flores (1992:230) reports that about 30m uphill from the Tetecaca outcrop sits a second smaller boulder with a carved niche. Local people in the neighborhood perceive it as a "paqarina/origin place from which all the animals and plants people eat emerge" (Paz Flores 1992:230). He assures the reader that the faithful who arrive from rural areas pray like Catholics at the foot of the cross, but in their hearts they petition the paqarina. Thus the sacred potency of the rock and the cross and associated Andean and Christian practices of reverence have become fused.

The upper section of Tetecaca has experienced very different kinds of uses. It is here where people from the neighborhood have often gathered and sat on the carvings to converse and socialize; some individuals have eaten snacks and left trash, and others have gone to the bathroom. After forms of vandalism increased in 2003 and 2004, tall concrete fence posts were erected between the upper section of Tetecaca and the road in June 2005. The fence was complete a year later, and local authorities are taking measures to protect the site and restrict access. Local residents, on the other hand, may perceive the fence as an act of taking a popular spot by the government.

Quite different settings of sacred rocks possibly worked by the Inka and still in use today exist in remote areas. Inge Bolin (1998) reports from Chillihuani located in the province of Quispicanchis of the Departamento Cusco. Bolin lived with the herders of Chillihuani at an altitude of close to 5,000 meters above sea level and has written about their customs, rituals, and views. Of special interest is her account of the lagoon *Waqraqocha*, which is shrouded in myths and stories and has a personality that aids and punishes human behavior (Bolin 1998:212–228). Near this important body of water, Bolin encountered three *mesarumis* or tables of stone. The first one she describes as being very large; it is associated with *Llaqtayoq Mach'aqway*, a great snake and the protector of Chillihuani, who used to sunbathe on this rock. Bolin illustrates the second one (1998:222): it is a small flat stone in front of a boulder with a flattened surface and a rounded projection which seems to have been worked. Her consultants identified it as "the guardian of the lagoon" (1998:221). The third mesarumi sits on the shore of the lagoon half immersed in water.

The cultural context of the mesarumis is radically different from the Cusco cases with respect to the fact that at the high altitude of the herding community of Chillihuani little outside contact or interruptions of people's traditional lifestyles have occurred. As Bolin (1998:221) describes, the ritual practice of letting coca leaves fall on the flat stone surface and reading the patterns displayed as they come to rest probably has a long history, and there is no need to invent traditions. The Chillihuani examples demonstrate that rural people continue to interact with selected rocks as empowered places of contact and manifestation of supernatural beings. Such practices probably constitute an unbroken tradition since early pre-Inka times,[4] only redefined during the Late Horizon when the Inka state overlaid a politicized imperial rock art style. The rural Andean customs of stone rituals have never been eradicated.

Overcoming Power Hierarchies Between Scholars and Andean Subjects/Objects of Study

First and foremost, any academic professional study in the Western sense sets up a hierarchical power relationship between the writer-scholar and the researched object-people, of which I have presented awareness throughout this book: the researcher places himself/herself in a position signaling higher education and a wider intellectual vision than the objects and subjects of research who are seen as confined in these respects and by implication as inferior. As Americans and Europeans, we take it for granted that science and technology are preparing a better future, and art history, archaeology, and anthropology are vehicles of this kind of progress in their own right. Individual researchers discover new empirical data which replace old ones until they arrive at the final answer or ultimate truth which we consider a universal; value hence it advances humankind.

Andean people have seen the world from a totally different perspective. In their eyes, all living beings as well as inorganic objects are endowed with a vital energy or animate essence which dynamically interacts between its avatars. These life forces have to be kept in a state of fluid equilibrium in order for all beings to prosper. The faraway future beyond the next agricultural cycle is of little concern. Traditionalists focus more on the here and now and relations within the community and with the natural world. In this sense, the personal discrepancies in the explanations given by Porfirio Huanca Sanchez and Ramon Calle of the rock art site Copacati (chapter 3) do not matter; what does matter is that both recognize this Inka site and its rock carvings as liminal places where vital energies can be accessed. On this level, continuities exist with the precontact past, and the power of stone perseveres. How these powers and life forces are approached and related to in ritual is a personal matter and preference of each ritual specialist. When we juxtapose this mindset to our Western thinking, we find no final answer or absolute truth since these animate essences are alive, dynamic, and they change. A ritual intercession successful today is not guaranteed to have the same outcome tomorrow.

It goes without saying that if ritual specialists have to continually test how to interact with the life forces of the cosmos, it is a daunting challenge for us to understand how they work more specifically. Instead of defined hierarchies, we are dealing with dynamic flows of powers. Ultimately, it is impossible for us to grasp animate essences to the extent a Quechua or Aymara traditionalist can, but the patient and extensive fieldwork of a number of respected anthropologists provides some important insights which I will briefly review here. The following observations refer to animism as a worldview as opposed to the narrower spirit forces in rocks and mountains the previous chapters concentrated on.

Two of the most perceptive and sensitive ethnographers are Catherine Allen and Frank Salomon. Allen has written extensively about the knowledge she gained from her consultants in Sonqo, a Quechua-speaking community in southern Peru (1998, 2002). She, as well as Salomon (1998), defines traditional Quechua selfhood as a "web of socioritual connections" (Allen 1998:23) which encompass relations with relatives, community members, sacred places in the landscape, wak'as, as well as with subservient beings, such as domestic animals and household utensils (see also Van Vleet 2008). In order to thrive, an adult is expected to maintain these relations in some state of balance according to the Andean system of social, economic, and ritual reciprocity. Part of this balancing act is internal and psychological. Allen (1998:23–24) adds interesting observations of how dissolution of this internal balance causes disease and even foreshadows death (see also Allen, unpublished manuscript). Salomon's methodology is linguistic and largely based on the Huarochiri manuscript.[5] He (1998:9–11) analyzes the ontological culture-specific meanings of Quechua verbs for aspects of being and describes the path of life all beings take as a progression: from a young, tender, fleshy, and fast-changing age; to a firmer and more resistant, static, and slow-changing adulthood; and finally, to the desiccated but enduring state of mummies and seeds who have left life but whose vitality has been preserved in a sacred space called *Uma Pacha* and continues to interact and to nourish the living. The most permanent of all beings are the hardest and driest: that is, the seeds and mummies of powerful ancestors; and in the natural setting, the geological features of mountains and stone. The interactions among beings situated at differing points in this progression in time and space are dynamic and fluid, circulating and ever renewing. He and Allen emphasize that the balancing act of life forces occurs on various scales beginning with internal balance within the self, to community relations, and extends into the physical landscape which is culturally transformed into a macro-replica or sinsign (in Peircean terms, see chapter 2) of the human internal and psychological balanced structure of animated essences. Allen (1998:23) cites the specific example of the duality of the self, which holds fascinating implications when extended to the organization of imperial Inka space and the marking of rock art sites (see chapter 2).

Ziolkowski (1966:30–34) discusses the equilibrium of life forces under the concept *hucha*. According to Gonzalez Holguin (in Ziolkowski 1966:30), the Quechua term *hucha* designates "sin" and "dispute, business (deal)." Ziolkowski richly illustrates with linguistic and ethnographic examples that in Andean thinking sin is conceptualized as a disruption in the balance of vital energies caused by failing to fulfill the obligations of reciprocity or *ayni*. Such an energy imbalance may have catastrophic consequences for the individual as well as the community, and sinners are required to do penance and make offerings in order to restore the initial order. When this established order does not produce the desired results, it can be elevated to a higher level, as did the kuraka *Caque Poma* (Ziolkowski 1966:33).[6]

The examples cited illustrate that traditional Andean power relations have never been strictly hierarchical but always fluid and dynamic, in a constant state of tension which had to be balanced through ritual. The major historical interlude was the Inka, who set up a well-documented political hierarchy.[7] Perhaps this is a reason why the Inka have received more scholarly attention than other Andean cultures, since their power concepts resemble ours and are easier for us to grasp.

Lila Abu-Lughod poignantly discusses the dilemma of how to collaborate ethically as well as productively with indigenous people and outlines strategies of how to overcome the self–other paradigm in her chapter with the provocative title "Writing against Culture" (1991; explained in the introduction). One such strategy is writing "ethnographies of the particular" (Abu-Lughod 1991:158–159) foregrounding local actions by individuals living their particular lives, inscribed in their bodies and words, as opposed to notions of cultural homogeneity and timelessness (Abu-Lughod 1991:150). This is what I have practiced in this book; see, for example, the explanations of rock art sites at Copacabana offered by Porfirio Huanca Sanchez and Ramon Calle (chapter 3).

As the discussion shifts from precontact to postcontact and contemporary empowered stony sites, the theoretical framework by necessity tilts more toward particular individual action as opposed to long-term cultural principles, since more localized ethnographic observations become available. Nevertheless, the following will reveal compelling continuities in the collective memory and social understanding of places centered upon certain stones and rock formations in the Andean landscape.

Relations with Stony Places Constructed in the Contemporary National and Global Worlds

Relations of power and the political landscape in Adam T. Smith's terminology, spun around sculpted rock outcrops today, of course, go well beyond the simple Western scholar–indigenous and local people dichotomy. They encompass relationships between Cusquenos and rural people, between neighborhoods and the Ministerio de Cultura, neighborhoods and city government, local residents and tourists, as well as the Ministerio de Cultura and related local branches and the government in Lima.

John Janusek (2004:3–25) sets this complexity into perspective in a cogent overview of the theoretical literature focusing on identity and power, grounded in practice theory which aims to weigh the impact of culture versus action on the processes of changing and reproducing human societies. Practice theory is the general perspective from which I have approached the carved rocks: Smith's model of political landscapes is more strongly oriented toward the stable pole of culture whereas phenomenology looks toward individual agency. We will now look at two more models aligning the binaries of culture and action under the umbrella of practice theory to help us better comprehend present-day activities around Inka carved rocks.

The first is the "figured world" concept coined by Dorothy Holland, William Lachicotte Jr., Debra Skinner, and Carole Cain (1998:49–65). They define "figured worlds" as "as if" realms collectively formed by a group of people. As they are socially and culturally constructed, particular characters are foregrounded, significance is assigned to selected acts, and particular outcomes are desired. Such a "figured, as-if" realm constitutes a simplified world populated by a calculated number of agents who engage in a limited set of relevant activities motivated by certain cultural forces (Holland et al. 1998:52). Holland et al. emphasize that figured worlds enter relations with larger, institutionalized power structures, which are based on rank and hierarchy (1998:57–58); and that they unfold around cultural artifacts, which the authors define broadly as possessing material and ideal aspects (1998:60–63).

A fitting example of a "figured, as-if" world are offerings (pagos) conducted by contemporary shamans, some of which take place at sculpted rock outcrops. As outsiders, we can hire our own shaman; and on this occasion, we can physically experience and mentally perceive the pago to pachamama, the earth. At other times, we find the burned leftovers at or near carved boulder sites and are left to imagine the process of the ceremonial offering (see chapter 2). In late May 2005, a traditionalist shaman from Q'ero in the Ausankati region came to our hotel room in Cusco and performed a pago to pachamama and the apus so that we would remain unharmed on our upcoming trek. He first spread his carrying cloth on the floor, and on top a large piece of paper, and began to assemble small objects to be offered to pachamama on the left half of the paper and to the apus on the right half. Pieces of gilded foil and small sticks wrapped with gilded foil as well as candy and sweet wafers were for pachamama, whereas silver foil and sticks wrapped with silver foil and salty crackers were designated for the apus. In the end, he folded the entire offering, packed it up, and said he would go and burn it at the Templo de la Luna (Temple of the Moon), which is Laqo (see chapter 2).[8]

I do not know whether our shaman really traveled up to Laqo that night and burned our offering for pachamama and the apus to feed on. During my first visits in 1998 and 1999, there was a large fireplace south of the south façade of Laqo. It was covered with rocks while not in use. Most likely, because of the increased number of visitors and vendors as well as the current excavations by the Ministerio de Cultura, this fireplace is no longer there, and the site to make pagos has been moved to the east of Laqo. The hillside northeast and east of Laqo is covered and crowned with boulders and outcrops, some of which are carved. Nearby to the east, there is a large natural niche in an outcrop with a flat bottom. On this table- or altar-like surface, offerings are burned (personal observations 2001, 2005).

We may imagine that similar offerings were performed by the Inka in pre-contact times. Holland et al. emphasize that figured worlds and their cultural artifacts as they explain them are not static but "constantly changing through the improvisations of actors" (1998:63). This is what I have shown throughout this book; and reviewing the cases described above, new sets of power relations, of dominance and hierarchy, may be sketched out:

a) Center–periphery distances contract: shamans from Q'ero travel to Cusco more frequently driven by economic pressure as well as opportunities. Here they sell their services to tourists from around the world, again for economic reasons.
b) Sacred–secular merge and are reinvented: offerings to the earth and the snow mountains are conveniently relocated in Cusco hotel rooms. Nevertheless, in the eyes of the outsider, the shaman emerges as a figure of great spiritual power and mystery who will go out at night and feed pachamama and the apus by burning his pago at a dark rock outcrop.
c) Inka/Quechua and Catholic traditions clash and are brought into a circuit at rock art sites, such as Tetecaca (see above, this chapter).
d) The sacred and the secular intensify in Cusco, the center, which crystallizes as the economic power. Memory agents of the sacred alluded to by the carved rocks, the shamans, and Inka heritage in general, project an image of Cusco and Machu Picchu as profoundly spiritual and otherworldly

> places in the global media throughout the world and fuel tourism, which in turn feeds Cusco's as well as the national economy.

Spirituality and economy appear to merge as the contemporary binary, yet at the same time tightly interwoven power players. Holland et al. (1998:63) relate these discourses to identity development.

The second model from practice theory is the process of inventing traditions discussed by Eric Hobsbawm (1983). Hobsbawm (1983:1) defines "invented tradition" as "a set of practices normally governed by overtly or tacitly accepted rules and of a ritual or symbolic nature, which seek to inculcate certain values and norms of behavior by repetition, which automatically implies continuity with the past." Hobsbawm (1983:2–3) clearly sets apart tradition from custom and from convention or routine: custom accommodates more variance than tradition, and the functions of convention/routine are technical rather than ideological. While all societies invent traditions, it occurs more frequently when rapid transformations alter the social context to which "old" traditions had responded, or when the institutional structures promulgating old traditions lose their authority and influence. As a result, people make new ones which address their changed needs (Hobsbawm 1983:4–14).

Multilayered examples of such invented traditions are again the pagos discussed above as well as the yearly Inti Raymi celebration which was initiated in Cusco and is now being copied and adapted by many communities in the southern Andes, including Copacabana (see chapter 3). The Inti Raymi was the Inka June solstice festival partially documented in the chronicles. Although the organizers claim historical correctness in the display of the colorful costumes and performances, the Inti Raymi has been reinvented to serve new economic, social, as well as national needs: it is a major tourist attraction and high-priced VIP-seating tickets which include gourmet lunch boxes quickly sell out. On the other hand, the Inti Raymi draws thousands of nationals, Peruvians who boast the Inka as an important foundation of their past and of the present notion of a national identity. The administration in Lima pressured by heart-wrenching social needs, continues to fund archaeological work through the Ministerio de Cultura, which aims to reconstruct Inka sites and advertise them as Peru's unique cultural heritage. In Bolivia, the solstice celebration at Copacabana is certainly a variant of the Inti Raymi, yet the word "Inka" is rarely mentioned. From the Bolivian perspective, the Inka have been appropriated by Peru, and authorities bolster their festivals as expressions of local Aymara (invented! as adapted from Inka customs) traditions. Carved rocks participate as cultural artifacts around which many performances are staged.

The issue of "authoring" mixed and new identities fitting into the world of the twentieth and twenty-first centuries is probably quite different at Chillihuani, a village with little outside contact as discussed above. The cultural context and the rituals conducted at the mesarumis are probably not invented traditions but practices passed down through generations through a process in which they have been reproduced but also modified following evolving needs. Holland et al. (1998:169–191) focus on the agent-based aspect of "authoring" the self by drawing on multiple theories from linguistics, anthropology, and psychology. They treat the self as "a position from which meaning is made, a position that is 'addressed' by and 'answers' others and the physical and cultural environment. In answering, the self 'authors' the world—including itself and others" (Holland et al. 1998:173). This viewpoint is fully consistent with phenomenology. We might say the Chillihuani herders continue to author themselves and their world in a space with little outside disturbance.

CONCLUSIONS

This book has attempted to balance a theoretical framework grounded in human agency with new empirical results documenting the cultural mechanisms through which Inka politics, economy, history and memory, ritual, and identity have been performed at carved rocks.

Saqsaywaman as Materialized Summary

The three main findings presented here are beautifully visualized in the authoritative Inka architectural complex whose appearance and perception continue to change due to ongoing work by archaeologists from the Ministerio de Cultura: Saqsaywaman. Here I can only touch upon the complex interwoven issues, leaving a possible monograph as a promising and fruitful future project. Saqsaywaman is a monumental complex conjoining architecture, rock outcrops, and rock art, and overlooking Cusco on one of the hills to the north of the city. The popular literature refers to Saqsaywaman as "Fortaleza" or fortress because the Spanish captured this complex and used its massive walls for their protection when Manqo Inka attempted to reconquer Cusco in the 1530s (see introduction). It is noteworthy that Saqsaywaman takes center stage in the Cusco coat of arms granted by the Spanish crown in 1540 (Dean 2010:148–149). It is portrayed like a European castle tower composed of geometric stone blocks and is thus effectively colonized.

Inka Saqsaywaman had many diverse functions mentioned by the Spanish writers. It would go well beyond the scope of this summary to evaluate individual statements, but a very potent remark comes from Garcilaso de la Vega (1989:469 [1539–1616:B. 7, Ch. 29]): "for it (Saqsaywaman) was considered a house of the Sun for arms and warfare, just as the temple was a house of the Sun for prayer and sacrifice" (also cited in Harvey and Harvey 2005:12). Garcilaso attributes the planning and procurement of building stones to Pachakuti Yupanki, the largest share of the construction to Thupa Inka Yupanki, and dates the completion of Saqsaywaman during the reign of Wayna Qhapaq.

The complex of Saqsaywaman may be divided into three sectors (see Gasparini and Margolies 1980:283, fig. 276): the northern sector with the reservoir where tall bedrock formations and rock art abound; the southern sector in which the architectural constructions of the towering zigzag walls and numerous building foundations are concentrated; and the flat, open intermediate space serving as performance area.

In the northern sector, the bedrock consists of unusual highly undulating compositions of diorite and limestone. Following van de Guchte's (1999:151) aesthetic of alterity, such conspicuous geophysical features would surely have attracted the interest of the Inka, and they may have been one factor in choosing the building site. This northern sector is perforated by caves, graced with rock sculptures, and the reservoir is surrounded by a complex system of now interrupted canals and probably served once as a water storage basin. Some of the rock sculptures form short grids which could have been used to count offerings. On a conceptual level visualized by the stonework, the northern sector would be understood as the realm of natural, untamed forces where limited human interactions or measures of control are evident. Piedra Cansada, the monumental building stone which refused to be moved any further (chapter 2), sits a short distance to the north; its travel story has brought to life such outside contacts with peripheral regions.

The southern sector contains the architecture Saqsaywaman is so famous for: enormous and irregular stone blocks have been shaped and precisely fitted in Cusco-style masonry to create three ascending rows of zigzag walls. Foundations of various building types occupy the hilltop. In 2004, archaeologist Sabino Quispe Serrano working for the Ministerio de Cultura gave me a tour of the excavations begun in 2000. Killke structures were encountered below Inka walls in the southern sector, indicating that the Killke people were drawn to this hillside location with the undulating diorite formations before the Inka. The Inka buildings under excavation in 2004 were double-room structures sharing the longitudinal wall. Some of the compounds did not have a formal access and therefore were dark and exhibited stairs leading to an upper level, prompting investigators to link them with the Chingana in Garcilaso de la Vega's description of Saqsaywaman. Other compounds were identified as ritual quarters. The Inka deposited a block of diorite in the entrance to one of them, reinforcing symbolic connotations of stone materials. In another ritual quarter, one section was a food and/or chicha preparation area; a different section yielded special objects, such as small crystals, pieces of shell, and one game piece. Most likely, the Inka themselves filled this room to conceal it from the eyes of the Spanish. A canal once ran along the outermost wall on the hilltop collecting drainage water from other channels and diverting it down and onto terraces. This canal was clearly utilitarian but could have shared overlapping ritual aspects alluding to a water temple Garcilaso mentions as being part of Saqsaywaman. Ministerio de Cultura archaeologists have identified a second set of zigzag walls downhill looking toward Cusco. This new southern set of zigzag walls is distinguished from its northern counterpart by masonry of smaller stones (Sabino Quispe Serrano, personal communication).

On a conceptual level, the southern sector exhibits the largest quantity of stone blocks which were quarried and fitted together with meticulous precision and without the use of mortar to form retaining and freestanding walls. In other words, here we observe the highest degree of stone manipulation and control over nature.

The intermediate, flat, and open space between the northern and southern sectors likely functioned as a performance area where large numbers of people could gather for state-sponsored events. Today the main reconstructed ceremonies of the Inti Raymi celebration take place on this bounded esplanade.

The three sectors are physically and conceptually connected by the mechanisms discussed throughout this book: sight lines offer important visual links between the Throne of the Inka/Sabacurinca and the intermediate and southern sectors as well as extending beyond to distant mountains and nevados personifying apus including Ausangate/Ausankati (chapter 2). The three sectors are situated physically near each other on the gently sloping hillside so that any human being can quickly move from one to the other.

Conceptually, the northern and southern sectors constitute variants of the opposing yet interwoven poles we have encountered in many places discussed in the chapters of this book: the natural versus the man-made, untamed versus civilized, free versus controlled, periphery versus center. Yet Inka lived and visual culture makes clear that these dichotomies are not composed of neatly separated categories but that both poles crosslink and interweave; for example, the block of diorite deliberately obstructing the entrance to one of the ritual quarters marks one of the instances in which the untamed emerges in a highly controlled context.

Structurally, one could see the northern and southern sectors as materializations of a duality with two visually very dissimilar halves. The southern sector would clearly be the Hanan part due to its dominant and controlled formal language. It faces and confronts the capital city while the northern sector is not visible from within Cusco. The elevation distinctions are reversed since the Hurin or northern sector sits geographically higher. One might postulate a second duality between the two sets of zigzag walls. The northern three rows of walls constructed of immense stone blocks and oriented toward the esplanade are the dominant half; the southern zigzag walls show smaller stones although they directly face the city.

Geophysically, structurally, as well as conceptually, the flat open space of the esplanade acts as the mediating ground between the sectors and halves. Multiple levels of interaction and discourse were performed through rituals and ceremonies, a practice which continues today with the Inti Raymi. In this manner, Saqsaywaman binds together bedrock, Inka architecture, rock art, and state rituals and functions as a lead agent in constructing the image of an Inka heritage in Cusco, Peru, and the global world.

Dean (2010:144–152; 1998) has argued that Cusco authorities and native Andeans have reconstructed Saqsaywaman as a site of an encounter with the sacred rather than of military defeat. These notions began with accounts of apparitions of Christian saints who assisted the Spanish in overpowering Manqo Inka's fighters. Later, such numinous forces were transferred to the stonework of Saqsaywaman where contemporary shamans continue to leave offerings (or pagos). My understanding differs somewhat in that I suggest that there has always been a fundamental Andean belief in animated essences of stone which the Spanish never extirpated. In the Afterword, I will illustrate this with two cases from the recent past.

NOTES

1. Certainly, reciprocity did not determine all relations between the Inka and conquered peoples. The well-known policy of forcefully relocating thousands of mita workers into newly acquired territories was an authoritarian measure solely to provide administrative and economic benefits to the state. In military matters, reciprocal relations were offered to subject tribes; and if they were rejected, a brutal and all-out war was conducted.

2. Drawing his data from the iconography of ceramic vessels, he postulates that the performers in sexual acts, funerary rites, and sacrificial scenes may be classified into three broad types of subjects: human beings and eventual sacrificial victims inhabit the World of the Living; the living-dead, skeletal, and mutilated beings are housed in the World of the Dead; and individuals with supernatural attributes, sacrificial victims, as well as animals populate the Afterworld (Bourget 2006:225). Rituals enable passage from one world to another and entail multiple levels of inversion: for

example, the passage from the World of the Dead into the Afterworld implies that the deceased regains a certain form of life. Bourget interprets his findings that the vast majority of actors on the sex pots who perform nonreproductive sex belong to the Worlds of the Living and Dead while those who have reproductive sex are supernaturals and others from the Afterworld as strategies of inverted fertility which serve to link the living with the beings endowed with supernatural attributes, sacrificial victims, and animals in a dualistic relationship.

3. This line of reasoning does not exclude the likelihood that in pre-Inka societies stone shrines were connected with ayllu-based power structures (huancas personifying ancestors) or that authoritative religious cults, such as Chavin, used stony cave-like architectural spaces. What is Inka-specific are the style of rock carving and the strategic placement of the sculpted wak'as consistent with an imperial vision of landscape.

4. New research in the Norte Chico region, which encompasses the Rio Huaura, Rio Supe, Rio Pativilca, and Rio Fortaleza valleys on the central coast of Peru, has identified over twenty sites with monumental stone architecture. Complex social structures were supported by inland agriculture supplemented by a coastal maritime economy. Certain sites displayed modified stone slabs (*wank'a*): for example, Caral, Caballete, Huaricanga (Haas, Creamer, and Ruiz 2004; Matthew Piscitelli, personal communication 2013).

5. The Huarochiri manuscript is the only collection of narratives of a precontact religious tradition written in Quechua. It originated in the central highlands and was edited by Father Francisco de Avila in 1608.

6. For some time, Caque Poma had hoped to be promoted to a higher administrative position. So when the Inka ruler requested that he send girls from his area to Cusco for the qhapaq ucha ceremonies and sacrifices, he sent his own daughter. Such an act communicated that he gave more than he was obligated to: it created an unbalanced state of hucha and positioned the Inka ruler as his debtor. The only way to restore reciprocity relations was for the Inka to promote the kuraka to his desired status. In terms of hucha, the kuraka committed the "sin" of sending an excessive gift which could only be compensated for with the higher position requiring what we may view as a sort of "business deal" from our Western twenty-first century perspective. Ziolkowski (1966:34) defines "hucha" as any activity against the established rules of reciprocal energy transmission which resulted in an energy imbalance.

7. I am not asserting that the Inka were the only Andean state-level society. The Chimu practiced a well-defined and ethnographically documented political hierarchy. In the Middle Horizon, Wari and Tiwanaku were clearly states, but internal power relations are still not well understood in present scholarship. The same could be said for Moche and Nasca.

8. Many local people know Laqo as the "Temple of the Moon." The reason for this name lies in one of the caves which terminates at a rock sculpted in the form of a table or altar. Above is an opening to the outside through which moonlight shines on the altar at certain times in the lunar cycle.

Afterword

I would like to conclude with two recent facets of identity expression grounded in Inka stonework and associated worldviews. The first is localized in Cusco, and the second has been fomenting in many Latin American countries during the past decade.

In the mid-twentieth century, the discourse over Bingham and the intervention of U.S. institutions in Peruvian cultural heritage kindled passionate feelings of regionalism in Cusco. Traditional Cusquenos perceived that outside entities in the name of universal and ever-advancing science threatened local knowledge and wisdoms, and they resorted to propagate their local cultural heritage and promote a regional economy. At the same time, Cusco regionalism has had to be counterbalanced with the dynamics of the city's interests on the national scene and now with the global world which has brought tourism as the primary economic factor.

Cusquenos have found themselves in an ongoing identity formation process. On the first and most important level, multiple strategies were launched to recapture and defend Cusco as a "sacred city" whose conceptual space included Machu Picchu: for example, in 1913 and still in the wake of the Bingham debate, the intellectual elite founded the Instituto Histórico del Cusco with the mission to study and publicize local history, to preserve ancient sites, and conduct archaeological investigations. Yet on a second level, Cusco is also constructed as an "ideal city" which, during Inka times, integrated other Andean cultures and now partly wants to and is coerced to open to the world for economic reasons (Lopez Lenci 2007:310). It is most intriguing that one of the Cusco intellectuals and writers, Uriel García, has proposed to condense the complex history of Cusco and Machu Picchu in the materiality of stone "in which the natural, the naive, and the sublime merge":

> In the fog of deeds transcribed by fantastic tales begins the history of Cuzco. And the material element in which the naïve fantasy of the indian incarnates itself is *stone*, and stone is the concretion of history, it survives triumphantly the hardships of destructive weather, it keeps the subtle secret of the soul of the indian, *it is the synthesis of the complicated life of the Inka*, and it is in *stone* that centuries have accumulated traces of their infinite course. . . . (cited in Lopez Lenci 2007:308; translated and emphasis added by the author)

Uriel García states that elements of a stone ideology have been carried over into the present: rock art is, of course, no longer the power tool of a particular state government, but it now lies at the heart of an ideology of national identity. The materiality of stone continues to be the tinkuy between the Inka past, colonial history, and the present; between man-made constructions and their natural environment; and between traditional Quechua worldview and a twenty-first century mind-set which primarily reproduces Inka stonework as a national icon in the tourism industry for economic purposes. Stone constructions and wak'as also feature in the works of many Cusco artists. As Andean people have always known that the static and timeless qualities of stone lend it its power. Lopez Lenci (2007:317–18) optimistically projects these ideas into the future by means of Andean cyclical time and the traditional concept of paqarina perceived as a contemporary and active agent: following the cyclical movement of history, a new Manqo Qhapaq will emerge from a new paqarina situated in Cusco, who will found a new American culture balancing the Hispanic and indigenous heritage.

The second case concerns the new visibility of sentient beings of nature, such as mountains, rivers, and the land, in the political arenas of Peru, Bolivia, and Ecuador. Marisol de la Cadena (2010) poignantly unpacks the notion of indigeneity as the presence of earth beings demanding a place in politics. She analyzes indigeneity as being partially connected with government institutions, forming an entity that is more than one but less than two (de la Cadena 2010:347). As such, indigeneity and its proponents threaten to rupture traditional politics–science, culture–nature dichotomies derived from the European Enlightenment era and grounded in the notions that human beings are ontologically different from things (including elements of nature) and as such the only actors worthy and capable of engaging in the antagonisms of politics.

De la Cadena (2010: 350–357) analyzes mountains as sites of equivocation where views from different worlds, rather than differing perspectives about the same world, collide. In recent anti-mining demonstrations, environmentalists and Quechua traditionalists have formed uneven alliances, often leading to successful outcomes. Although both groups fight against pollution and for the preservation of cultural heritage lands, Andean traditionalists view mountains, such as Ausangate/Ausankati and Machu Picchu, as sentient earth beings. They are living entities with likes and dislikes, who would feel the pain and amputation from present open-sky mining technologies, and who wield power to interfere in the lives of communities. De la Cadena (2010:357) positions mountains like Ausangate/Ausankati in the global context of "emergent forms of life," which establish networks of socionatural relations which make life locally possible. Their destruction, on the other hand, does not only cause environmental damage, but at the same time threatens survival of the local people. Thus the movement of indigeneity with loud voices and global public visibility has come to undermine the Western paradigm of nature–culture, science–politics and has opened doors for new relationships to be forged between diverse worlds which would be interconnected without being made commensurable. It appears that mountains and stone as their existential building block will assume new roles as agents and actors in emerging global and pluriversal power constellations. In such a scenario, mountains like Ausangate/Ausankati would count as beings, contend with political issues, and their multiple ontologies would engage in a contemporary political landscape[1] that bursts the Eurocentric nature–culture, science–politics straitjacket. On December 21, 2010, Bolivia followed suit when President Evo Morales Ayma signed Law 071, the Law of the Rights of Mother Earth (Ari 2014: appendix 2).

NOTE

1. T. A. Smith's model of *political landscapes* (2003) is well suited to compare and contrast Inka and contemporary Andean landscapes, which has been practiced throughout this book by sketching out and reconstructing numerous scenarios of power relations between objects and agents.

Bibliography

Abu-Lughod, Lila. 1991. "Writing against Culture," in *Recapturing Anthropology. Working in the Present*, edited by Richard Fox, pp. 137–162. Santa Fe: School of American Research Press.

Albornoz, Cristobal de. 1967/1580(?). "Un Inedit de Cristobal de Albornoz: La Instruccion para Descubrir Todas las Guacas del Peru y Sus Camayos y Haciendas," edited by Pierre Duviols. *Journal de la Societe des Americanistes*. n.s. vol. 56, no.1: 7–39. Paris.

Alcina Franch. 1976. *Arqueologia de Chinchero. 1: La Arquitectura*. Mision Cientifica Espanola en Hispanoamerica II. Madrid: Ministerio de Asuntos Exteriores.

Allen, Catherine. "The sadness of jars: separation and rectification in Andean understandings of death." Unpublished manuscript.

———. 1998. "When utensils revolt. Mind, matter, and modes of being in the pre-Columbian Andes," in *RES* 33, pp. 18–27.

———. 2002. "The Incas Have Gone Inside: Pattern and Persistence in Andean Iconography." *RES* 42, pp. 180–203.

Angles Vargas, Victor. 1988. *Historia del Cusco Incaico*. 3 vols., published by the author in Cusco.

Ansion, Juan. 1990. "Como Calculaban los Incas," in *Quipu y Yupana. Coleccion de Escritos*, edited by Carol Mackey, Hugo Pereyra, Carlos Radicati, Humberto Rodríguez, and Oscar Valverde, pp. 257–266. Lima: Consejo Nacional de Ciencia y Tecnología.

Ari, Waskar. 2014. *Earth Politics. Religion, Decolonization, and Bolivias's Indigenous Intellectuals*. Durham: Duke University Press.

Arkush, Elizabeth. 1994. *Pilgrims and Emperors: Small Inca Ceremonial Sites in the Southwest Lake Titicaca Basin, Peru*. Paper presented at the 64th Annual Meeting of the Society for American Archaeology, Chicago.

———. 2005. "Inka Ceremonial Sites in the Southwest Titicaca Basin," in *Advances in Titicaca Basin Archaeology–1*, edited by C. Stanish, A. Cohen, and M. Aldenderfer, pp. 209–242. Los Angeles: Cotsen Institute of Archaeology, University of California at Los Angeles.

Arriaga, Pablo Jose de. 1968/1621. "La Extirpacion de la Idolatria del Piru," in *Cronicas peruanas de interes indigena*, edited by Francisco Esteve Barba. Biblioteca de Autores Espanoles (continuacion), vol. 209, pp. 191–275. Madrid: Ediciones Atlas.

Ascher, Marcia. 2002. "Reading Khipu: Labels, Structure, and Format," in *Narrative Threads: Accounting and Recounting in Andean Khipu*, edited by Jeffrey Quilter and Gary Urton, pp. 87–102. Austin: University of Texas Press.

Ascher, Marcia and Robert Ascher. 1997 [1981]. *Code of the Qhipu*. New York: Dover Publications.

Ashmore, Wendy and A. Bernhard Knapp (editors). *Archaeologies of Landscape. Contemporary Perspectives*. Malden, MA: Blackwell Publishers.

Aveni, Anthony F. 1980. *Skywatchers of Ancient Mexico*. Austin: University of Texas Press.

———. 1981. "Horizon Astronomy in Incaic Cuzco," in *Archaeoastronomy in the Americas*, edited by R. A. Williamson, pp. 305–319. Ballena Press Anthropological Papers 22. Los Altos, CA: Ballena Press/Center for Archaeoastronomy.

Bandelier, Adolph F. 1910. *The Islands of Titicaca and Koati*. New York: Hispanic Society of America.

Bastien, Joseph W. 1978. *Mountain of the Condor: Metaphor and Ritual in an Andean Ayllu*. American Ethnological Society: Monograph 64 (series editor Robert F. Spencer). St. Paul: West Publishing Company. Reissued in 1985 by Waveland Press.

———. 2006. "Mountain-Body Metaphor of Kallawaya Andeans Expressed in Mesas," in *Mesas & Cosmologies in the Central Andes*, edited by Douglas Sharon, pp. 125–135. San Diego Museum Papers 44. San Diego: Museum of Man.

Bauer, Brian S. 1991. "Pacariqtambo and the mythical origins of the Inca." *Latin American Antiquity* 2:7–26.

———. 1992. *The Development of the Inca State*. Austin: University of Texas Press.

———. 1998. *The Sacred Landscape of the Inca. The Cusco Ceque System*. Austin: University of Texas Press.

———. 2004. *Ancient Cusco*. Austin: University of Texas Press.

Bauer, Brian and Alan Covey. 2002. "Processes of State Formation in the Inca Heartland (Cuzco, Peru)." *American Anthropologist* 104(3): 846–864.

Bauer, Brian S. and David S. P. Dearborn. 1995. *Astronomy and Empire in the Ancient Andes*. Austin: University of Texas Press.

Bauer, Brian, Mary Futrell, Lisa Cipolla, Alan Covey, and Joshua Terry. 2004. "Excavations at Inca Sites on the Island of the Sun," in *Archaeological Research on the Islands of the Sun and Moon, Lake Titicaca, Bolivia: Final Results from the Proyecto Tiksi Kjarka*, edited by Charles Stanish and Brian Bauer, pp. 43–91. Monograph 52, Cotsen Institute of Archaeology. Los Angeles: University of California.

Bauer, Brian S. and Charles Stanish. 2001. *Ritual and Pilgrimage in the Ancient Andes. The Islands of the Sun and Moon*. Austin: University of Texas Press.

Bejar Luksic, Luis Fernando, Karen Durand Caceres, and Alexei Vranich. 2010. "Investigating the *Huacas* of the Empire: The Archaeological Project 'Balcon del Diablo,' Cuzco, Peru." Poster presentation at the 50th Anniversary Conference of the Institute of Andean Studies, University of California Berkeley, California.

Bell, Bernard and Jessica Christie. Ms. New Fieldwork from the Cerro San Miguel Ridge, Machu Picchu.

Bengtsson, Lisbet. 1998. *Prehistoric Stonework in the Peruvian Andes. A Case Study at Ollantaytambo*. Gotarc Series B, No.10, Etnologiska studier 44. Goteborg: Goteborg University, Department of Archaeology.

Berlo, Janet and Ruth Philipps. 1998. *Native North American Art*. Oxford: Oxford University Press.

Betanzos, Juan de. 1996 (1557). *Narrative of the Incas*, translated and edited by Roland Hamilton and Dana Buchanan. Austin: University of Texas Press.

Bingham, Hiram. 1922. *Inca Land: Explorations in the Highlands of Peru*. Boston: Houghton Mifflin. Cambridge: The Riverside Press.

———. 1979 *Machu Picchu A Citadel of the Incas*. New York: Hacker Art Books.

———. 1997 *Lost City of the Incas*. Lima: Condor Books, Special Book Services.

Boero Rojo, Hugo and Sonia de Boero Rojo. 1987. *El Imperio del Sol*. Editorial Hispania.

Bolin, Inge. 1998. *Rituals of Respect. The Secret of Survival in the High Peruvian Andes*. Austin: University of Texas Press.

———. 2006. "The Power of the Enqaychu and the Cosmos of Andean Pastoralists," in *Mesas and Cosmologies in the Central Andes*, edited by Douglas Sharon, pp. 113–123. San Diego Museum Papers 44. San Diego: Museum of Man.

Bourget, Steve. 2006. *Sex, Death, and Sacrifice in Moche Religion and Visual Culture*. Austin: University of Texas Press.

Bouysse-Cassagne, Therese and Olivia Harris. 1987. "Pacha: En Torno al Pensamiento Aymara." *Tres reflexiones sobre el pensamiento andino*, pp. 11–59. La Paz: Hisbol.

Bray, Tamara (editor). 2014. *The Archaeology of Wak'as: Explorations of the Sacred in the Pre-Columbian Andes*. Boulder: University Press of Colorado.

Burger, Richard. 1992. *Chavin and the Origins of Andean Civilization*. London: Thames and Hudson.

Burger, Richard, Morris Craig, and Ramos Matos Mendieta (editors). 2007. *Variations in the Expression of Inka Power*. Washington, DC: Dumbarton Oaks Research Library and Collection.

Burger, Richard L. and Lucy C. Salazar (editors). 2003. *The 1912 Yale Peruvian Scientific Expedition Collections from Machu Picchu, Human and Animal Remains*. Yale University Publications in Anthropology, no. 85. New Haven: Yale Univesity Press.

———. 2004. *Machu Picchu: Unveiling the Mystery of the Incas*. New Haven: Yale University Press.

Cabello Valboa, Miguel. 1951. *Miscelanea Antartica. Una Historia del Peru Antiguo*. Universidad Nacional Mayor de San Marcos, Instituto de Etnologia, Lima.

Caceres Silva, Carmela and Carmen Gladys Jurado Carrasco. 2000. *"Iconografia de Chinchero": Una introducción a su estudio*. Tesis, Departamento Academico de Antropologia, Arqueologia y Sociologia. Universidad Nacional de San Antonio Abad del Cusco.

"Ancient Ruins found in Bolivia." *Vancouver Sun*, July 10, 2008. http://www.canada.com/vancouversun/news/story.html?id=b9522ae5-c6db-4614-82e0-3b1a55d64777.

De la Cadena, Marisol. 2010. "Indigenous Cosmopolitics in the Andes: Conceptual Reflections Beyond 'Politics.'" *Journal of the Society of Cultural Anthropology*, vol. 25, no. 2, pp. 334–370.

Cardenas Martin, Mercedes. 2001. "El Espacio en el Modelo Urbano Inca: los Centros Administrativos Provinciales." *Boletin de Lima* vol. XXII, no. 123, pp. 15–29.

Carrion Cachet, Rebecca. 1955. El Culto al Agua en el Antiguo Peru. *Separata de la Revista del Museo Nacional*, II, 2, Lima.

Chavez, Karen Mohr. 2001. "La culture Chiripa." *Dossiers d'Archeologie*, no. 262, pp. 24–29.

Chavez, Sergio. 2002. "Identification of the Camelid Woman and Feline Man Themes, Motifs, and Designs in Pucara Style Pottery," in *Andean Archaeology II, Art, Landscape, and Society*, edited by Helaine Silverman and William Isbell, pp. 35–69. New York: Kluwer Academic/Plenum Publishers.

Chavez-Ballon, Manuel. 1970. "Ciudades Incas: Cuzco, Capital del Imperio." *Wayka* 3: 1–15.

Christie, Jessica Joyce. 2002. *Samaipata—Imperial Rock Art at the Eastern Border of the Inca Empire*. Paper presented at the Annual Meetings of the Society for American Archaeology (SAA), Denver.

———. 2003. "The Sacred Inca Rocks of Sayhuite," in *Rock Art Papers*, edited by Ken Hedges, vol. 16, pp. 133–142. San Diego Museum Papers 41. San Diego: Museum of Man.

———. 2005a. "The Carved Rocks of Machu Picchu," in *Rock Art Papers*, edited by Ken Hedges, vol. 17, pp. 139–154. San Diego Museum Papers 43, San Diego: Museum of Man.

———. 2005b. "Rock Art as a Political Strategy in the Inca Empire," in *Rock Art Papers*, edited by Ken Hedges, vol. 17, pp. 123–137. San Diego Museum Papers 43. San Diego: Museum of Man.

———. 2005c. *Piedras Talladas Inka y Yupanas: Una Posible Esencia Compartida*. Paper presented at the International Symposium on Archaeology in the South Central Andes, the Catholic University, Arequipa, Peru.

———. 2006a. "Houses of Political Power among the Ancient Maya and Inka," in *Palaces and Power in the Americas: From Peru to the Northwest Coast*, edited by Jessica Christie and Patricia Sarro, pp. 353–396. Austin: University of Texas Press.

———. 2006b. "Inca Copacabana: A Reconstruction from the Perspective of the Carved Rocks." *Anthropos* 101: 179–201.

———. 2007a. "Did the Inka Copy Cusco? An Answer Derived from an Architectural-Sculptural Model." *Journal of Latin American and Caribbean Anthropology*, vol.12, no.1, pp. 164–199.

———. 2007b. *Possible Late Forms of the Inka Usnu in the Vilcabamba Region*. Paper presented in the symposium *Inka Architecture of Power* at the annual Conference of the Society for American Archaeology, Austin.

———. 2008. "Inka Roads, Lines, and Rock Shrines: A Discussion of the Contexts of Trail Markers." *Journal of Anthropological Research*, vol. 64, no. 1, pp. 41–66.

———. 2009. *Inka Carved Rocks and Khipus: Were Some Sculpted Grids Counting Devices Using the Same Structure as Khipus?* Paper presented at the annual Conference of the American Rock Art Research Association, Bakersfield, CA.

———. 2011. *Exporting Inkaness: An Argument for Replication in Architecture and Sculpture in the Ayacucho Region*. Paper presented at the annual SAA Conference, Sacramento, CA.

———. 2012. "Thoughts on Inka Visual Representation at Q'enqo and Other Rock Art Sites." *Ethnohistory* 59(3): 597–630.

———. 2013a. *Inka Iconographic Rocks*. Paper presented at the International Rock Art Congress IFRAO/ARARA in Albuquerque.

———. 2013b. *Negotiating Inka Borders in the North: The Case of Ingapirca, Ecuador*. Paper presented at the Society for Amazonian and Andean Studies Conference at the University of Central Florida in Orlando.

———. 2014. *Inka Stone Ideology in Peripheral Regions of the Empire*. Paper presented at the annual Conference of the Society for American Archaeology in Austin.

Christie, Jessica and John Staller. 2004. *Inka Pacariqtambo—It Depends which Eyes are Looking at It*. Paper presented in the session *Landscapes of Origin in the Americas* organized and cochaired by Christie and Staller at the annual Conference of the American Anthropological Association, Atlanta.

Cieza de Leon, Pedro de. 1959 (1553). *The Incas*, edited by Victor Wolfgang von Hagen. Norman: University of Oklahoma Press.

Coben, Lawrence. 2006. "Other Cuzcos: Replicated Theaters of Inka Power," in *Archaeology of Performance*, edited by Takeshi Inomata and Lawrence Coben, pp. 223–259. Lanham, MD: AltaMira Press.

Cobo, Bernabe. 1956 (1653). *Historia del Nuevo Mundo*, Book 12, tomo III. Edited by Luis A. Pardo. Publicaciones Pardo-Galimberti, Cuzco.

———. 1990 (1653). "Historia del Nuevo Mundo, Books I and II" in *Inca Religion and Customs*. Edited and translated by Roland Hamilton. Austin: University of Texas Press.

Conklin, William. 1996. "Structure as Meaning in Ancient Andean Textiles," in *Andean Art at Dumbarton Oak*, edited by Elizabeth Hill Boone, vol. 2, pp. 321–328. Washington, DC: Dumbarton Oaks Research Library and Collection.

Conklin, William J. and Jeffrey Quilter (editors). 2008. *Chavin: Art, Architecture, and Culture*. Los Angeles: Cotsen Institute of Archaeology, University of California.

Cornejo Bouroncle, Jorge. 1946. "Huakaypata; la Plaza Mayor del Viejo Cuzco." *Revista universitaria*, pp. 85–116. Cuzco: Editorial de Cultura Andina.

Covey, Alan. *Inka Agricultural Intensification in the Imperial Heartland and Provinces*. Paper presented at the 68th Annual Meeting of the Society for American Archaeology, Milwaukee.

———. 2006. *How the Incas Built their Heartland. State Formation and the Innovation of Imperial Strategies in the Sacred Valley, Peru*. Ann Arbor: University of Michigan Press.

———. 2011. "Landscapes and Languages of Power in the Inca Imperial Heartland (Cuzco Peru)." *The SAA Archaeological Record* 11(4): 29–32.

Cummins, Tom. 2007. "Queros, Aquillas, Uncus, and Chulpas: The Composition of Inka Artistic Expression and Power," in *Variations in the Expression of Inka Power*, edited by Richard Burger, Craig Morris, and Ramiro Matos Mendieta, pp. 267–311. Washington, DC: Dumbarton Oaks.

D'Altroy, Terence. 2002. *The Incas*. Malden, MA: Blackwell Publishers.

David, Bruno and Meredith Wilson (editors). 2002. *Inscribed Landscapes. Marking and Making Place*. Honolulu: University of Hawai'i Press.

Dean, Carolyn. 1998. "Creating a Ruin in Colonial Cusco: Sacsahuaman and What Was Made of It." *Andean Past* 5: 161–183.

———. 2006a. What is the Trouble with the Term Art ? *Art Journal*.

———. 2006b. "Rethinking Apacheta." *Nawpa Pacha* 28: 93–108. Berkeley, CA: Institute of Andean Studies.

———. 2007. "The Inka Married the Earth: Integrated Outcrops and the Making of Place." *Art Bulletin* 89, no. 3: 502–18.

———. 2010. *A Culture of Stone. Inka Perspectives on Rock*. Durham: Duke University Press.

Dearborn, David. 2000. "The Inca Rulers of the Andes, Children of the Sun," in *The History of Non-Western Astronomy. Astronomy across Cultures*, edited by Helaine Selin. Dordrecht: Kluwer Academic Publishers.

Dearborn, David and R.E. White. 1982. "Archaeoastronomy at Machu Picchu," in *Ethnoastronomy and Archaeoastronomy in the American Tropics*, edited by Anthony Aveni and Gary Urton, pp. 249–259. Annals of the New York Academy of Sciences, vol. 385. New York: New York Academy of Sciences.

Diaz, Julio Palomino. 1994. *Intiwatanas y Numeros*. Municipalidad del Qosqo.

DiscoveryNews.com. 2004. Incan Counting System Decoded? www.discoverynews.com. Accessed February 3, 2004.

Duviols, Pierre. 1976. "La Capacocha." *Allpanchis phuturinqa* 9: 11–58.

———. 1979a. "La Dinastia de los Incas ¿Monarquia o Diarquia?: Argumentos Heuristicos a Favor de una Tesis Estructuralista." *Journal de la Societe des Americanistes* 66. Paris.

———. 1979b. "Un Simbolisme de L'occupation, de l'Amenagement et de L'exploitation de L'espace." *L'Homme; Revue Francaise d'Anthropologie* 19(2): 7–31.

———. 1980. "Algunas Reflexiones Acerca de la Tesis de la Estructura Dual del Poder Incaico." *Historica* 4(2, December): 183–196. Lima: Fondo Editorial, Pontificia Universidad.

Elorrieta Salazar, Fernando E. and Edgar Elorrieta Salazar. 1996. *Sacred Valley of the Incas: Myths and Symbols*. Cusco: Sociedad Pacaritanpu Hatha.

Escalante Moscoso, Javier F. 1997. *Arquitectura Prehispanica en los Andes Bolivianos*. La Paz: Produccion Cima.

Fejos, Paul. 1944. *Archeological Explorations in the Cordillera Vilcabamba Southeastern Peru*. Viking Fund Publications in Anthropology, no. 3. New York.

Fonseca Santa Cruz, Javier. 2009. Informe: "Proyecto de Investigacion Arqueologica de Espiritu Pampa – Vilcabamba." Cusco: Instituto Nacional de Cultura Región Cusco.

Frame, Mary. 2001. "Beyond the Image: The Dimensions of Pattern in Ancient Andean Textiles," in *Abstraction: The Amerindian Paradigm*, edited by Cesar Paternosto, pp. 113–136. Brussels: Societe des Expositions du Palais des Beaux-Arts de Bruxelles.

Franco Inojosa, J. M. 1937. "Informe de—sobre las Ruinas de Janan Kosko.—II (7a Entrega)." *Revista del Museo Nacional* 6(2): 201–231. Lima: Imprenta del Museo Nacional.

———. 1939. Ruinas de Chamankalla, Quillarumi, Inkarakay y Salakaka. Arqueología del Cusco. *Revista del Museo Nacional* 8(2): 250–264. Lima.

Friedman, Jonathan. 1994. *Cultural Identity and Global Process*. London: Sage Publications.

Garcilaso de la Vega, Inca. 1963 (1605). *Comentarios Reales de los Incas*, edited by Carmelo Saenz de Santa Maria. Madrid: Biblioteca de Autores Espanoles.

———. 1989 (1539–1616). *Royal Commentaries of the Incas, and General History of Peru*, edited by H. V. Livermore. Austin: University of Texas Press.

Gasparini, Graziano and Luise Margolies. 1980. *Inca Architecture*. Bloomington: Indiana University Press.

Gisbert, Theresa. 1997. "Angeles y Dioses en Copacabana," in *Arqueologia, Antropologia e Historia en los Andes. Homenaje a Maria Rostworowski*, edited by Rafael Varon Gabai and Javier Flores Espinoza, pp. 617–641, Instituto de Estudios Peruanos, Banco Central de Reserva del Peru.

Glowacki, Mary and Michael Malpass. 2003. "Water, Huacas, and Ancestor Worship: Traces of a Sacred Wari Landscape." *Latin American Antiquity* 14(4): 431–448.

Gonzalez Holguin, Diego. 1952 (1608). *Vocabulario de la Lengua General de Todo el Peru Llamada Lengua Qquichua o del Inca*. Lima: Edicion del Instituto de Historia, Imprenta Santa Maria.

———. 1989 (1608) *Vocabulario de la Lengua General de Todo el Peru Llamada Lengua Qquichua o del Inca*. Universidad Nacional Mayor de San Marcos, Editorial de la Universidad.

Grieder, Terence. 1978. *The Art and Archaeology of Pashash*. Austin: University of Texas Press.

———. 1982. *Origins of Pre-Columbian Art*. Austin: University of Texas Press.

Guaman (Huaman) Poma de Ayala, Felipe. 1956 (1615). *La nueva cronica y buen gobierno*, edited by Luis Bustios Galvez. Part I. Lima: Editorial Cultura, Ministerio de Educacion Publica del Peru.

———. 1987 (1615). *Nueva cronica y buen gobierno*, edited by John Murra, Rolena Adorno, and Jorge Urioste. 3 vols. Madrid: Historia 16.

———. 1993 (1615). *Nueva cronica y buen gobierno*, edited by Franklin Pease and Jan Szeminski. Mexico D. F.: Tomo I. Fondo de Cultura Economica.

Guillen Naveros, Jorge W. 2006. *Informe Final del Proyecto de Investigacion Arqueologica Sector "Salonniyoq— Templo de la Luna" Saqsaywaman*. Cusco: Instituto Nacional de Cultura.

———. 2007?. *Proyecto de Investigacion Arqueologica Salonniyoq o Templo de la Luna. Parque Arqueologico nacional de Saqsaywaman*. Cusco: Instituto Nacional de Cultura.

Haas, Jonathan, Winifried Creamer, and Alvaro Ruiz. 2004. "Dating the Late Archaic occupation of the Norte Chico región in Peru." *Nature* 432, no. 23/30: 1020–1023.

Hamilton, Roland. 2008. Cobo, Bernabe (1580–1657) in *Guide to Documentary Sources for Andean Studies, 1530–1900*, edited by Joanne Pillsbury, pp. 152–155. Norman: University of Oklahoma Press.

Harvey, Katherine and Armando Harvey. 2005. *Saqsaywaman. Centro Cosmico Andino*. Puno: Editorial Pacifico.

Heffernan, Ken. 1996. "The Mitimaes of Tilka and the Inka Incorporation of Chinchaysuyu," in *Tawantinsuyu* 2: 23–36.

Helms, Mary W. 1993. *Craft and the Kingly Ideal: Art, Trade, and Power*. Austin: University of Texas Press.

———. 1998. *Access to Origins: Affines, Ancestors, and Aristocrats*. Austin: University of Texas Press.

Hemming, John and Edward Ranney. 1982. *Monuments of the Incas*. New York Graphic Society. Boston: Little, Brown and Company.

Hobsbawm, Eric. 1983. "Introduction: Inventing Traditions," in *The Invention of Tradition*, edited by Eric Hobsbawm and Terence Ranger, pp. 1–14. Cambridge: Cambridge University Press.

Holland, Dorothy, William Lachicotte Jr., Debra Skinner, and Carole Cain. 1998. *Identity and Agency in Cultural Worlds*. Cambridge: Harvard University Press.

Houston, Stephen D. and Tom Cummins. 2004 (2008). "Body, Presence, and Space in Andean and Mesoamerican Rulership," in *Palaces of the Ancient New World*, edited by Susan Toby Evans and Joanne Pillsbury, pp. 359–398. Washington, DC: Dumbarton Oaks Research Library and Collection.

Huaycochea Nunez de la Torre, Flor de Maria. 1994. *Qolqas Bancos de Reserva Andinos. Almacenes Inkas Arqueologia de Qolqas*. Cusco: Universidad Nacional de San Antonio Abad del Cusco.

Hyslop, John. 1990. *Inka settlement planning*. Austin: University of Texas Press.

Inomata, Takeshi. 2006. "Politics and Theatricality in Mayan Society," in *Archaeology of Performance*, edited by Takeshi Inomata and Lawrence Coben, pp. 187–221. Oxford: Altamira Press.

Janusek, John Wayne. 2004. *Identity and Power in the Ancient Andes*. New York: Routledge.

Jimenez de la Espada, Don Marcos. 1965 (1881–1897). *Relaciones Geograficas de Indias. Peru*, edited by Jose Urbano Martinez Carreras, vol.I. Madrid: Biblioteca de Autores Espanoles.

Julien, Catherine. 2000. *Reading Inca History*. Iowa: University of Iowa Press.

Kellett, Lucas. 2006. "Public Archaeology in an Andean Community." *The SAA Archaeological Record* 6(2): 8–11.

Kendall, Ann. 1985. *Aspects of Inca Architecture: Description, Function, and Chronology*. London: BAR International Series 242.

Kendall, Ann, Rob Early, and Bill Sillar. 1992. "Report on Archaeological Field Season Investigating Early Inca Architecture at Juchuy Coscco (Q'aqya Qhawana) and Warq'ana, Province of Calca, Dept. of Cuzco, Peru," in *Ancient America. Contributions to New World Archaeology*, edited by Nicholas Saunders, pp. 189–255. Oxbow Monograph 24. Oxford: Oxbow Books.

Kirchhoff, Paul. 1949. "The Social and Political Organization of the Andean Peoples," in *Handbook of South American Indians*, vol. 5, *The comparative ethnology of South American Indians*, edited by J. Steward. Bureau of American Ethnology Bulletin 143, pp. 293–311. Washington, DC: U.S. Government Printing Office.

Knapp, Bernhard and Wendy Ashmore. 1999. "Archaeological Landscapes: Constructed, Conceptualized, Ideational," in *Archaeologies of Landscape: Contemporary Perspectives*, edited by Wendy Ashmore and Bernhard Knapp, pp. 1–30. Malden, MA: Blackwell Publishers.

Kolata, Alan and Carlos Ponce Sangines. 1992. "Tiwanaku: The City at the Center," in *The Ancient Americas: Art from Sacred Landscapes*, edited by Richard Townsend, pp. 317–333. Art Institute of Chicago. Munich: Prestel Verlag.

Krickeberg, Walter. 1949. *Felsplastik und Felsbilder bei den Kulturvoelkern Altamerikas*. Band I, Palmen Verlag vormals. Berlin: Dietrich Reimer.

Kubler, George. 1962. *The Art and Architecture of Ancient America; the Mexican, Maya, and Andean Peoples*. Baltimore: Penguin Books.

Laurencich Minelli, Laura. 2000. "The Inca 'Quipu,'" in *The Inca World*, edited by Laura Laurencich Minelli, pp. 189–192. Norman: University of Oklahoma Press.

Lechtman, Heather. 1996. "Cloth and Metal: The Culture of Technology," in *Andean Art at Dumbarton Oaks*, edited by Elizabeth H. Boone, vol. 1, pp. 33–43. Washington, DC: Dumbarton Oaks Research Library and Collection.

Lee, Vincent. 2000. *Forgotten Vilcabamba*. Sixpac Manco Publications.

———. 2008. Paper presented at the annual Meeting of the Institute of Andean Studies, Berkeley, California.

Legnani, Nicole Delia. See under Titu Cusi Yupangui.

Llanos, Luis. 1936. "Informe sobre Ollantaitambo. Trabajos arqueologicos en el Departmento del Cuzco bajo la direccion del Dr. Luis E. Valcarcel." *Revista del Museo Nacional, II Semestre* 5(2): 123–156. Lima.

Locke, L. Leland. 1912. "The Ancient Quipu, a Peruvian Knot Record." *American Anthropologist* 14: 325–332.

Lopez Bejarano, Jose Maria. 2009. *Sacred Shrines and Pilgrimage Routes in the Lake Titicaca Basin: Results of an Archaeological Survey of the Copacabana Peninsula, Bolivia*. Paper presented at the Conference Surveying Andean Legacy Archaeological Research along the Inka Road System, Smithsonian, National Museum of the American Indian, Washington, DC.

Lopez Lenci, Yazmin. 2007. *El Cusco, Paqarina Moderna*. Cusco: Instituto Nacional de Cultura.

Lumbreras, Luis G. 1974. *The Peoples and Cultures of Ancient Peru*. Translated by Betty J. Meggers. Washington, DC: Smithsonian Institution Press.

MacCormack, Sabine. 2008. "Ramos Gavilan, Alonso (ca.1570–ca.1639)," in *Guide to Documentary Sources for Andean Studies, 1530–1900*, edited by Joanne Pillsbury, pp. 559–563. Norman: University of Oklahoma Press.

MacLean, Margaret Greenup. 1986. *Sacred Land, Sacred Water: Inca Landscape Planning in the Cuzco Area*. PhD dissertation, University of California, Berkeley. University Microfilms, Ann Arbor.

Malville, J. McKim. 2008. *Cosmological Motifs of Peruvian Huacas*. Paper presented at the meeting of the South American Explorers Club, Westcliff, CO.

Malville, J. McKim, Michael Zawaski, and Steven Gullberg. In press. *Cosmological Motifs of Peruvian Huacas*.

Mantilla, Roberto. 1972. Arquitectura rupestre en Copacabana. *Arte y Arqueologia* 2: 61–69. La Paz: Instituto de Estudios Bolivianos, Facultad de Humanidades.

McEwan, Colin and Maarten van de Guchte. 1992. "Ancestral Time and Sacred Space in Inca State Ritual," in *The Ancient Americas: Art from Sacred Landscapes*, edited by Richard Townsend, pp. 358–371. Chicago: Art Institute of Chicago.

McEwan, Colin, Frank Meddens, Gabriel Ramon, Francisco Ferreira and Cirilio Vivanco Pomacanchari. 2010. *Re-cognizing and Marking the Andean Landscape: Ushnus, Apachetas, Sayhuas, and Wankas*. Paper presented at the 75th Anniversary Meeting of the Society for American Archaeology, St. Louis, MO.

McEwan, Gordon. 2005. *Pikillacta: the Wari Empire in Cuzco*. Iowa City: University of Iowa Press.

Meddens, Frank M. 1997. "Function and Meaning of the Usnu in Late Horizon Peru." *Tawantinsuyu* 3: 5–14.

———. 2006. "Rocks in the Landscape: Managing the Inka Agricultural Cycle." *The Antiquaries Journal* (86): 36–65.

Meddens, Frank M., Nick Branch, Millena Frouin and Colin McEwan. 2010. *Locating Usnus: The Inca Rationale for the Placement of Sacred Platforms*. Paper presented at the 75th Anniversary Meeting of the Society for American Archaeology, Saint Louis, MO.

Meddens, Frank M., Colin McEwan, and Cirilio Vivanco Pomacanchari. 2010. "Inca 'Stone Ancestors' in Context at a High-Altitude *Usnu* Platform." *Latin American Antiquity* 21(2): 173–194.

Meyers, Albert. 1998. "Las Campanas Arqueologicas en Samaipata, 1994–1996. Segundo Informe de Trabajo." *Sociedad de Investigacion del Arte Rupestre de Bolivia (SIARB)*, Boletin 12:59–86. La Paz.

———. 2007. "Toward a Reconceptualization of the Late Horizon and the Inka Period: Perspectives from Cochasqui, Ecuador, and Samaipata, Bolivia," in *Variations in the Expression of Inka Power*, edited by Richard Burger, Craig Morris, and Ramiro Matos Mendieta, pp. 223–254. Washington, DC: Dumbarton Oaks.

Meyers, Albert and Cornelius Ulbert. 1997. "Inka Archaeology in Eastern Bolivia: Some Aspects of the Samaipata Project." *Tawantinsuyu* 3: 79–85.

Miller, George R. 2003. "Food for the Dead, Tools for the Afterlife: Zooarchaeology at Machu Picchu," in *The 1912 Yale Peruvian Scientific Expedition Collections from Machu Picchu, Human and Animal Remains*, edited by Richard L. Burger and Lucy C. Salazar, pp. 1–63. New Haven: Yale University Publications in Anthropology, no.85.

Millones, Luis. 2008. "Albornoz, Cristobal de (ca. 1529–ca. 1610)," in *Guide to Documentary Sources for Andean Studies, 1530–1900*, edited by Joanne Pillsbury, pp. 21–25. Norman: University of Oklahoma Press.

Molina (el Cusqueno), Cristobal de. 1943 (1574). "Relacion de las Fabulas y Ritos de los Incas," in Las Cronicas de los Molinas, edited by Francisco Loayza, pp. 1–84. Los Pequenos Grandes Libros de Historia Americana, Serie I, Tomo IV. Lima.

———. 1959 [1575]. *Ritos y Fabulas de los Incas*. Buenos Aires: Editorial Futuro.

Morris, Craig. 2004. "The Architecture of Tahuantinsuyo," in *The Incas. Art and Symbols*, by Franklin Pease, Craig Morris, Julian Santillana, Ramiro Matos,

Paloma Carcedo/Luisa Vetter, Carmen Arellano, Lucy Salazar/Vuka Roussakis, pp. 1–59. Lima: Banco de Credito del Peru.

Morris, Craig and Donald Thompson. 1985. *Huanuco Pampa: An Inca City and its Hinterland*. London: Thames & Hudson.

Muriel, C.C., director. n.d. Ceques y Huacas, Primera Parte, Camino al Chinchaysuyu. *Cuadernos Andinos*. Ano 1, #5, Qosqo, Peru.

Murra, John. 1980. *The Economic Organization of the Inka State*. Greenwich, CT: JAI Press Inc.

Murua, Martin de. 1962 (c. 1611–1615). *Historia General del Peru, Origen y Descendencia de los Incas* . . . Introduction and notes by Manuel Ballesteros-Gaibrois. Madrid.

———. 2004 (1590). *Codice Murua. Historia y genealogía de los reyes Incas del Peru (Codice Galvin)*, 2 vols., edited by Juan M. Ossio. Madrid: Testimonio Compania Editorial.

Nair, Stella Elise. 2003. *Of Remembrance and Forgetting: The Architecture of Chinchero, Peru from Thupa 'Inka to the Spanish Occupation*. Unpublished PhD dissertation, Department of Anthropology, University of California, Berkeley.

———. 2010. *In Search of the Kallanka: Modern Myths, Colonial Tales and Other Architectural Fables*. Paper presented at the 50th Annual Meeting of the Institute of Andean Studies, University of California, Berkeley.

Niles, Susan. 1987. *Callachaca*. Iowa City: University of Iowa Press.

———. 1992. "Inca Architecture and Sacred Landscape," in *The Ancient Americas: Art from Sacred Landscapes*, edited by Richard Townsend, pp. 347–357. Chicago: Art Institute of Chicago.

———. 1999. *The Shape of Inca History*. Iowa City: University of Iowa Press.

Nowack, Kerstin. 1998. *Ceque and More: A Critical Assessment of R. Tom Zuidema's Studies on the Inca*. Markt Schwaben, Germany: A. Saurwein.

Ossio, Juan M. 2004. "Introduccion," in *Codice Murua. Historia y genealogia de los Reyes Incas del Peru (Codice Galvin)*, 2 vols., by Martin de Murua, edited by Juan M. Ossio, pp. 7–72. Madrid: Testimonio Compania Editorial.

Pachacuti Yamqui Salcamaygua, Juan de Santa Cruz. 1993 (1613). *Relacion de Antiguedades Deste Reyno del Piru*, edited by Pierre Duviols and Cesar Itier. Cusco: Centro de Estudios Regionales Andinos "Bartolome de Las Casas."

Paerssinen, Martti. 1992. *Tawantinsuyu: The Inca State and its Political Organization*. Helsinki: Societas Historica Finlandiae.

Pardo, Luis A. 1946. "La Metropoli de Paccarictampu." *Revista de la Seccion Arqueologica de la Universidad Nacional del Cuzco* 2: 3–46.

Paternosto, Cesar. 1996. *The Stone and the Thread. Andean Roots of Abstract Art*. Austin: University of Texas Press.

Paz Flores, M. Percy. 1992. "El Senor de Teteqaqa se da Buen Trato (La Cruz mas Festejada de la ciudad)," in *El Qosqo. Antropologia de la Ciudad*, edited by Hiroyasu Tomoeda and Jorge A. Flores Ochoa, pp. 226–248. Qosqo: Ministerio de Educacion del Japon y Centro de Estudios Andinos Cuzco.

Pease Garcia Yrigoyen, Franklin. 1994. *Los Incas. Una Introduction*. Biblioteca "Lo que debo saber" Vol. 1, Pontificia Universidad Catolica del Peru. Lima: Fondo Editorial.

———. 1995. *Las Cronicas y los Andes*. Pontificia Universidad Catolica del Peru, Instituto Riva-Agüero. Lima: Fondo de Cultura Economica.

———. 2004. *Los ultimos Incas del Cuzco*. Instituto Nacional de Cultura, Peru.

Peirce, Charles S. 1985. "Logic as Semiotic: The Theory of Signs," in *Semiotics. An Introductory Anthology*. Edited by Robert E. Innis, pp. 1–23. Bloomington: Indiana University Press.

Pereyra Sanchez, Hugo. 1990. "La Yupana, Complemento Operacional del Quipu," in *Quipu y Yupana. Coleccion de Escritos*, edited by Carol Mackey, Hugo Pereyra, Carlos Radicati, Humberto Rodríguez, and Oscar Valverde, pp. 235–255. Lima: Consejo Nacional de Ciencia y Tecnología.

Pillsbury, Joanne (editor). 2008. *Guide to Documentary Sources for Andean Studies, 1530–1900*, 3 volumes. Norman: National Gallery of Art and University of Oklahoma Press.

Pizarro, Pedro. 1978 (1571). *Relacion del Descubrimiento y Conquista de los Reinos del Peru*. Lima: Pontificia Universidad Catolica del Peru.

Polo de Ondegardo, Juan. 1916 (1571). "Relacion de los Fundamentos Acerca del Notable Dano que Resulta de no Guardar a los Indios sus Fueros," edited by Horacio H. Urteaga, in Vol. 3, *Informaciones Acerca de la Religión y Gobierno de los Incas*. Colección de Libros y Documentos Referentes a la Historia del Peru, ser.1, 45–189. Lima: Sanmarti.

Porras Barrenechea, Raul. 1963. *Fuentes Historicas Peruanas (Apuntes de un Curso Universitario)*. Lima: Instituto Raul Porras Barrenechea, Universidad Nacional Mayor de San Marcos.

Portugal Ortiz, Max. n.d. *Proyecto de Trabajo Arqueologico en la Peninsula de Copacabana*. Manuscript in the Archives of the Instituto Nacional de Arqueologia in La Paz.

Portugal Ortiz, Maks, and Dick Edgar Ibarra Grasso. 1957. *Copacabana: El Santuario y la Arqueologia de la Peninsula e Islas del Sol y la Luna*. Cochabamba, Bolivia: Editorial Atlantic.

Portugal Zamora, Maks. 1977. Estudio Arqueologico de Copacabana. *Arqueologia en Bolivia y Peru*, vol. 2, pp. 285–323.

Presta, Ana Maria and Catherine Julien. 2008. Polo Ondegardo (ca. 1520–1575) in *Guide to Documentary Sources for Andean Studies, 1530–1900*, edited by Joanne Pillsbury, pp. 529–535. Norman: University of Oklahoma Press.

Protzen, Jean-Pierre. 1985. "Inca Quarrying and Stonecutting." *Journal of the Society of Architectural Historians*, vol. 44, no. 2, pp. 161–182.

———. 1993. *Inka Architecture and Construction at Ollantaytambo*. Oxford: Oxford University Press.

———. 2000. "Inca Architecture," in *The Inca World. The Development of Pre-Columbian Peru, A.D.1000–1534*, edited by Laura Laurencich Minelli, pp. 193–217. Norman: University of Oklahoma Press.

———. 2010 *Inca Architecture: A Perspective*. Paper presented at the 50th Annual Meeting of the Institute of Andean Studies, University of California, Berkeley.

Protzen, Jean-Pierre and Stella Nair. 1997. "Who Taught the Inca Stonemasons Their Skills?" *Journal of the Society of Architectural Historians* 56(2): 146–167.

Quilter, Jeffrey and Gary Urton (editors). 2002. *Narrative Threads*. Austin: University of Texas Press.

Radicati di Primeglio, Carlo. 1990. "Tableros de Escaques en el Antiguo Peru," in *Quipu y Yupana. Coleccion de Escritos*, edited by Carol Mackey, Hugo Pereyra, Carlos Radicati, Humberto Rodríguez, and Oscar Valverde, pp. 219–234. Lima: Consejo Nacional de Ciencia y Tecnología.

Ramos Gavilan, Alonso. 1887. *Historia de Copacabana y de la Milagrosa Imagen de su Virgin*. Edited by Rafael Sans.Lima: Impresa por J. Enrique del Campo.

———. 1988 (1621) *Historia del celebre Santuario de Nuestra Senora de Copacabana*. Edited by Ignacio Prado Pastor. Lima: Grafico P. L. Villanueva.

Reinhard, Johan. 1991. *Machu Picchu. The Sacred Center*. Lima: Nuevas Imagenes S.A.

———. 1992. "Interpreting the Nazca Lines," in *The Ancient Americas. Art from Sacred Landscapes*, edited by Richard Townsend, pp. 291–301. The Art Institute of Chicago. Munich: Prestel Verlag.

———. 2007. *Machu Picchu. Exploring an Ancient Sacred Center*. Fourth Revised Edition. Los Angeles: Cotsen Institute of Archaeology, University of California.

———. *Relaciones Geograficas de Indias* (RGI). 1965 (1557–1586). 3 vols. Biblioteca de Autores Espanoles (continuacion), vols. 183–185. Madrid: Ediciones Atlas.

Rivera Sundt, Oswaldo. 1978. "Arqueologia de la Peninsula de Copacabana." *Pumapunku* 12: 69–86.

———. 1984. "La Horca del Inka." *Arqueologia Boliviano* 1: 91–101. La Paz.

Rodriguez Limache, Oscar C. 2007. *Informe Final: Del Proyecto de Investigacion Arqueologica, Saqsaywaman Sector Qochapata*. Cusco: Instituto Nacional de Cultura.

Rojas C., J. Heriberto. 2006. *El Complejo Arqueologico de Ingapirca*. Tomo III. Azogues: Offset Color Cuenca.

Rostworowski de Diez Canseco, Maria. 1953. *Pachacutec Inca Yupanqui*. Lima.

———. 1999 *History of the Inca Realm*. Cambridge: Cambridge University Press.

Rowe, John H. 1946. "Inca Culture at the Time of the Spanish Conquest," in *The Andean Civilizations*, edited by Julian Steward, pp. 183–330. Handbook of South American Indians, vol. 2. Bureau of American Ethnology Bulletin 143, Washington, DC: U.S. Government Printing Office.

———. 1967. "What Kind of Settlement Was Inca Cusco?" *Nawpa pacha* 5: 59–75. Berkeley: Institute of Andean Studies.

———. 1980. "Relacion de las guacas del Cuzco (1653)," in "Account of the Shrines of Ancient Cuzco," translated and edited by J. H. Rowe. *Nawpa Pacha* 17(1979): 2–80.

———. 1985. "La Constitucion Inca del Cuzco." *Historica* 9(1): 35–73.

———. 1990a. "A Machu Picchu a la Luz de Documentos de Siglo XVI." *Historica* XIV(1): 139–154. Lima: Pontificia Universidad Catolica del Peru, Departamento de Humanidades.

———. 1990b. "El plano mas antiguo del Cuzco." *Historica* 24(2): 367–377.

———. 1997. "Las tierras reales de los incas," in *Arqueologia, antropologia, e hsitoria en los Andes, homenaje a Maria Rostworowski*, ed. Rafael Varon Gabai and Javier Flores Espinoza, 277–287. Instituto de Estudios Peruano, Lima.

Salazar, Lucy C. 2004. "Machu Picchu: Mysterious Royal Estate in the Cloud Forest," in *Machu Picchu: Unveiling the Mystery of the Incas*, edited by Richard Burger and Lucy Salazar, pp. 21–47. New Haven: Yale University Press.

Salazar, Lucy C. and Richard L. Burger. 2004. "Lifestyles of the Rich and Famous: Luxury and Daily Life in the Households of Machu Picchu's Elite," in *Palaces of the Ancient World*, edited by Susan Toby Evans and Joanne Pillsbury, pp. 325–357. Washington, DC: Dumbarton Oaks Research Library and Collection.

Salles-Reese, Veronica. 1997. *From Viracocha to the Virgin of Copacabana*. Austin: University of Texas Press.

Sallnow, Michael J. 1987. *Pilgrims of the Andes: Regional Cults in Cusco*. Washington, DC: Smithsonian Institution Press.

———. 1990. "Pilgrimage and Cultural Fracture in the Andes," in *Contesting the Sacred: The Anthropology of Christian Pilgrimage*, edited by John Eade and Michael J. Sallnow, pp. 137–153. London: Routledge.

Salomon, Frank. 1991. "Introductory Essay: The Huarochiri Manuscript," in *The Huarochiri Manuscript*, edited by Frank Salomon and George Urioste, pp. 1–38. Austin: University of Texas Press.

———. 1998. "How the *Huacas* Were: The Language of Substance and Transformation in the Huarochiri Quechua Manuscript." *RES* 33:7–17.

Salomon Frank and George Urioste (editors). 1991. *The Huarochiri Manuscript*. Austin: University of Texas Press.

Santos Escobar, Roberto. 1980. *Probanza de los Inkas Aukaylli de Copacabana (1619–1716)*. Manuscript in the Instituto Nacional de Arqueologia, La Paz.

Sarmiento de Gamboa, Pedro. 1942 (1572). *Historia de los Incas*. Buenos Aires: Emece Editores.

Seddon, Matthew Thomas. 1998. *Ritual, Power, and the Development of a Complex Society: The Island of the Sun and the Tiwanaku State*. PhD dissertation, Department of Anthropology, University of Chicago.

———. 2004. "Excavations at the Site of Chucaripupata: A Tiwanaku IV and V Temple and Domestic Occupation," in *Archaeological Research on the Islands of the Sun and Moon, Lake Titicaca, Bolivia: Final Results from the Proyecto Tiksi Kjarka*, edited by Charles Stanish and Brian Bauer, pp. 93–137. Monograph 52, Cotsen Institute of Archaeology. Los Angeles: University of California.

Seddon, Matthew and Brian Bauer. 2004. "Excavations at Tikani," in *Archaeological Research on the Islands of the Sun and Moon, Lake Titicaca, Bolivia: Final Results from the Proyecto Tiksi Kjarka*, edited by Charles Stanish and Brian Bauer, pp. 83–91. Monograph 52, Cotsen Institute of Archaeology. Los Angeles: University of California.

Segovia, Bartolome de. 1968 (1553). "Relacion de Muchas Cosas Acaecidas en el Peru . . ," in *Cronicas Peruanas de Interes Indigena* (often attributed to Cristobal de Molina [el Almagrista]). Biblioteca de Autores Espanoles vol. 209. Madrid: Ediciones Atlas.

Sharman, Russell Leigh. 2006. "Re/Making La Negrita: Culture as an Aesthetic System in Costa Rica." *American Anthropologist* 108(4): 842–853.

Sherbondy, Jeanette. 1982. *The Canal Systems of Hanan Cuzco*. PhD dissertation, University of Illinois, Urbana-Champaign. University Microfilms, Ann Arbor.

———. 1992. "Water Ideology in Inca Ethnogenesis," in *Andean Cosmologies through Time*, edited by Robert Dover, Katharine Seibold, and John McDowell, pp. 46–66. Indianapolis: Indiana University Press.

Silverman, Helaine. 1990. "Beyond the Pampa: the Geoglyphs of the Valleys of Nazca." *National Geographic Research*, 6(4): 435–56.

———. 2002. "Touring Ancient Times: The Present and Presented Past in Contemporary Peru." *American Anthropologist* 104(3): 881–902.

Smith, Adam T. 2003. *The Political Landscape. Constellations of Authority in Early Complex Polities*. Berkeley: University of California Press.

Socualaya Davila, Carlo. 2007. *Proyecto de Investigacion Arqueologica de los Sectores Chincana Grande, y Qocha. Parque Arqueologico nacional de Saqsaywaman*. Cusco: Instituto Nacional de Cultura.

Squier, Ephraim George. 1877. *Peru: Incidents of Travel and Exploration in the Land of the Incas*. London: Macmillan.

Staller, John. E. n.d. "Forms of Interaction and their Significance to Archaeological Patterning in a Sacred Landscape." *Journal of World Prehistory*.

Staller, John and Jessica Christie. 2006. *Political Authority and Sacred Space in the Ancient Andes: Exploring the Antiquity of the Ushnu Concept*. Paper presented at the 2006 Conference of the Society for American Archaeology, Puerto Rico.

Stanish, Charles. 2003. *Ancient Titicaca: the Evolution of Complex Society in Southern Peru and Northern Bolivia*. Berkeley: University of California Press.

Stanish, Charles and Brian Bauer (editors). 2004. *Archaeological Research on the Islands of the Sun and Moon, Lake Titicaca, Bolivia: Final Results from the Proyecto Tiksi Kjarka*. Monograph 52, Cotsen Institute of Archaeology. Los Angeles: University of California.

Steele, Paul and Catherine Allen. 2004. *Handbook of Inca Mythology*. Santa Barbara, CA: ABC-CLIO.

Stone-Miller, Rebecca. 2002. *Art of the Andes from Chavin to Inca*. New York: Thames & Hudson.

Taylor (see Ethnohistory article).

Taylor, Gerald. 1974–1976. "Camay, camac, et camasca dans le manuscript quechua de Huarochiri." *Journal de la Societe des Americanistes* vol.63, series 2, pp. 231–244.

Tilley, Christopher. 1994. *A Phenomenology of Landscape Places, Paths and Monuments*. Oxford: Berg Publishers.

Tilley, Christopher and Wayne Bennett. 2004. *The Materiality of Stone. Explorations in Landscape Phenomenology: 1*. Oxford: Berg Publishers.

Titu Cusi Yupangui. 2005 (1570). *A 16th Century Account of the Conquest*, edited by Nicole Delia Legnani. Cambridge: Harvard University Press.

Topic, John. 2002. "From Stewards to Bureaucrats: Architecture and Information Flow at Chan Chan, Peru." *Latin American Antiquity* 14(3): 243–274.

Trimborn, Hermann. 1967. *Archaeologische Studien in den Kordilleren Boliviens*. Berlin: Verlag von Dietrich Reimer.

Ubbelohde-Doering, Heinrich. 1952. *Kunst im Reiche der Inca*. Tuebingen: Verlag Ernst Wasmuth.

Uhle, Max. 1910. Zur Deutung der Intihuatana. *Verhandlungen des XVI. Internationalen Amerikanisten-Kongresses*, Wien September 1908, pp. 371–388. Vienna: A. Hartleben's Verlag.

Urbano, Henrique Oswaldo. 2008. Molina, Cristobal de (ca. 1529–1585) in *Guide to Documentary Sources for Andean Studies, 1530–1900*, edited by Joanne Pillsbury, pp. 427–428. Norman: University of Oklahoma Press.

Urton, Gary. 1981. *At the Crossroads of the Earth and Sky. An Andean Cosmology*. Austin: University of Texas Press.

———. 1990. *The History of a Myth. Pacariqtambo and the Origin of the Inkas*. Austin: University of Texas Press.

———. 1999. *Inca Myths*. Austin: University of Texas Press.

———. 2002. "An Overview of Spanish Colonial Commentary on Andean Knotted-String Records," in *Narrative Threads*, edited by Jeffrey Quilter and Gary Urton, pp. 3–25. Austin: University of Texas Press.

———. 2003. *Signs of the Inka Khipu*. Austin: University of Texas Press.

———. 2004. *Three Matching Khipu Accounts from Lake of the Condors*. Paper presented at the 44th Annual Meeting of the Institute of Andean Studies, Berkeley, California.

Valcarcel, Luis. 1935a II.—"Sajsawaman Redescubierto (1). Los Trabajos Arqueologicos del Cusco. 2a Entrega." *Revista del Museo Nacional*, 211–233. Lima: Imprenta del Museo Nacional.

———. 1935b. Sajsawaman Redescubierto (III). Trabajos arqueologicos en el Dep. del Cusco, 3a Entrega. *Revista del Museo Nacional, I Semestre* 4 (1): 1–34, Lima.

———. 1935c. Sajsawaman Redescubierto (IV). Los trabajos arqueologicos en el Dep. del Cusco. 4ª Entrega. *Revista del Museo Nacional, II*: 163–233. Imprenta del Museo Nacional, Lima.

———. 1964. *Machu Picchu el mas famoso monumento arqueologico del Peru*. Biblioteca de America/Libros del Tiempo Nuevo, Eudeba Editorial Universitaria, Buenos Aires.

Van Dommelen, Peter. 1999. "Exploring Everyday Places and Cosmologies," in *Archaeologies of Landscape*, edited by Wendy Ashmore and A. Bernard Knapp, pp. 277–285. Malden, MA: Blackwell Publishers.

Van de Guchte, Maarten J. D. 1984. "El Ciclo Mitico Andino de la Piedra Cansada." *Revista Andina* 2(2): 539–556.

———. 1990 "'Carving the World': Inca Monumental Sculpture and Landscape." Unpublished PhD dissertation, Department of Anthropology, University of Illinois, Urbana-Champaign.

———. 1996. "Sculpture and the Concept of the Double Among the Inca Kings." *RES* 29/30: 256–268.

———. 1999. "The Inca Cognition of Landscape: Archaeology, Ethnohistory, and the Aesthetic of Alterity," in *Archaeologies of Landscape*, edited by Wendy Ashmore and Bernhard Knapp, pp. 149–168. Malden, MA: Blackwell Publishers.

Van Dyke, Ruth M. 2007. *The Chaco Experience: Landscape and Ideology at the Center Place*. Santa Fe: School for Advanced Research Press.

Van Vleet, Krista E. 2008. *Performing Kinship: Narrative Gender and the Intimacies of Power in the Andes*. Austin: University of Texas Press.

Verano, John W. 2003. Human Skeletal Remains from Machu Picchu: A Reexamination of the Yale Peabody Museum's Collection, in *The 1912 Yale Peruvian Scientific Expedition Collections from Machu Picchu, Human and Animal Remains*, edited by Richard L. Burger and Lucy C. Salazar, pp. 65–117. New Haven: Yale University Publications in Anthropology, no. 85.

Villanueva Urteaga, Horacio. 1971. "Documentos sobre Yucay en el Siglo XVI." *Revista del Archivo Historico del Cuzco* 13 (1970): 1–148, Cuzco.

Wassen, Henry. 1931. *The Ancient Peruvian Abacus*. (reprint from *Comparative Ethnographical Studies* 9: 189–205), Goeteborg.

———. 1940. "El Antiguo Abaco Peruano Segun el Manuscrito de Guaman Poma." *Etnologiska Studier*, 11: 1–30, Goeteborg.

———. 1990. "El Antiguo Abaco Peruano segun el Manuscrito de Guaman Poma," in *Quipu y Yupana. Colección de Escritos*, edited by Carol Mackey, Hugo Pereyra, Carlos Radicati, Humberto Rodríguez, and Oscar Valverde, pp. 205–218. Lima: Consejo Nacional de Ciencia y Tecnología.

Wiener, Charles. 1880. *Perou et Bolivie. Recit de Voyage*. Paris: Librairie Hachette.

Wright, Kenneth, Alfredo Valencia Zegarra, Ruth Wright, and Gordon McEwan. 2000. *Machu Picchu. A Civil Engineering Marvel*. Reston, VA: American Society of Civil Engineers Press.

Wright, Ruth and Alfredo Valencia Zegarra. 2001. *The Machu Picchu Guide Book*. Boulder: Johnson Books.

Ziolkowski, Mariusz. 1966. *La Guerra de los Wawqui. Los objetos y los mecanismos de la rivalidad dentro de la elite Inka, siglos XV–XVI*. Coleccion Biblioteca Abya-Yala 41, Ediciones. Quito: Abya Yala.

Zuidema, R. Tom. 1964. *The ceque system of Cuzco: The social Organization of the capital of the Inca*, translated by E. M. Hooykaas. International Archives of Ethnography, supplement to vol. 50. Leiden, E. J. Brill.

———. 1977. "The Inca Calendar," in *Native American Astronomy*, edited by Anthony F. Aveni, pp. 219–259. Austin: University of Texas Press.

———. 1980. "El Ushnu." *Revista de la Universidad Complutense*, pp. 317–362. Madrid: Universidad Complutense de Madrid.

———. 1981. "Comment." *Latin American Research Review* 16(3): 167–170.

———. 1982a. "Bureaucracy and Systematic Knowledge in Andean Civilization," in *The Inca and Aztec States 1400–1800, Anthropology and History*, edited by George Collier, Renato Rosaldo, and John Wirth, pp. 419–458. New York: Academic Press.

———. 1982b. "The Sidereal Lunar Calendar of the Incas," in *Archaeoastronomy in the New World*, edited by Anthony F. Aveni, pp. 59–107. Cambridge: Cambridge University Press.

———. 1990. "Ceques and chapas: An Andean Pattern of Land Partition in the Modern Valley of Cuzco," in *Circumpacifica: Festschrift fur Thomas S. Barthel*, edited by Bruno Illius and Matthias Laubscher. Sonderdruck: Peter Land.

Index

About the Author

Jessica Joyce Christie was born in Germany and has been an American citizen since 1989. She is associate professor in the art history concentration of the School of Art and Design at East Carolina University, Greenville, North Carolina, where she teaches courses about Andean, Mesoamerican, and Native North American visual culture. Her research and teaching have been strengthened by extended fieldwork at archaeological sites and with indigenous communities in all three culture regions.